ADVANCES IN
CORROSION SCIENCE AND TECHNOLOGY
VOLUME 7

ADVANCES IN CORROSION SCIENCE AND TECHNOLOGY

Editors:

M. G. Fontana
Fontana Corrosion Center, Department of Metallurgical Engineering
The Ohio State University, Columbus, Ohio

and

R. W. Staehle
Institute of Technology
University of Minnesota, Minneapolis, Minnesota

A Continuation Order Plan is available for this series. A continuation order will bring delivery of each new volume immediately upon publication. Volumes are billed only upon actual shipment. For further information please contact the publisher.

ADVANCES IN
CORROSION SCIENCE AND TECHNOLOGY
VOLUME 7

Edited by
Mars G. Fontana and Roger W. Staehle

PLENUM PRESS · NEW YORK AND LONDON

The Library of Congress cataloged the first volume of this title as follows:

Advances in corrosion science and technology. v. 1-

New York, Plenum Press, 1970-
  v. illus. 24 cm.
Editors. v. 1—M. G. Fontana and R. W. Staehle.
1. Corrosion and anti-corrosives—Collected works. I. Fontana, Mars Guy,
1910—   ed. II. Staehle, R. W., 1934—   ed.
TA418.74.A3   620.1'1223   76—107531

Library of Congress Catalog Card Number 76-107531
ISBN 0-306-39507-X

© 1980 Plenum Press, New York
A Division of Plenum Publishing Corporation
227 West 17th Street, New York, N.Y. 10011

Printed in the United States of America

PREFACE

This series was organized to provide a forum for review papers in the area of corrosion. The aim of these reviews is to bring certain areas of corrosion science and technology into a sharp focus. The volumes of this series are published approximately on a yearly basis and each contains three to five reviews. The articles in each volume are selected in such a way as to be of interest both to the corrosion scientists and the corrosion technologists. There is, in fact, a particular aim in juxtaposing these interests because of the importance of mutual interaction and interdisciplinarity so important in corrosion studies. It is hoped that the corrosion scientists in this way may stay abreast of the activities in corrosion technology and *vice versa*.

In this series the term "corrosion" is used in its very broadest sense. It includes, therefore, not only the degradation of metals in aqueous environment but also what is commonly referred to as "high-temperature oxidation." Further, the plan is to be even more general than these topics; the series will include all solids and all environments. Today, engineering solids include not only metals but glasses, ionic solids, polymeric solids, and composites of these. Environments of interest must be extended to liquid metals, a wide variety of gases, nonaqueous electrolytes, and other nonaqueous liquids. Furthermore, there are certain complex situations such as wear, cavitation, fretting, and other forms of degradation which it is appropriate to include. At suitable intervals certain of the review articles will be updated as the demands of technology and the fund of new information dictate.

Another important aim of this series is to attract those in areas peripheral to the field of corrosion. Thus, physicists, physical metallurgists, physical chemists, and electronic scientists all can make very substantial contributions to the resolution of corrosion problems. It is hoped that these reviews will make the field more accessible to potential contributors from these other areas. Many of the phenomena in corrosion are so complex that it is impossible for reasonable progress to be made without more serious and enthusiastic interdisciplinary interest.

This series, to some extent, serves as a "dynamic" handbook. It is well known that preparing a handbook is a long, tedious process and parts become out of date by the time the final volume is published. Furthermore, certain subjects become out of date more quickly than others. Finally, in a handbook it is never possible to prepare the individual discussions with sufficient detail and visual material to be properly useful to the reader. It is hoped that the format of this series serves to overcome some of these difficulties.

In addition to the discussion of scientific and technological phenomena the articles in this series will also include discussions of important techniques which should be of interest to corrosion scientists.

M. G. FONTANA
R. W. STAEHLE

CONTENTS

CORROSION CREEP AND STRESS RUPTURE

J. K. Tien and J. M. Davidson*

Henry Krumb School of Mines
Columbia University
New York, New York

INTRODUCTION

Corrosive attack by, say, the oxidizing combustion environment in a power turbine or in a coal gasification unit, coupled with high temperature, can allow an otherwise low stress to result in premature failures of structures. Generally speaking, high-temperature· plastic deformation can be designed against through enhancing creep resistance, prolonging stress rupture life, and increasing the material's stress rupture ductility. Unfortunately, even without the complication of environmental effects, the current state of knowledge of creep and stress rupture is at best qualitative. [1-7] Further, the known effects of environments on creep and stress rupture have yet to be reviewed, assessed, and compacted.

We view such compaction as one mandate for this treatise. The other goal is the critical assessment and construction of fundamental models to begin the process of explaining the major observations in these hyphenated areas of corrosion–creep and corrosion–stress rupture. In this interpretive review, we attempt to broaden our overview sufficiently to provide the necessary educational base for those whose backgrounds are concentrated in either of these areas but who wish to expand into the hyphenated areas.

We begin, then, with rather elementary analysis of high-temperature creep and stress rupture of structural alloys, and document the various observed effects of environmental interactions. This is followed by a brief tutorial on high-temperature corrosion and a discussion of the many ways in which this corrosion could interact with the mechanical properties of alloys, which leads directly to a discussion specifically on corrosion creep and corrosion stress rupture. Throughout the chapter we stay at high temperatures, generally above $0.5T_m$, the absolute melting temperature of the

* J. M. Davidson is now with The International Nickel Company, Inc., The Inco Research and Development Center, Sterling Forest, Suffern, New York 10901.

alloy of interest. Accordingly, we refrain from reviewing such complicated effects as stress corrosion cracking, liquid metal embrittlement, or surfactant degradation of fracture strength, all of which have been competently reviewed elsewhere.[8,9]

ENGINEERING ASPECTS OF CREEP AND STRESS RUPTURE

Creep in the engineering sense refers to slow and time-dependent plastic deformation caused by prolonged loading. At lower temperatures, say, less than one-half the absolute melting temperature T_m, substantial plastic deformation occurs only when the applied stress is near the yield strength. Creeping plastic flow, however, can take place with time at stresses much below the yield strength at high temperatures greater than $0.5T_m$. A simplified physical argument for this phenomenon is that as temperature increases, there will be more thermal energy to help the applied mechanical energy to cause dislocations to move and grain boundaries to slide, and hence, creep. Creep, therefore, can be defined as thermally assisted time-dependent plastic deformation of materials.

Typical creep curves are shown schematically in Fig. 1. In the general case, a creep curve (A) consists of four continuous stages. First is the instantaneous regime, which represents the elastic strain at the loading stress. Next is the primary or transient creep regime, which is characterized by an initially high and then decreasing creep rate. This is followed by the secondary or steady-state creep stage, which is a regime of relatively constant creep rate. The final stage, namely, tertiary creep, represents the regime of runaway creep rate, leading eventually to failure or stress rupture of the system. Quaternary and quinary stages of creep are sometimes reported. However, as we will see, these higher stages of creep are repetitions of earlier stages caused by environmental effects. Depending on the stress and temperature, one or more of these stages may be missing. For example, secondary creep is practically nonexistent at low temperatures ($<0.3T_m$) and low stress (curve B). At very high temperatures ($>0.8T_m$) and high stresses, tertiary creep is predominant (curve C).

Creep at low temperatures and low stress (curve B) is of little practical interest except, for example, in high-precision instruments, because it results in a transient flow which soon dies away almost completely so that the dimensions of the system become stable and the applied load can be supported safely for very long times. Creep at high stresses and high temperatures,

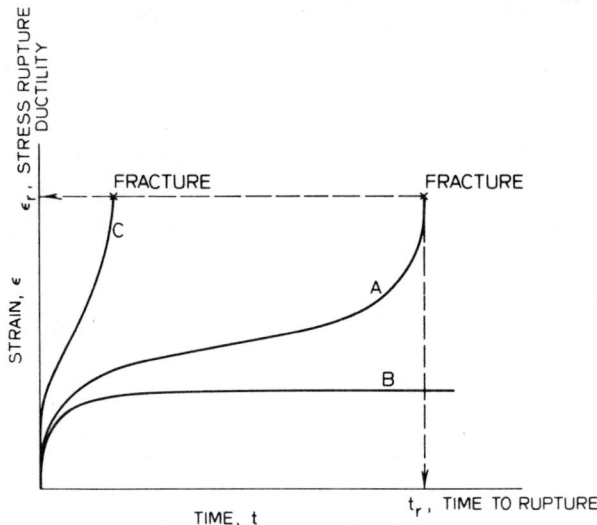

Fig. 1. Schematics of typical creep curves (see the text for an explanation). (After Tien et al.[1])

although rarely courted intentionally in practice for static load application, can be of importance in high-temperature cyclic fatigue applications in which a significant "creep component" can occur between cycles.[10] Furthermore, creep at high stresses and high temperatures has recently been suggested to control hot crack growth in some structural alloys.[11]

Most elevated-temperature structures are designed so that they spend their service lives in the steady state, if not the primary stage of creep (i.e., under conditions where the creep behavior of the material follows that schematically shown in curve A). In fact, a quantity known as "creep strength" is often referred to in structural design. This particular quantity is more or less governed by the steady-state or minimum creep rate, since it is defined as the stress necessary to result in an allowable strain (usually about 2–5%) after either 100 or 1000 hrs of elapsed loading time, longer time conditions generally being determined by extrapolation techniques such as that due to Larson and Miller.[12] Hence, at stresses where the dominant mode of deformation is the time-dependent creep, creep resistance means a low steady-state creep rate, or conversely, a high creep strength, provided that primary strains are small.

Phenomenological equations have been devised that characterize the entire creep curve or parts thereof. One of the more frequently used expressions describes the steady state creep rate, $\dot{\varepsilon}_s$, portion of the creep curve as

$$\dot{\varepsilon}_s = A\,[\sigma_A/E(T)]^n(\mathbf{b}/d)^p \exp\,(-Q_c^*/RT) \tag{1}$$

where A is a constant of proportionality, σ_A the applied stress, n the apparent stress sensitivity, p the sensitivity of the creep rate to grain size d, Q_c^* the activation energy for this process, E Young's modulus, $\mathbf{b}$ the Burgers vector, and RT has its usual meaning.

Although this expression has been found to suitably characterize the steady-state creep behavior of a wide variety of metals and alloys, it has recently been found[13] that in particle-strengthened creep-resisting alloys (such as superalloys and dispersion-strengthened alloys), an internal stress, σ_i, that opposes dislocation generation and movement must be included, where

$$\dot{\varepsilon}_s = A'\,[\sigma_A - \sigma_i)/E(T)]^{n'}(\mathbf{b}/d)^{p'} \exp\,(-Q_c^*/RT) \tag{2}$$

where the primed parameters are internal stress corrected.

Again from an engineering viewpoint, and referring to Fig. 1, the major stress rupture properties are the stress rupture life t_r and the stress rupture elongation (or ductility) ε_r. It has been experimentally determined for a variety of metals and alloys[3] that a relationship

$$t_r \cdot (\dot{\varepsilon}_s)^\beta = B \tag{3}$$

exists between t_r and $\dot{\varepsilon}_s$. The parameters B and β, which are constants for single stress and temperature conditions, can, as we will see, be highly dependent on environment. Interestingly, if β is unity, Eq. (3) reflects a strain-controlled criterion for stress rupture (i.e., B is then a function of stress rupture ductility). It appears then that for a given system and environment, it can be argued that as the creep rate is manipulated through alloy or microstructural design, the stress rupture life of the system will also be affected in a manner defined by Eq. (3) [i.e., as $\dot{\varepsilon}_s$ is lowered, t_r increases].

CLASSIFICATION OF CORROSION CREEP AND STRESS RUPTURE

Before beginning the task of compacting the environmental creep and stress rupture literature, we introduce for bookkeeping purposes the terms creep resistance factor, F_C, stress rupture life factor, F_L, and stress rupture ductility factor, F_D, where

$$F_C = (\dot{\varepsilon}_s^A - \dot{\varepsilon}_s^E)/\dot{\varepsilon}_s^A \tag{4}$$

$$F_L = (t_r^E - t_r^A)/t_r^A \tag{5}$$

$$F_D = (\varepsilon_r^E - \varepsilon_r^A)/\varepsilon_r^A \tag{6}$$

The superscript A denotes behavior in air and E denotes behavior in an environment other than air. Hence, positive F's mean that air impairs the property more so than the environment in question under the same applied stress and temperature conditions. Conversely, negative F's imply the inverse; that is, the environment in question is more deleterious than air with respect to the property in question.

Air is used as the standard for comparison only because historically most of the creep and stress rupture tests have been performed in air. We are strict in defining air as laboratory air. The definition does not encompass partial vacuum or artificial levels of pure oxygen. This is still, however, an unfortunately vague standard, as air may be wet or dry, and contain varying amounts of pollutants and other airborne particles.

Based on the F's, we take the first step toward compacting the information on corrosion creep and stress rupture by offering the classification scheme shown in Table 1 for use in this chapter, and perhaps, as a standard scheme for future use. This table contains all possible combinations or types of creep and stress rupture behavior patterns. For example, Type IA exhibits an

Table 1. Classification Scheme for
Corrosion Creep and Stress Rupture
Behaviors for Environments Other
than Laboratory Air Relative to
Laboratory Air[a]

Type	F_C	F_L	F_D
IA	−	−	−
IB	−	−	+
IIA	+	+	−
IIB	+	+	+
IIIA	−	+	−
IIIB	−	+	+
IVA	+	−	−
IVB	+	−	+

[a] The F terms are defined by Eqs. (4)–(6).

air-strengthening behavioral pattern, implying that the environment of interest is more corrosive than air with respect to creep resistance ($-F_C$), stress rupture life ($-F_L$), and stress rupture ductility ($-F_D$). In all four types of behaviors, the subtype designations A and B refer, respectively, to negative and to positive stress rupture ductility factors.

CRITICAL REVIEW OF BEHAVIORAL TYPES

There is no lack of references concerning environmental effects on creep and stress rupture in the literature.[14-66] There is, however, a lack of systematic studies on the subject. In our search for information, the classification scheme just introduced proved indispensable in sorting the scattered information. In the critical review that follows, attempts are made to base the environment–microstructure property correlations on the few available systematic studies.

Type I Behavior

This behavioral pattern is one of the few that has been investigated systematically in a rather complete parametric study,[14] which was concerned with air and vacuum ($\sim 10^{-5}$ torr) as the environments and a $\gamma'[Ni_3(Al, Ti)]$-strengthened nickel-base superalloy as the alloy of interest. Although constitutionally complex, the microstructure of this often-used class of high-temperature materials is fairly straightforward, consisting of about 30–40 vol % of fine ($\sim 0.1 \mu m$) and periodically spaced γ' particles within the grains, with the grain boundaries strengthened against sliding by discrete particles of refractory metal and chromium carbides sandwiched in a continuous γ' layer that follows the grain boundaries. A schematic representation of a typical microstructure of the material studied is shown in Fig. 2. It should be noted that the grain size of the system studied was 300 μm. As will become evident, grain size is very important in environmental creep.

The Type IA behavior is clearly illustrated in Figs. 3 and 4, which compare typical creep curves of this superalloy in air and in a vacuum of 10^{-5} torr at 760°C ($\sim 0.65 T_m$) and at 982°C ($\sim 0.75 T_m$), respectively. The values of the creep parameters, σ_i, n, and Q_c^*, and the stress rupture parameters, β and B, are tabulated in Table 2 for vacuum and air conditions at 760°C as well as 982°C. Also, although the trends cannot be simply expressed, the primary creep strains were observed to be higher in vacuum than in air (Figs. 3 and 4).

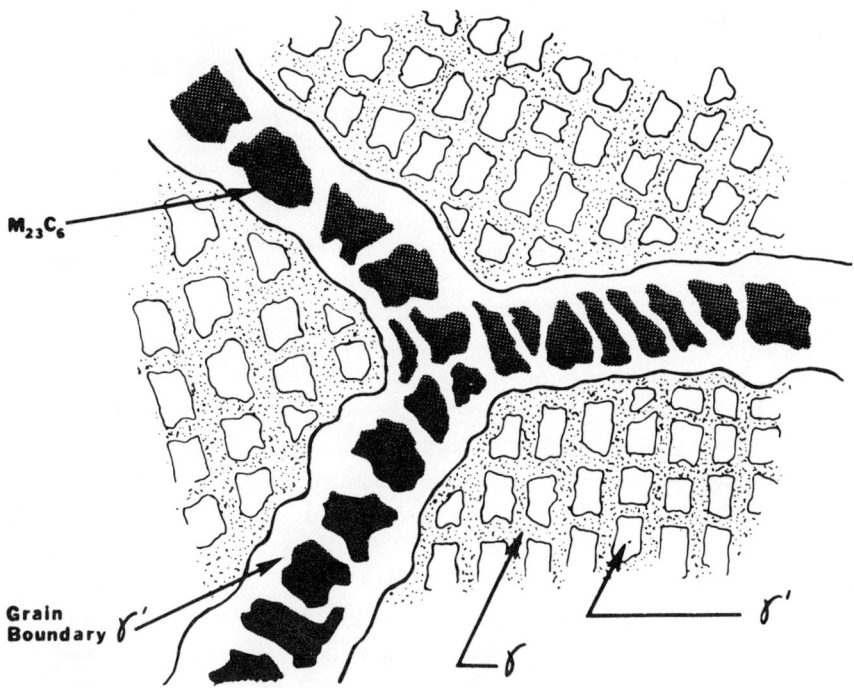

Fig. 2. Schematic illustration of a typical nickel-base superalloy microstructure showing homogeneous distribution of γ' precipitates in periodic cube arrays in a γ solid solution, and grain boundary carbides sandwiched between γ'.

As can be seen from Eqs. (2) and (3) and the data bank in Table 2, for this system air strengthened the alloy during steady-state creep. This strengthening appeared to be manifested in creep mainly through the apparent stress exponent n, and higher values for the internal stress, σ_i. The activation energies for creep remain, however, fairly invariant with test environment.

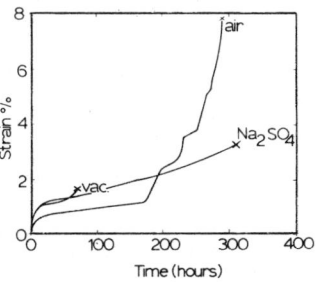

Fig. 3. Comparison of creep curves in air, vacuum (10^{-5} torr), and air with surface deposits of sodium-sulfate of polycrystalline superalloy of 300-μm grain size at 760°C and 483 MN/m² tensile stress. Vacuum: $F_C = 0.85$, $F_L = -0.61$, and $F_D = -0.81$. Na$_2$SO$_4$: $F_C = -0.26$, $F_L = +0.06$, $F_D = -0.66$. (After Aning.[14])

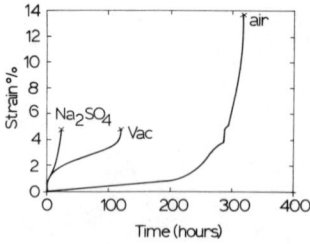

Fig. 4. Comparison of creep curves in air, vacuum $(10^{-5}$ torr), and air with surface deposits of sodium sulfate of polycrystalline superalloy of 300-μm grain size at 982°C and 108 MN/m^2 tensile stress. Vacuum: $F_C = -0.36, F_L = -0.63, F_D = -0.63$. Na$_2SO_4$: $F_C = -9.44, F_L = -0.93, F_D = -0.67$. (After Aning.[14])

The stress rupture behavior of the superalloy system with 300-μm grain size in vacuum and in air is also functionally compared in Table 2 and with the use of Eq. (3). Air was observed to prolong stress rupture lives at both temperatures (i.e., β is higher in vacuum than in air). The effect of air or vacuum environment on stress rupture ductility of the superalloy is clearly shown in Figs. 3 and 4. At both temperatures and over the entire stress range studied (440–800 MN/m^2 at 760°C and 90–120 MN/m^2 at 982°C), stress rupture ductility was lower in vacuum than in air.

In addition to the detailed study discussed above, Type IA behavior has often been reported in other studies. It has been observed in air vis-à-vis vacuum, ordinarily of 10^{-4}–10^{-6} torr, in other superalloys like Hastelloy-X,[15,16] in a single-crystal superalloy,[17] in large-grained (1–2 mm grain size) Udimet-700,[18–21] and in nickel and various other nickel-base alloys,[22–30] magnesium alloys,[31] and in silver-containing trace impurities.[32,33] Although the low oxidizing test environment is usually vacuum, Type IA behavior has also been reported for such nonoxidizing, or experimentally very low oxidizing environments as argon[34,35] and helium.[36–38] Also, it has recently been shown

Table 2. Experimental Creep Parameters of Udimet-700[a] Tested in Air and Vacuum at 760°C and 982°C[14]

Parameter	760°C		982°C	
	Air	Vacuum	Air	Vacuum
n	7.64	4.86	6.75	6.32
Q_c^* (KJ/mol)	313.8	304.6	367.9	323.9
β	1.31	3.31	0.68	1.13
B	4.5×10^{-5}	1.7×10^{-20}	4.26	1.36×10^{-4}
σ_i (MN/m^2)	245	117	45	38

[a] Grain size: 300 μm.

for vanadium alloys that Type IA behavior becomes more pronounced with ultrahigh vacuums ($\sim 10^{-9}$ torr).[39]

NaCl environments have been observed to lead to creep weakening with respect to air of a cobalt-base alloy and of a thoria dispersion-strengthened nickel–chromium alloy.[40] Also, deposits of sodium sulfate,[14] lead oxide,[41] oil ash,[42] vanadium ash with and without nickel sulfide additions,[43] and atmospheres of SO_2[43,44] have been observed to lead to Type IA behavior when compared to air.

Type IB behavior, where there is air strengthening but where the ductility is enhanced in the absence of air, has been observed in the creep and stress rupture behavior of certain commercial steels (DM 45, AISI 304, S-816) in air vis-à-vis nitrogen and vacuum (10^{-4} torr).[24] Another example of Type IB behavior is that of 304 stainless steel in pure oxygen as compared with air.[35] This result illustrates that pure oxygen might not necessarily affect creep properties in a fashion similar to air.[32,33,35,45]

Again, the difference between A and B subtypes rests solely with the relative effects of an environment and of air on stress rupture ductility. There are indications in the literature, where, for example, vacuum can result in either a lowering of ε_r, Type IA, or enhancement of ε_r, Type IB, on the same system,[33] or in many cases ε_r's were just not reported.[46-48] There are no quantitative explanations for this apparent controversy. However, it must be noted that ε_r is very sensitive to the microstructure of the system, especially to the presence of nonintentional inclusions which may vary from specimen to specimen.[5]

The decrease in creep rate, and subsequent increase in rupture life, Type IA or B behavior, has also been observed in metals and alloys whose surface scales are not oxides but rather coatings of other metals. For example, such behavior has been observed in copper-plated zinc single crystals and polycrystals and similarly plated single crystals of nickel.[49] Type I behavior has also been observed in a Ni–20Cr alloy coated with a ceramic film,[50] and cadmium with hydroxide and plastic coatings.[51]

Type II Behavior

The type II behavior is the inverse of Type I behavior (i.e., creep behavior characterized by air weakening instead of strengthening). Interestingly, Type II behavior has been observed in superalloy in vacuum of 10^{-5} torr when either the grain size became small, less than 300 μm, or when the number of grains per specimen cross section became sufficiently high, say, greater than five, or both.[18-21] These data, however, are complicated by the fact that only grain size was varied, rather than varying the grain size and specimen cross-sectional area independently. It is difficult to distinguish from this set of studies whether the behaviors were A or B type, since the stress rupture ductilities in air or vacuum were found to be comparable.

In contrasting the coarser-grain superalloy behavior, which is discussed above as Type I, with this set of Type II behavior in the same alloy, one comes to the conclusion that there can be a transition between the Type I and Type II behavior with grain size or density of grain boundaries, or both[18-21] (Fig. 5).

In addition to these superalloy observations, Type II behavior has also been observed in low carbon steels,[48] mild steels,[52] stainless steels,[45,53-56] Incoloy 800,[35] and in nickel and other nickel-base alloys.[23-27,57,58] These studies also establish that the transition between Types I and II can be temperature- and stress-dependent,[23-27,53] as illustrated in Fig. 6. It was observed with the nickel–chromium alloys that at lower temperatures, say, less than 700°C, and at higher stresses, the creep and stress rupture behavior in vacuum was Type II, whereas at higher temperatures and lower stresses, the behavior was Type I.[23-27] There is also the observation that steam strengthens with respect to air during creep.[59] Nitrogen has also been observed to strengthen as a consequence of the formation of nitride scales in the absence of oxygen.[34,35] Finally, in thin stainless steel wires, vacuum of 10^{-4} torr has been seen to be beneficial as compared to air.[60]

Summary of Corrosion Creep and Stress Rupture Behavior

The compacted phenomenological knowledge discussed above certainly does not leave one with the impression that all the major corrosion creep and

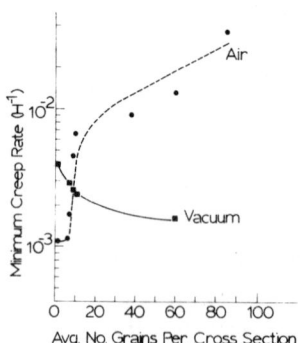

Fig. 5. Minimum creep rate of monocrystalline and polycrystalline superalloy tested in air and in vacuum (10^{-5} torr). Ordinate reflects change in grain size with constant specimen gauge diameter. (After Sessions et al.[21])

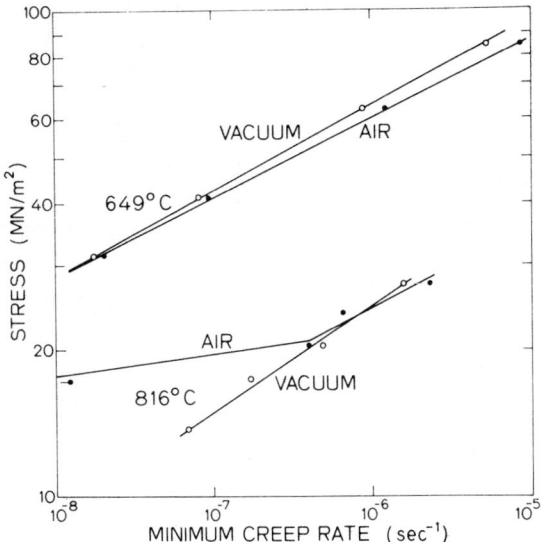

Fig. 6. Minimum creep rate of fine-grained nickel (grain size ~ 200 μm) in air and vacuum (10^{-5} torr) at 649 C and 816 C. (After Shahinian and Achter.[26])

stress rupture effects have been observed or uncovered, since systematic studies on the subject are still rare. In fact, Type III behavior has not been reported in the mainstream literature and Type IV behavior has been observed only rarely. The behavior was observed not in metals but in the Si_3N_4- and SiC-type ceramics.[61] However, perhaps as a consequence of the systematic classification that took place during the compaction described above, certain trends do emerge, and we describe these next.

Creep Rate and Stress Rupture Life The comparative environmental studies show that these properties appear to vary inversely with each other, which is consistent with the premise of a strain- or pseudo-strain-controlled stress rupture rationale of the type shown in Eq. (3). The environmental effects themselves, on the other hand, appear to be related to whether adherent oxide scales are present, noting that the absence of oxide scales can result not only from vacuum tests but also from tests in aggressive environments that can dynamically destroy scales. Further, the external oxide scale effects appear to become less significant when grain size becomes finer, or other internal effects predominate.

Stress Rupture Ductility. This property depends strongly on system and presumably on the initial microstructure and the hot-environment-produced microstructural changes.

These phenomenological trends suggest that rationales for the behavioral types rest with the effects of oxide scales, and other corrosion-related microstructural and alloy chemical changes, on hot plastic deformation, crack initiation, and propagation. In the following two sections, we present the compacted knowledge on scale formation and on the effects of scales and other hot-environment-caused microstructural and chemical changes on mechanical behavior, including information from the literature in areas other than creep and stress rupture. The discussion that follows these sections will agglomerate the overall knowledge, and rationales will be presented and the uncertainties noted for future research.

HIGH-TEMPERATURE CORROSION

What follows is by no means meant to be exhaustive, but merely sufficient to introduce the reader without a corrosion background to some of the fundamental concepts governing hot gas metal reactions. Of particular interest here are the different morphologies and spatial distributions of the reaction products, which directly determine the environmental interactions with mechanical properties, as documented and explained in the section below on oxidation- and corrosion-induced microstructural and chemical effects and mechanical behavior.

Emphasis is accordingly on thermodynamics, as it is generally capable of predicting what will form during the reactions, and where in relation to the alloy surface these reactions are likely to occur.

We begin with simple oxidation which can be thermodynamically treated as any other chemical reaction. In general terms the oxidation of a metal Me to MeO, according to the reaction

$$Me_{(s)} + \tfrac{1}{2}O_{2(g)} \rightleftharpoons MeO_{(s)} \tag{7}$$

is governed by the Gibbs free-energy change ΔG accompanying the reaction

$$\Delta G = \Delta G^{\circ}_{MeO} + RT\ln\frac{a_{MeO}}{(a_{Me})(a_{O_2})^{1/2}} \tag{8}$$

where a represents the activities of the three phases, and ΔG is the standard free energy of formation of the particular compound. The spontaneous forward reaction (oxidation) is expected for $\Delta G < 0$, and the spontaneous reverse reaction (reduction) is expected for $\Delta G > 0$. The special case for $\Delta G = 0$ defines the equilibrium condition between the metal and its oxide. By equation a_{O_2} to the partial pressure of O_2, P_{O_2}, which at high temperatures and low partial pressures is a good approximation, the oxygen partial pressure for Me–MeO at temperature T can be determined from

$$P_{O_2}(\text{Me–MeO}_{\text{equilibrium}}) = \exp\left(+2\Delta G^{\circ}_{\text{MeO}}/RT\right) \tag{9}$$

From ΔG versus T plots such as those introduced by Richardson and Jeffes,[67] we can determine the oxidation or reduction tendencies of metals or metal oxides at various partial pressures of oxygen and temperatures.

This oxidation need not necessarily occur in pure oxygen gas, for equilibrium P_{O_2}'s can also be achieved at high temperatures in gases such as carbon dioxide:

$$CO_2 \rightleftharpoons CO + \tfrac{1}{2}O_2 \tag{10}$$

or in the presence of water vapor:

$$H_2O \rightleftharpoons H_2 + \tfrac{1}{2}O_2 \tag{11}$$

Oxidation reduction reactions for metals in these gas mixtures, as well as equilibrium ratios for (CO/CO_2) and (H_2/H_2O) for Me–MeO equilibrium can be determined just as easily from the ΔG° versus T plots.[67]

These simple thermodynamic considerations can also be extended to the oxidation of alloys. During the initial stage of alloy oxidation, all oxide species that meet the thermodynamic stability criteria [Eq. (8)] can form (Table 3). The amount of each oxide species initially formed during this rapid uptake of oxygen has been shown to be roughly proportional to the concentration of the oxide-forming elements at the alloy surface[68,69] (Fig. 7).

The formation of these surface oxide nuclei of various compositions is only a transient phenomenon, however. As the oxidation reaction progresses with time, the necessary and sufficient conditions for the initial oxidation (oxide-phase stability) may no longer be sufficient to ensure the continued growth of the oxide phases. This is a consequence of displacement reactions of the form

$$\text{Me}_{(i)}O + \text{Me}_{(j)} \rightleftharpoons \text{Me}_{(j)}O + \text{Me}_{(i)} \tag{12}$$

which may occur as a result of relative oxide stability and long-range diffusion. The more thermodynamically stable oxides (more negative ΔG°) may displace the less stable oxides (less negative ΔG°), resulting in a layering of the oxide

Table 3. Compositions (Weight %) of Various Structural Alloys, the Initial Possible Oxides, and the Stable Oxides That Have Been Observed to Form on Them[74]

Alloy	Compositions (wt %)							Initial possible oxides	Stable oxide
	Fe	Cr	Ni	C	Ti	Al	Other		
410	Bal.[a]	12.0		0.15				Cr_2O_3, FeO	Cr_2O_3
304	Bal.	19.0	10.0	0.08				Cr_2O_3, FeO, NiO	Cr_2O_3
OR-1	Bal.	13.0	0.5	0.5	0.8	3.0		Cr_2O_3, TiO_2, Al_2O_3, FeO	Al_2O_3
18-SR	Bal.	18.2	0.4	0.05	0.4	2.0	1.1Si, 0.18Mn	Cr_2O_3, TiO_2, Al_2O_3, FeO	Al_2O_3
RA 330	Bal.	19.0	35.0	0.05			1.25Si, 1.5Mn	Cr_2O_3, FeO, MnO, SiO_2	Cr_2O_3
Hastelloy-X	18	22.0	47.8	0.10			9Mo, 1.5Co, 0.5W, 0.5Mn	Cr_2O_3, NiO, FeO	Cr_2O_3
Inconel 601	16	23.0	60.5	0.05		1.35		Cr_2O_3, NiO, FeO, Al_2O_3	Cr_2O_3

[a] Bal. = balance of the composition (wt%).

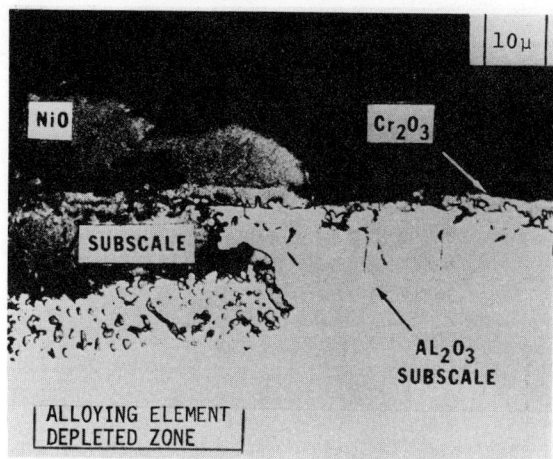

Fig. 7. Transient oxidation microstructure of Ni–10Cr–1Al oxidized for 20hr at 1000 C in 0.1-atm oxygen, showing transient formation of thermodynamically possible oxides NiO, Cr_2O_3, and Al_2O_3. (After Pettit and Tien.[71])

phases. Because the partial pressure of oxygen is expected to decrease away from the oxide gas interface, increasingly more stable oxides generally lie nearer the alloy oxide interface.

With time and a sufficient metal supply, the most stable oxide phase is expected, thermodynamically, to displace all other oxides above it, the final steady-state stage of oxidation being reached when this oxide covers the alloy substrate. Isothermal stability diagrams are commonly used to represent this stable oxide as a function of oxygen and alloying element activity.[70] If this stable oxide is also slow-growing (i.e., protective) it will obviously afford more oxidation resistance to the alloy than will a fast-growing oxide—a definite consideration in alloy design for oxidation resistance. Such a scale (Al_2O_3) is shown in Fig. 8. At elevated temperatures, protectivity as measured by the diffusivity of oxygen through the scale is generally highest for Al_2O_3, then Cr_2O_3, and then the nickel and iron oxides.[71-74]

The kinetics of oxidation are also generally related to the oxygen partial pressure through an exponential of the form

$$\text{oxidation rate} \propto P_{O_2}^{1/n} \qquad (13)$$

where n ranges from 2 to 6, depending, of course, on the oxide species, its morphology, and the defect structure.[73]

Fig. 8. Steady-state oxidation microstructure of Ni–20Cr-4Al oxidized for 20 hr at 1200 C in 0.1 atm oxygen showing only thermodynamically most stable oxide Al_2O_3. (After Pettit and Tien.[71])

As mentioned, formation of the most stable steady-state metal oxide requires a sufficient supply of that metal at the alloy surface. This necessary and sufficient supply is often minute (e.g., only 1 ppm Al is necessary in nickel for the sole formation of Al_2O_3 on the NiAl alloy in air between 900 and 1300°C[70]). Nonetheless, kinetics as well as internal oxidation can prevent the formation of this stable oxide. Again, in the case of Ni–Al alloys, it has been demonstrated that if the diffusion of aluminum within the alloy is slower than in the oxide, aluminum may be depleted to the extent that other oxide phases, such as NiO or $NiAl_2O_4$, are stable.[70] In alloys with dilute concentrations of the least noble element, nuclei of this most stable oxide might be overrun by a less stable oxide and incorporated as discrete particles into the oxide scale.[75] With such an alloy, the formation of internal oxide particles of the least noble element is also likely, as long as the solute concentration is below a critical limit.[76] This internal oxidation is further promoted by low diffusion coefficients of the solute in the alloy and a high oxygen partial pressure in the gas.[76] However, in gas mixtures with a very low oxygen activity, the inability of alloys to form protective adherent scales often results in internal corrosion.[36-38] The size, shape, and distribution of these internal oxides vary with the systems and condition of interest, although smaller particles tend to form with the more stable internal oxides, and all particles tend to be more highly concentrated at grain boundaries.[77,78]

It is worth mentioning here that loss of oxide protectivity might also occur due to the repeated spallation of these stable scales, thereby allowing accelerated oxidation of the less stable oxides on the bare alloy underneath.[72] Vaporative losses might also impair the formation of a thin protective scale, particularly with Cr_2O_3-bearing alloys,[79] although, interestingly enough, vaporization of Cr_2O_3 on some Ni–Cr–Al alloys has been observed to actually promote protectivity by allowing the formation of a thin single Al_2O_3 scale on these alloys.[69]

When we are investigating the high-temperature reactions between metals or alloys and gases that not only oxidize but also sulfidize, carburize, etc., it is necessary, as in the case of mixed oxide phases on alloys, to resort to isothermal stability diagrams that show the thermodynamically stable corrosion product for varying partial pressures of the different gases (see, e.g., Ref. 80). Because most metals have a greater affinity for oxygen than other oxidants, corrosion products formed on metals and alloys in any hot environment that contain even a small percentage of oxygen should be predominantly oxides.[72] For this reason, much of the later sections will be concerned with the effects of oxides on creep and stress rupture, which should by no means suggest that carbides, sulfides, and nitrides might not affect these properties in a similar manner. As in the case of mixed oxides forming on alloys in oxygen, thermodynamic predictions of stable phases might be modified by kinetic or spatial considerations. Furthermore, many industrial environments develop particularly pugnacious condensed phases that might often be coupled with localized physical attack of protective oxide scales, so that alloy corrosion behavior is far from what is thermodynamically expected.

In the presence of deposited ash such as Na_2SO_4 on the alloy surface, as is common in jet(gas)-turbine and coal gasification applications, oxidation often becomes accelerated or even catastrophic—a phenomenon generally referred to as "hot corrosion."[81–93] Vanadium pentoxide,[85,86] in the presence or absence of Na_2SO_4, has also been shown to accelerate oxidation and flux away oxide scales when it is in liquid form. Although some workers consider the sulfur in Na_2SO_4 to be the deleterious species,[84,90] others point to the activity of the Na_2O in the Na_2SO_4 to be the critical factor in hot corrosion.[72,83,88] As mentioned, sodium chloride has also been suggested as a likely culprit in the degradation of oxide scales.[87]

Although the mechanism of hot corrosion attack is still controversial, it appears that the presence of a liquid phase from any ash deposit can destroy the integrity of adherent oxide scale barriers, resulting in porous, nonadherent surface phases that are often accompanied by internal sulfides and oxides.[81] A typical chronology of hot corrosion attack, elegantly shown in microstructural form, is given in Fig. 9.[93]

In hot carbonaceous atmospheres such as CH_4/H_2 and CO/CO_2, carbon may be dissolved interstitially into an alloy,[94] or, in the extreme of saturation,[95] may appear as carbides in the alloy. Similarly, in carbon-containing alloys, the

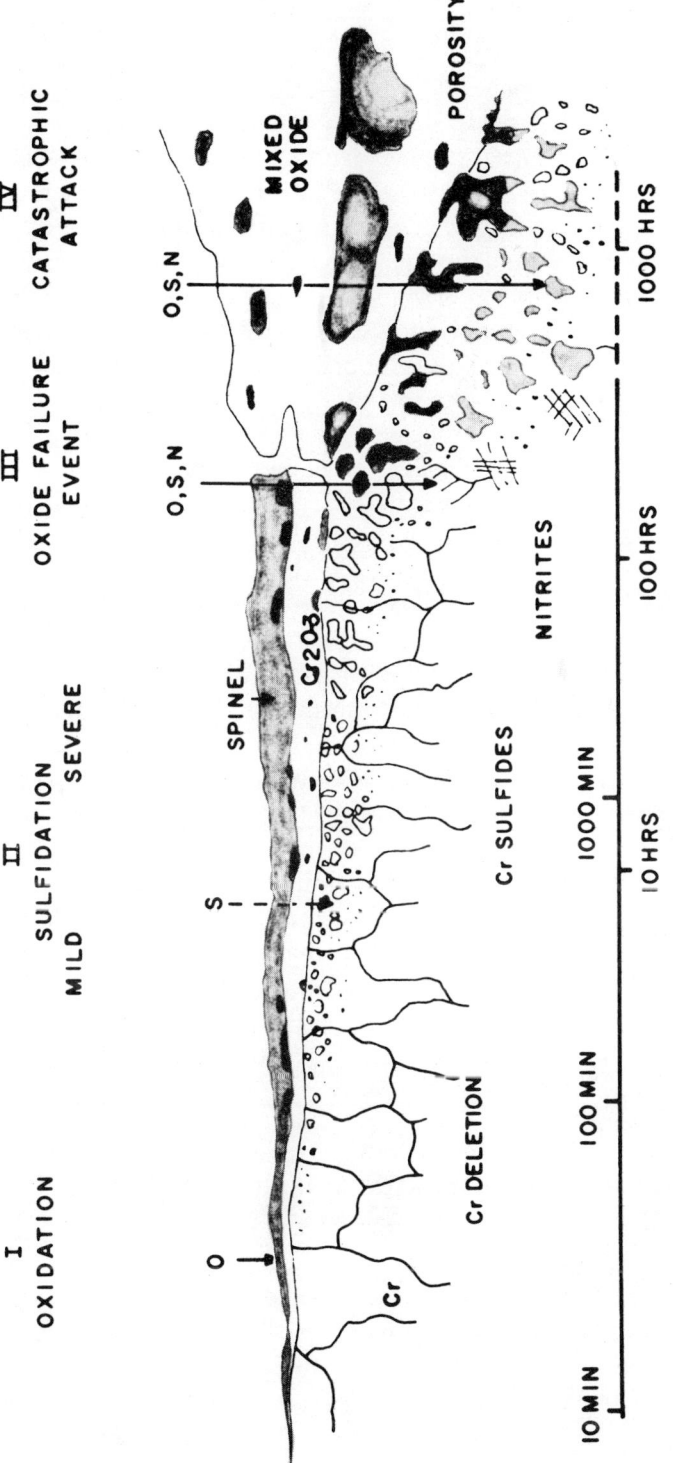

Fig. 9. Composite photomicrograph and schematic diagram illustrating the progression from "simple" oxidation (I) to severe hot-corrosion (IV) for a high-chromium nickel-base superalloy, approximately 250×. Reproduced at 100%. (After Sims.[93])

carbon activity of the environment may be such that alloy carbon is depleted by reacting to form gaseous oxides or hydrocarbons. A catastrophic form of carburization known as "metal dusting" has been reported for iron- and nickel-base alloys between 425 and 800°C.[96,97] This highly localized form of attack and pitting generally occurs at areas stripped of their protective oxide scales, which are carburized and subsequently mechanically[96] or chemically[97] broken down to dust composed of graphite, metal, and mixed oxides and carbides. The thermodynamics and kinetics of nitrogen dissolution in alloys as well as nitride precipitation[98] and the formation of surface nitride scales[99] have also been thoroughly investigated.

Finally, when an alloy is in a state of stress, as during creep applications, one should consider the possibility of applied stress affecting the thermodynamics and kinetics of oxidation or corrosion.[100-112] Stress has been suggested to affect corrosion rates through its influence on kinetic rate-law transition,[106] although such proposals are in dispute.[109] Furthermore, various theories[101] and experimental observations[35,108] indicate the possibility of enhanced corrosion kinetics due to the rupturing of surface scales of corrosion products by the applied stress. Recent work on the corrosion of heat-resisting CoCrAl and NiCrAl alloys, with and without yttrium additions, which promote oxide adherence,[111] have shown that although extreme compressive deformation can lead to ridging and cracking in the Al_2O_3 scale, the amount of subsequent spallation and reoxidation, and hence the oxidation kinetics, were not significantly altered.[110]

On the more microscopic scale, environmental penetration and oxidation kinetics have been observed to increase in localized areas where microcracks appear at the alloy surface,[14,18-21,103,110] and at grain boundary/external surface intersections.[14,107] As mentioned, grain boundaries can be rapid diffusion paths for the corrosive environment, and deep penetrations of oxidation or corrosion product phases are frequently observed down grain boundaries (e.g. Refs. 14, 18–21, and 107).

OXIDATION- AND CORROSION-INDUCED MICROSTRUCTURAL AND CHEMICAL EFFECTS AND MECHANICAL BEHAVIOR

The complexities and variants in structural and alloy chemistry changes produced in hot environments are perhaps best illustrated in Figs. 7–10. Certainly, the changes are not simply restricted to a few surface atomic layers, since the oxidation- or corrosion-affected zone is usually on the order of tens of

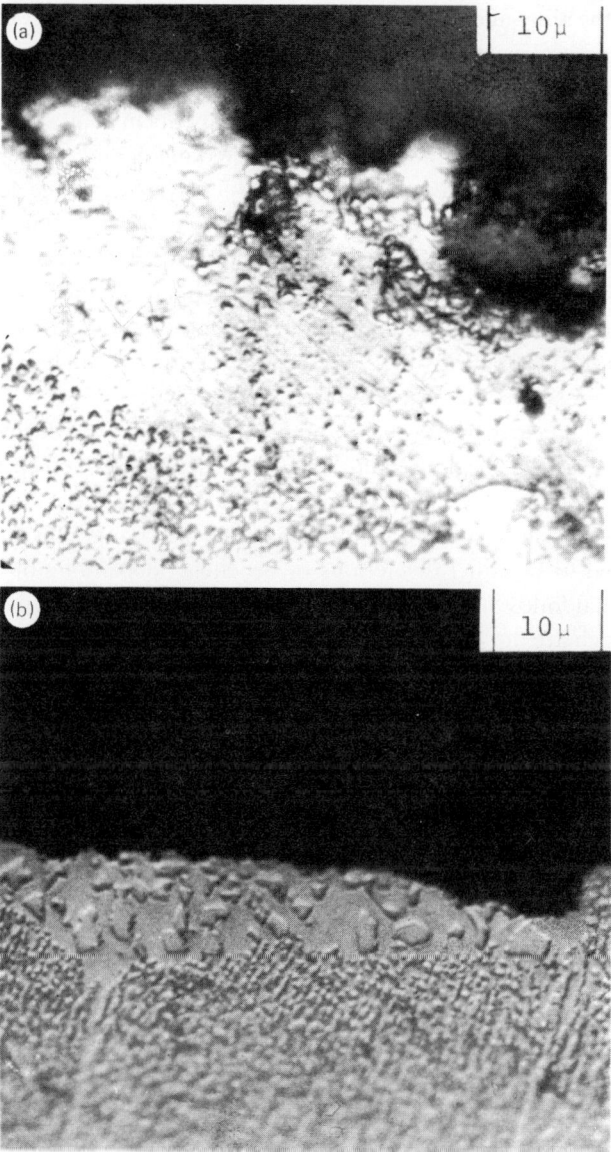

Fig. 10. Micrographic comparison of external surfaces of superalloy creep tested at 982°C, 108 MN^{-2} in (a) air and (b) vacuum (10^{-5} torr) for 315 and 110 hr, respectively. Note altered surface layer microstructures as well as the presence of oxide scale in the air case. (After Aning.[14])

microns and is basically determined by the diffusion penetration distance of oxygen into the alloy and of alloying elements undergoing selective oxidation. In the general case, as one proceeds inward from the gas–oxide interface, the microstructural and chemical changes encountered are (1) an external oxide scale, (2) a zone enriched with environmental elements in solid solution, (3) a subscale of oxide particles, and (4) a zone depleted of alloying elements which have been selectively oxidized, which may manifest itself as a zone depleted of strengthening precipitates. Further, as mentioned, oxidation may be preferential down chemically active short-circuit diffusion paths such as (5) grain boundaries and (6) crack faces. In the case of cracks, one can treat the crack surfaces as free surfaces, and all the changes itemized above can take place perpendicular to these surfaces and the crack tip, although gas activities might be significantly altered by the tip morphology.

It would be nearly impossible to predict how the various structural and chemical changes will affect mechanical behavior in general or creep and stress rupture in particular, even if these changes are quantitatively documented. This is because, synergistic effects aside, many of the changes discussed have been observed to be either deleterious or beneficial, depending on the alloy system. In what follows, we discuss, drawing from a broad spectrum of studies (many of which are not creep and stress rupture studies), the known effects of these microstructural and chemical changes on mechanical behavior.

Surface Coating, Scale, and Residual Stress Effects on Dislocation Production and Mobility

In the 1950s and the early 1960s, a considerable amount of work was done on the role of surfaces, including overlay coatings and oxide scales, on metal and alloy mechanical properties. Although much of this work did not involve creep or stress rupture, it is nonetheless of interest here since it speaks directly to the sensitivity of dislocation production and mobility to surfaces and scales. It serves no purpose in the context of this chapter for us to undertake an extensive review of all the experimental observations. The really interested reader is referred to the excellent reviews by Kramer and Demer,[113] Machlin,[114] Douglas,[115] or Westwood.[116] In what follows we select those investigations that bear more directly on the question at hand.

Roscoe[117] in 1934 was one of the first to note a surface effect on mechanical properties. He observed that the critical resolved shear stress of a cadmium single crystal was reduced by a factor of 2 when an external oxide scale developed on the crystal was removed. Since this work, numerous investigators have studied the role of oxide, ceramic, or metal coatings on the shear stress,[118-121] twinning stress,[122,123] the shape of the stress–strain curve,[119,121] microscopic deformation characteristics,[121,122] fracture,[124] internal friction[125] and abnormal strain recovery effects,[126] and even in a very few cases the fatigue and creep behavior of a wide variety of single crystals and polycrystals.[127,128]

The resulting theories and supporting experimental observations can be classified into two main categories:

Production of Dislocations

It can be easily shown that it would require twice the stress value to activate a double-ended Frank–Read type of dislocation source than a single-ended source.[129] It is proposed that single-ended sources should predominate at free and clean surfaces, since any mobile dislocation half-loop intersecting the free surface can potentially be a source.[114] Hence, clean surfaces are viewed as potent sources for dislocations. Conversely, this situation is neutralized when a coherent coating or scale covers the free surface.[114] In other words, a heavily oxidized surface is not considered a preferred region for the production of dislocations. That free surfaces are potent sources for dislocations, and coated surfaces are not, is confirmed, for example, through interpretations of the initial plastic flow behavior of silver-coated versus noncoated copper[118] and oxidized versus unoxidized or polished aluminum single crystals.[127]

Mobility of Dislocations

Adherent scales or coatings are also shown to provide strength through resisting the escape of edge dislocations from the surface,[122] and motion of screw dislocations intersecing the surface.[114] Simple image-force analysis indicates that the opposing stress due to a surface scale is proportional to $[(\mu_S - \mu_A)/(\mu_S + \mu_A)]$,[130] μ_S and μ_A being the shear moduli of the scale and the alloy, respectively. Conceivably this stress can be an attractive stress if the elastic modulus of the scale is less than that of the substrate. However, this is usually not the case when the scales are oxides or other corrosion products. That this restraining stress may exist is supported, for example, through ambient temperature plastic deformation results from studies on copper-plated zinc crystals,[122] oxidized aluminum crystals,[121] and oxidized cadmium crystals[125] and polycrystals.[126] Although not yet documented, one would expect this effect also with respect to grain boundary sliding in that such sliding—or grain rotation—does result in surface steps.

Surface stresses may evolve during high-temperature corrosion through any combination of stress sources including thermal mismatch strains and oxide growth strains.[131] Thermal growth stresses can arise during temperature change because of differences in thermal expansion between the oxide and the metallic substrate and can be approximated as[74]

$$\sigma_o^T = \frac{-E_o(T_i - T_f)(\alpha_A - \alpha_o)}{[1 + (E_o/E_A)(t_o/t_A)](1 - v)} \tag{14}$$

$$\sigma_A^T = \frac{E_A(T_i - T_f)(\alpha_A - \alpha_o)}{[1 + (E_A/E_o)(t_A/t_o)](1 - v)} \tag{15}$$

where σ_o^T and α_o are the normal thermal stress in, and average expansion coefficient of, the oxide, respectively; σ_A^T and α_A are the normal thermal stress in, and average expansion coefficient of, the alloy substrate, respectively; T_i and T_f are the initial and final temperatures of the composite oxide and substrate, and v is the Poisson ratio, assumed to be invariant with material. Table 4 tabulates elastic moduli, α's, and $\Delta\alpha$'s for various often encountered oxide and metal systems.

Oxide growth stresses may arise because of volumetric differences between the oxide scale and the metallic substrate, although the magnitude of such stresses can be diminished by anisotropy factors and the outward flux of metallic cations.[132] If we assume perfect isotropy and oxide growth by the inward diffusion of oxygen, we might approximate the growth stresses in the same manner:

$$\sigma_o^G = \frac{-E_o[(V_R)^{1/3} - 1]}{[1 + (E_o/E_A)(t_o/t_A)](1 - v)} \tag{16}$$

$$\sigma_A^G = \frac{E_A[(V_R)^{1/3} - 1]}{[1 + (E_A/E_o)(t_A/t_o)](1 - v)} \tag{17}$$

where σ_o^G is the normal growth stress in the oxide scale and σ_A^G the normal growth stress in the alloy substrate, and V_R is the ratio of the unconstrained atomic volume of oxide to that of the alloy, also known as the Pilling–Bedworth ratio.[133]

The sense of the stresses in the oxide and in the alloy is of interest since it will determine whether these stresses will add to or subtract from the superimposed applied stress. As can be seen from Eqs. (16) and (17), except in the MgO/Mg and LiO/Li cases, which are rarely found in commercial systems, the growth stress is always compressive in the oxide and tensile in the substrate. From Eqs. (14) and (15) and Table 4, we see that the thermal stress is usually compressive in the oxide and tensile in the substrate during down-quench, and vice versa during up-quench. The NiO/Fe system is one important exception to this generality. Hence, depending on the sense of the applied (mechanical) stress and the sense of the temperature change during a mission, the growth or thermally induced residual stresses in the substrate can be either deleterious, in the case that they add to the applied stress, or beneficial, in the case that they subtract from the applied stress, to creep resistance.

If we use Eqs. (16) or (17) and calculate σ_A^G's, we find that for all V_R's and except in cases of vanishing oxide thicknesses, the calculated values are on the order of hundreds of thousands, if not millions of psi, which should invariably fracture the alloy substrates. Such stresses may indeed affect surface cracking, but massive alloy fractures due to these stresses have not been reported. In fact, the yielding of these unrealistically high values by the standard growth stress expressions strongly suggests that the expressions may not be altogether realistic. At high temperatures the growth strains can conceivably be accommodated by localized flow in the alloy or even in the oxide itself, which would diminish the stresses to the level of the at-temperature flow stress in either phase. One major problem with these growth stress expressions is that they implicitly assume that the interface between oxide and alloy substrate is

Table 4. Young's Modulus (E), Thermal Expansion Coefficient (α), and $\Delta\alpha$ between Oxide–Oxide and Oxide–Metal Combinations[74,187,188]

Young's modulus (10^3 MN/m^2)	Thermal expansion coefficient (10^{-6}/°C)		$\Delta\alpha(10^{-6}$/°C)					
			Al_2O_3	Cr_2O_3	FeO	NiO	Ni	Fe
353	8.1	Al_2O_3	0	−0.8	+4.1	+9.0	+9.5	+7.2
276	7.3	Cr_2O_3	+0.8	0	+4.9	+9.8	+10.3	+8.0
172	12.2	FeO	−4.1	−4.9	0	+4.9	+5.4	+3.1
317	17.1	NiO	−9.0	−9.8	−4.9	0	+0.5	−1.8
149	17.6	Ni	−9.5	−10.3	−5.4	−0.5	0	—
138	15.3	Fe	−7.2	−8.0	−3.1	+1.8	—	0

coherent (i.e., there is expitaxy or coherent matching of lattices across the interface, with the volumetric differences being accommodated by interfacial coherency strains or stresses). Although some degree of coherent matching may be possible in the very early stages of oxidation, a mature oxide scale, with its crystal structure and chemical nature so unlike that of metals, would most probably be separated from the alloy substrate by an incoherent interphase boundary. In such cases, σ_A^G can no longer be calculated from Eqs. (16) and (17). In fact, in the extreme the volumetric differences would be accommodated sharply across the now-incoherent interface—in which case the growth stresses, in oxides or substrates, would be small.

From this discussion we conclude that the magnitude of the oxidation residual stresses, due to either growth or thermal inducement, could be on the order of the at-temperature flow stress of that material, with the senses of the stresses given by the appropriate expression in Eqs. (14) through (17). These stresses should be considered to be superimposed onto any applied stress when discussing the effects of oxidation or oxide scales on mechanical behavior of alloy substrates, when the oxide scales remain integrally attached to the substrate.[134] For example, oxide growth stress-generated creep strains of up to 2% have been observed in stainless steel wires after applied stress free exposures at around 1000°C.[134]

The question of adherence can be important to corrosion creep and stress rupture, since there will certainly be no scale effects on mechanical behavior if the scales spall off the alloy substrates. We have discussed that hot corrosion is certainly a means for oxide scale detachment, if not complete spallation. Spallation can also be caused by the thermal stresses discussed above. The various mechanisms of oxide spallation, including those due to ductility exhaustion, creep, and fatigue modes of spallation, have recently been reviewed.[135] Experimentally, except in cases when rare earth alloy additions are added to promote oxide adherence,[111] oxides are seen to spall from, for example, Ni–20Cr–4Al-type alloys after only one cycle between 300°C and ambient temperature.[135]

In addition to the major scale effects noted above, it has been proposed that the properties of the substrate material near the surface can be affected by surface changes. For instance, adsorbed gas atoms can affect the elastic modulus of a thin ($\sim 30\text{-}\alpha$) near-surface layer and the lattice parameter of this layer.[114] This type of effect has been cited to explain aqueous adsorption effects on surface weakening in selected nonmetallic solids,[136] and is illustrated quite dramatically in Fig. 11 for the reduction in elevated-temperature rupture strength of silver wires, especially the thinner wires, in various gaseous atmospheres.[137]

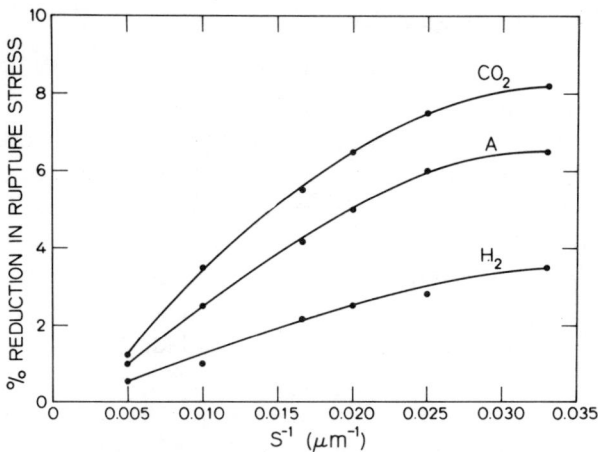

Fig. 11. Percent reduction in rupture strength of silver wires in carbon dioxide, argon, and hydrogen environments relative to vacuum at 300°C as a function of the reciprocal of specimen size, S^{-1}. Note the increased weakening for smaller specimen diameters. (After Forestier and Claus.[137])

Zone Enriched with Environmental Elements and Vacancies

Although corrosion reactions are normally confined to external alloy surfaces, particularly on the more-corrosion-resistant structural alloys, the diffusion of gaseous elements can certainly extend far below the surface to a distance on the order of $(Dt)^{1/2}$, D being the at-temperature diffusivity of the gaseous element into the alloy and t the time of environmental exposure. One must consider, then the possibility of these environmental elements affecting the mechanical behavior of the subscale regions.

Interstitial oxygen, formed at oxygen partial pressures too small to produce continuous surface oxides or internal oxidation, has been observed to improve the creep resistance of 304 stainless steel,[35] iodine–titanium, and impure silver.[32,33]

Somewhat on these same lines, it has been suggested[139,140] that the process of oxidation itself might induce solid-solution weakening by the injection of vacancies into the alloy substrate, as well as promoting the formation of deleterious voids. An example of void formation due to vacancy condensation has been observed in nickel aluminum alloys.[141] These Kirkendall vacancies and voids may very well enhance both the diffusional and dislocation aspects of creep, in a manner similar to high-density irradiation production of excess vacancies, which, in addition to being responsible for the notorious voids-related swelling effects,[142] have been observed to enhance creep rates.[143]

Internally Formed Particles

Perhaps somewhat more ostentatious and apparent in the subscale region is the reaction of the diffusing gaseous elements with the more highly oxidizable (in terms of ΔG^0) alloying elements to form internal oxide particles: an example is shown in Fig. 7. It has long been recognized[144-149] that internally formed corrosion particles can increase ambient temperature alloy hardness and tensile properties in a manner similar to the intentionally oxide dispersion-strengthened (ODS) alloys.[5,148] In addition to oxides,[150-154] resistance to creep due to internally formed corrosion-product particles has also been observed for internally precipitated carbides[34,150-152] and nitrides in alloys containing iron or chromium.[152,155-157]

Recent creep work has been performed on type 304 stainless steel that had been exposed to flowing liquid sodium at 600–700°C with only 0.4–0.8 ppm carbon, which fairly well simulates the heat transport system environment in prototype liquid metal fast breeder reactors (LMFBR).[62] This particular test environment was found to be carburizing, which not only increased the percentage of atomic carbon but also the precipitation and growth of carbides. This carburization led to improved creep resistance (i.e. lower creep rates and longer stress rupture lives), as compared to unexposed specimens, when tested in purified argon after exposure. Interestingly, it was also determined in this study that the apparent stress exponent, n, for steady-state creep increased from ~ 6 for the unexposed steel to 10.7 for the steel carburized at 600°C in the liquid sodium.

Unfortunately, whereas internally formed particles may enhance creep resistance and stress rupture lives, they correspondingly reduce stress rupture ductility.[5] For example, whereas the typical stress rupture ductility of a Ni–Cr solid solution alloy is about 45%,[27] it is only 1–9% when oxides are present as strengtheners.[5] Although the reason for this deleterious effect is not well known, it is believed to be related to enhanced work hardening and local hydrostatic tension, and hence earlier void or microcrack initiation at the oxide particle interface.[5,158]

Zones Depleted of Alloying Elements and Precipitates

The selective oxidation that occurs during scale and subscale formation can result in a subscale chemistry change which, in the extreme, can result in the dissolution of strengthening precipitates. In reducing environments, for instance, interstitial carbon may be lost through decarburization, or strengthening carbides might even be dissolved, thereby impairing the alloy creep properties,[58,103,159] as has been shown to be particularly important in decarburizing liquid sodium environments for nuclear applications.[160] The loss of near-surface precipitates is particularly noticeable in oxidizing environments in alloys where aluminum, a prime oxide former, is a major contributor to strength through the precipitation of aluminum intermetallic phases such as $Ni_3Al(\gamma')$, which is the major strengthening phase in the heat-resisting superalloys. The depletion of these γ' phases is extremely apparent during creep tests (e.g., Fig. 10a). It has also been observed,[14] that a vacuum environment at high temperatures can lead to a γ'-enriched zone (Fig. 10b) as a consequence of the evaporation of the more volatile elements such as chromium. The effect of such affected layers on mechanical behavior, however, has not been uniquely determined.

Grain Boundary Microstructural Changes

Except at very high temperatures, diffusion of gaseous species along grain boundaries can be significantly more rapid than intragranular diffusion.[161] Hence, the occurrence of the aforementioned solid-solution strengthening of alloys may occur farther into the alloy at grain boundary regions than elsewhere. Grain boundary solid-solution strengthening due to such diffusion has in fact been observed in internal friction experiments[162] and in impact tests.[153,163] In the latter tests,[163] a $(Dt)^{1/2}$ dependence of the diffusional strengthening was confirmed, as well as the loss of potency of grain boundaries as favored diffusional paths at very high temperatures.

Because the high-energy grain boundaries are preferential nucleation sites for internal oxidation and precipitation, one would expect a higher volume of area fraction of internally formed oxides, carbides, nitrides, etc., on boundaries than in the matrix, resulting in strengthening and enhanced resistance to grain boundary sliding creep.[5,18-21,140] An experimental study on impure silver supports this hypothesis[32,33]; it also emphasizes that this increased resistance to grain boundary sliding is bought at the price of decreased stress rupture ductilities because grain boundary precipitates can and do facilitate the formation of deleterious grain boundary voids,[33,55,164,165] and subsequent crack initiation, leading eventually to stress rupture.[140]

Many workers have observed complete oxide networks along grain boundaries in alloys tested in high temperatures in highly oxidizing conditions.[29,30,154] The previous analysis of grain boundary strengthening is no longer valid in this case, in which the grain boundaries are replaced by an oxide phase. Whether this replacement strengthens or weakens the alloy under creep conditions is open to question, since both effects are found.[29,30,154]

Finally, it is well established that grain boundaries can become particularly susceptible to decohesion when a liquid phase appears along the grain boundary—as in liquid metal embrittlement.[89,166] This effect could be particularly noticeable in hot corrosive environments in which low-melting-point eutectics are not uncommon (e.g., Ni_3S_2–Ni and CrS–Cr[91]). Further, it is known that adsorbed gases on boundaries which can, for example, diffuse preferentially down grain boundaries at lower temperatures where corrosion products are not expected to form can decrease surface energy and, presumably, the cohesive force across a boundary.[167,168] This effect may enhance grain boundary sliding and cracking.

DISCUSSION

Environmental Effects on Creep Rate

As might be readily discerned from the preceding sections, complete parametric analyses of corrosion creep phenomena are few. However, these few investigations, as well as comparison of results from environment related creep tests performed at different laboratories, might suggest some consistent corrosion creep trends related to environment, alloy systems, and microstructure. These trends are most clearly discussed in terms of the correlation between corrosion-affected microstructure and mechanical properties just documented. These effects can be summarized with respect to creep resistance in the form shown in Table 5. In principle, all these effects should be considered in the rationalization of creep behavior of a system in different environments, say, air vis-à-vis vacuum. However, depending on the system, it is conceivable that some effects may be less significant than others. For example, with reference to the investigations on Udimet-700,[14,18–21] in the coarser grained superalloy which exhibits Type I behavior, one can assume that the grain boundary effects are less important than the other effects. This, of course, is not true when one is considering Type II behavior where the systems can be fine grained.

Again, Type I behavior is typified by air strengthening. More specifically, for coarse grained nickel-base superalloy it appears that the environment affected creep rate mainly through the stress-dependence term and not so much through the temperature-dependence or activation-energy term. That environment did not affect the temperature dependence of creep is not surprising, since one would not expect a surface environment to result in a drastic change of the thermally activated processes of dislocation motion, which are usually rate-controlled by vacancy self-diffusion. That environment did affect the stress-dependent term is interesting. Although the precise meaning of n is presently not understood, particularly with reference to complex particle-strengthened alloy systems, many phenomenological theories for steady-state creep attribute the n to the combined stress sensitivity of mobile dislocation density, ρ_m, and dislocation mobility, M, such that

$$\rho_m M \propto [\sigma_A/E(T)]^n \tag{18}$$

Although there is as yet no first-principle understanding of ρM as it applies to creep, a lower n can certainly imply that dislocation generation and/or the stress-dependent aspect of dislocation mobility is inhibited.

This inhibition is perhaps more clearly understood in terms of internal back stress [Eq. (2)]. If we assume that the same dislocation creep mechanisms operate in both the air and vacuum systems, then by setting $n' = 4$, as

Table 5. Summary of Expected Corrosion Effects on Creep Resistance

Enhancing creep resistance $(-\sigma_C)$	Reducing creep resistance $(+\sigma_C)$
Adherent oxide scale	Removal of oxide scales
Residual stresses[a] due to up-quencing	Residual stresses[a] due to down-quenching and Oxide growth stresses[a]
Influx of gaseous interstitials into the substrate solid solution	Outflux of selected solid solution elements due to selective oxidation or evaporation
	Influx of Kirkendall vacancies with concomitant void formation due to oxidation process
Internally precipitated oxides or other corrosion products	Dissolution of strengthening precipitates due to selective oxidation or evaporation
Reduced grain boundary (sliding) creep due to oxides at boundaries	Enhanced grain boundary (sliding) creep due to oxygen at boundaries

[a] Exceptions noted in the section, "Oxidation- and Corrosion-Induced Microstructural and Chemical Effects and Mechanical Behavior."

predicted by recovery creep theories,[13] for all tests, σ_i values as listed in Table 2 are obtained. Apparently, oxidation in air increases this internal stress (e.g., at 760°C, the internal stress in air is 245 MN/m^2 while that in vacuum is 117 MN/m^2). By comparing these internal stresses, one can speculate that the average internal stress due to the surface oxide film is approximately 128 MN/m^2 for the polycrystalline alloy studied, which seems to indicate that the deformation of the region near the oxide–alloy interface could be very important in the creep tests in air for the grain size (300 μm) tested.

As suggested in the previous section, strengthening might also arise from internal effects, particularly dispersoid-type strengthening due to internal oxide (carbide, nitride, etc.) precipitates. Consistent with surface-film-strengthening mechanisms, the study[62] on the creep strengthening of stainless steel due to internal carburization found that n increased from ~ 6 for the uncarburized steel up to 10.7 for the carburized steel. Here again, the increase in n probably reflects a greater internal opposition stress, which has also been found in superalloys mechanically alloyed with oxide dispersions.[13]

One way of including such effects in the creep equation [Eq. (2)] would be to add an environmental dependent back stress term σ_c:

$$\dot{\varepsilon}_s = A''[(\sigma_A - \sigma_i \pm \sigma_c)/E(T)]^{n'}\exp(-Q_c^*/RT) \qquad (19)$$

σ_c will then be the algebraic sum of the different surface back-stress terms when more than one contribution is expected. A plus sign would imply environmental weakening; a minus sign is appropriate in the case of environmental strengthening.

Many of these σ_c terms have yet to be defined quantitatively, or even qualitatively. Nevertheless, we postulate that some effects should become predominant with various alloy systems, microstructures, environments, and states of stress. For example, in applying this particular means of rationalization to the major observed Type I behavior, namely, those studies involving the fairly coarse grained nickel-base[14,18-21] superalloy, we make the following observations with respect to the effects in Table 5. In such a strengthened system as this alloy (1033 MN/m^2 tensile strength even at 760°C[169]), we do not expect any of the solid solution effects to significantly affect the internal stress of the system; and, as mentioned, we choose to neglect the grain boundary effects. Instead, it is presumed that the major effect is expected to be associated with the σ_c's of the type defined in Eq. (19); i.e., air strengthening of this system is through opposition to dislocation generation and movement by the adherent oxide scale, which develops during creep in air and not during vacuum creep (Fig. 10a and b) or creep in hot salt environments.[14] The photomicrographs in Fig. 10, for example, further show that after either air or vacuum creep, the surface substrate layers are progressively becoming single-phase—γ in the case of air, probably through the selective oxidation of aluminum and titanium, and γ' in the case of vacuum, as a consequence of chromium evaporation. In either case neither of the surface layers are precipitation-strengthened; that is, these effects will tend to be self-cancelling when one is qualitatively trying, as we are doing now, to rationalize the vis-à-vis creep behavior of the system in air and in vacuum.

This is not to say that all environments other than vacuum will behave like air. As mentioned, helium environments containing trace impurities might also induce Type I behavior, because the oxygen content of this environment is often very small, or even immeasurable.[104] Air strengthening might also be eliminated through degradation of adherent oxide scales by NaCl environments[40] or hot corrosive ash deposits.[14,43] But in considering such environments, we must be careful not to confuse surface effects with any internal or otherwise localized effects that may be operative simultaneously. It has been suggested that the creep weakening often observed in helium test environments are, in fact, related to solid solution effects,[58] or the prevalent formation of internal corrosion products[104] that weaken the grain boundaries. Certainly, in hot corrosive environments, loss of creep resistance may be directly related to the localized nature of the corrosive attack as well as the subsurface penetration, particularly along grain boundaries, of molten salts, as depicted in Fig. 12. Support for this assumption might be found in work on the creep weakening of Inconel-700 in vanadium ash, which was found to be more significant when the ash contained a low-melting nickel/nickel sulfide eutectic.[43] Also, comparison of Figs. 3 and 4 reveal that hot corrosive creep weakening due to sodium sulfate deposits was found to be effective only above the melting point of the deposit.[14]

Again, Type II creep behavior is exemplified by air weakening and, as discussed, it was seen in the nickel-base superalloy with sufficiently fine grain size,[18-21] becomes and, as mentioned, in other nickel-base alloys when the test temperature was low or the stress was high.[23-30]

A major unknown quantity in the creep and stress rupture behavior of the usual polycrystalline structures, even in cases where the environment is not a variable, is the interplay between intragranular or dislocation creep and such grain-boundary-related creep modes as grain boundary sliding and diffusional creep. Such interplay, which is expected to include mutual accommodation processes,[170,171] will, of course, be grain size dependent. It is no surprise, then, that one of the major observations in corrosion creep and stress rupture is the

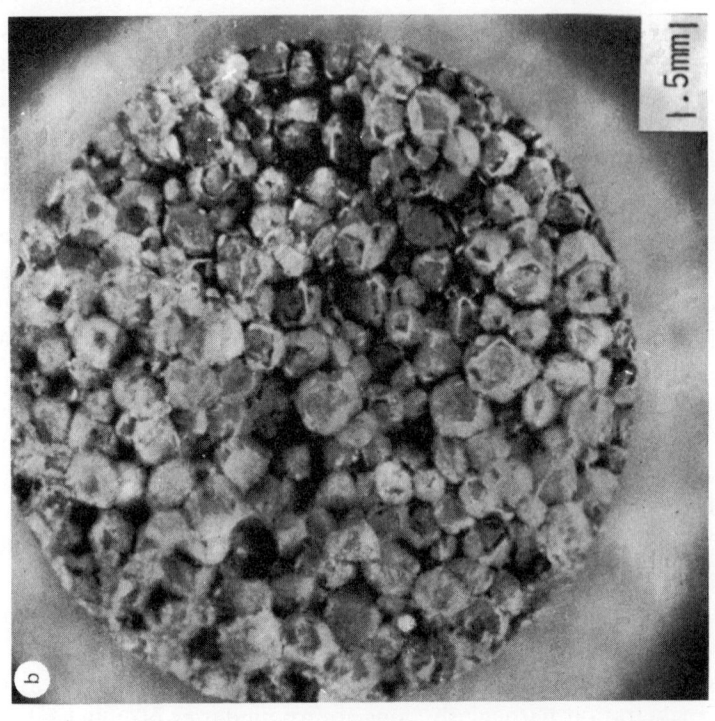

Fig. 12. Example of a thoroughly corroded nickel-base superalloy creep specimen tested at 870 C with Na_2SO_4 deposition in air: (a) side view; (b) top view of fracture surface. Note extreme intergranular mode of attack and fracture. (After Aning.[14])

grain size-dependent transition between Type I and Type II behavior. This becomes clearer if we again refer to Table 5. In the same material system, air strengthening of intragranular surfaces might be increasingly opposed by air weakening of surface-connected grain boundaries as grain size becomes finer, and intergranular (grain boundary sliding) creep becomes the dominant creep mode, leaving the grain boundary effects in Table 5 as the important ones. Further, as discussed, oxidation or air penetration along the boundaries could enhance grain boundary sliding through, say, a reduction of the cohesive forces.[29,30,55] The first premise is reasonable and has been seen for single-phase systems.[170]

The validity of the second premise, that air environments can enhance grain boundary sliding, is supported by recent work on the air versus vacuum creep behavior at 760°C of a high volume fraction γ' strengthened nickel-base superalloy for which both grain size and specimen size were independently varied.[172] For that system, in which grain boundaries were kept fairly free of strengthening carbides, air weakening was observed. Expectedly, in that study, specimens with coarse grains (275 μm) were found to be more creep resistant in air than were specimens with fine grains (100 μm). In contrast, for tests in vacuum, creep was found to be fairly insensitive to grain size. This is indeed consistent with the rationale of enhanced grain boundary sliding due to air penetration, which is also supported by grain boundary offset observations of larger grain boundary sliding contributions to total creep strain in air compared to vacuum. Interestingly, for specimens of the same alloy system aged to precipitate carbides along the grain boundaries, the air weakening behavior was no longer observed; instead, air strengthening occurred. These results can be rationalized based on the surface oxide scale strengthening which is expected to be effective when grain boundaries are constrained against sliding, as is the case with grain boundaries pinned by carbides or, as discussed, when grains are very coarse or when grain boundaries are altogether eliminated, as in single crystals.

As both grain size and specimen size were independently varied in that study,[172] the number of grains per cross-section parameter was also investigated, based on numerous grain size/specimen size combinations. No clear correlation between creep rate and the number of grains per cross section was noted in either air or vacuum. However, for nearly all of the grain size/environment combinations, thick specimens were observed to be more creep resistant than were thin specimens. For creep in air, this phenomenon can be attributed to the greater proportion of specimen grains, and hence grain boundary channels for air penetration, at the surface for thin specimens. This effect is in direct competition with surface oxide scale strengthening, which leads to greater creep resistance for thinner specimens.[115] On the other hand, for creep in vacuum, that thick specimens were more creep resistant than thin specimens is consistent with the existence of an intrinsically weak surface layer. That is, in vacuum (10^{-6} torr) environments, external specimen or component surfaces, with perhaps adsorbed gases but without oxide scales, can be inherently less creep resisting than the bulk material, perhaps because of the lack of geometric constraints at the alloy surface.

Many questions remain, even after these studies. Although air environments appear to enhance grain boundary sliding, the nature of the gaseous reaction along the boundaries, and the mechanism of enhanced sliding, remain unknown. Also, surface oxide scale strengthening and its profile away from the oxide–alloy interface have yet to be modeled in terms of oxide scale and alloy thicknesses, and oxide integrity and adherency. Furthermore, if one believes that the physical presence of an oxide scale can cause strengthening of surface grains, one must also consider the state of long-range growth and thermal stresses that this adherent scale would produce in the underlying substrate and the effects of these stresses on creep and stress rupture (see Table 5). If one considers the ideal situation of a shear stress at the alloy–oxide surface being transmitted to the alloy as a normal compressive or tensile stress, elementary mechanics predicts an inverse relationship between creep rate and specimen diameter. This oxide stress effect can also either assist or compete with the other surface effects. Certainly selected up-quench and down-quench thermal treatments of creep specimens prior to creep can be a means of determining the effects of thermal stresses on the creep and stress rupture of oxide-covered systems. Also, transmission electron microscopy can help in determining whether there are long-range stresses, and whether they are of sufficient magnitude to cause significant dislocation activity.

The often-observed temperature- and stress-dependent transitions between Types I and II behaviors are less easily rationalized, and are most probably related to numerous variable parameters pertaining to oxide type and morphology, creep mode, and alloy system. For instance, at higher temperatures, especially in the high-chromium or -aluminum oxidation-resistant alloys, one would expect thicker oxide scales, and hence air strengthening. In fact, this temperature transition has been suggested to reflect a switchover from oxide scale strengthening at higher temperatures, to gaseous adsorption weakening, particularly at crack tips, at lower temperatures.[23-27] However, a change in temperature might influence environmental creep indirectly by altering the dominant creep mode.[1-6] This is also likely with changes in applied stress.[1-6] In fact, very high stress states might preclude the existence of steady-state creep so that we are, in essence, observing environmental influences on tertiary creep or stress rupture behavior instead. It is unfortunate, then, that most of the environmental creep studies reported in the literature, and for that matter most engineering creep studies,[1-6] do not include continuous strain monitoring in the determination of stress rupture lives.

Environmental Effects on Stress Rupture

In all the complete corrosion creep studies discussed in this chapter, however, environmentally reduced steady-state creep rates were accompanied by longer stress rupture lives, whereas shorter lives resulted from higher creep rates. Regardless of the mode of failure, then, an inverse creep rate/rupture life relationship, as described in Eq. (3), apparently remains valid even when environmental effects are considered.

Stress rupture ductility, however, does not seem to follow any clear pattern; both increased and decreased ductility have been observed for both air strengthening and air weakening. However, through comparison of some of the more complete corrosion creep studies, a metallographic hint of the predominant modes of stress rupture in various environments is obtained.

For example, in the Type I creep behavior of the coarse grained superalloy, tertiary creep and certainly stress rupture occurred in air as a consequence of the propagation of one or two of the externally initiated cracks (Fig. 13a), whereas in vacuum (Fig. 13b), stress rupture occurred as a consequence of the linking of the many triple-point voids that formed in the interior of the specimens. The air cracks initiated at grain boundary/surface intersections, where the precipitates had been depleted due to oxidation, and then propagated along the grain boundaries. Another interesting feature of the tests performed in air is that the tertiary stage was characterized by steps (Figs. 3 and 4). These seem to be due to the fact that cracks get initiated easily at the surface-connected boundaries leading to the steep portion of the step, but get blunted and probably stopped by the oxidized and γ'-denued layer in front of the crack, as shown in Fig. 14, leading to the relatively flatter portion of the step. This "stop-and-go" propagation of the cracks tends to prolong the tertiary life of the specimen. This effect is also apparently dynamic, for preoxidized specimens creep-tested in vacuum did not show this "stop-and-go" behavior, although creep rate was reduced by the oxide scale. On the other hand, in vacuum-tested samples, internal triple-point cracks initiated and propagated much earlier during the

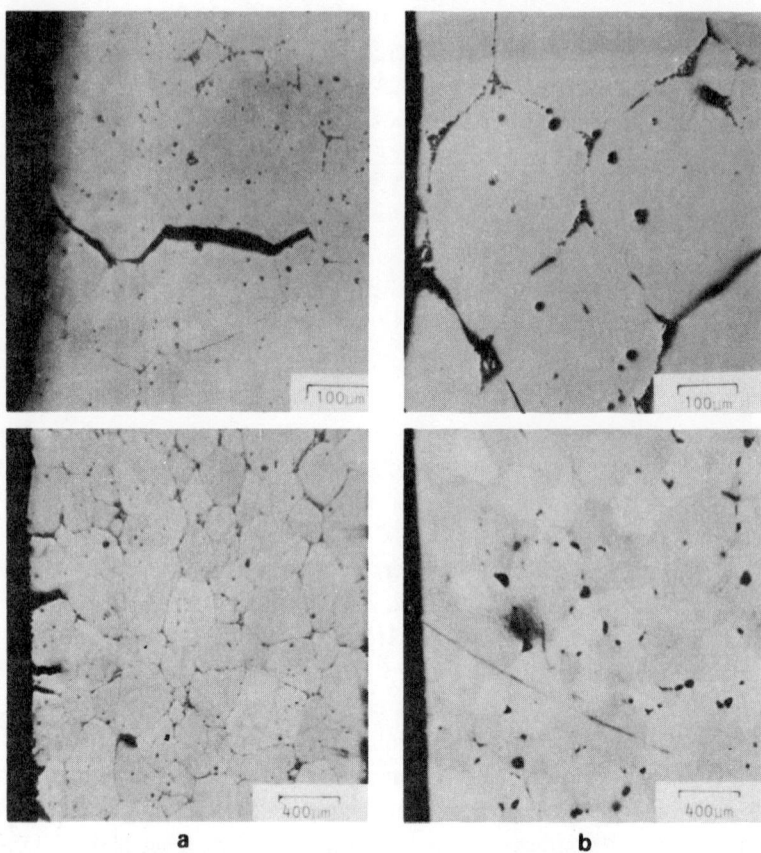

Fig. 13. Micrographic comparison of stress rupture modes of coarse grained nickel-base superalloy in (a) air and (b) vacuum (10^{-5} torr) after stress rupture at 982°C, 15.6 ksi, and 315 hr, showing predominance of surface cracking in air and internal triple-point cracking in vacuum. (After Aning.[14])

steady-state creep region as compared to air tests (Figs. 3 and 4). These sharp and unobstructed cracks propagated much faster, resulting in much shorter tertiary creep times and ductilities.[14]

In tests on finer grained superalloy ($<250\,\mu$), essentially the same metallographic results were obtained: a greater number of surface-intersecting grain boundary cracks in air, and a greater number of internal triple-point voids in vacuum.[18] For this finer grained material, however, the greater number of surface cracks in air, as opposed to the coarse grained superalloy, led to "premature" stress rupture and shorter rupture lives in air.[18]

It is worth mentioning here that highly oxidizing environments such as air[55] and pure oxygen[32] have also been noted to increase internal cavitation in

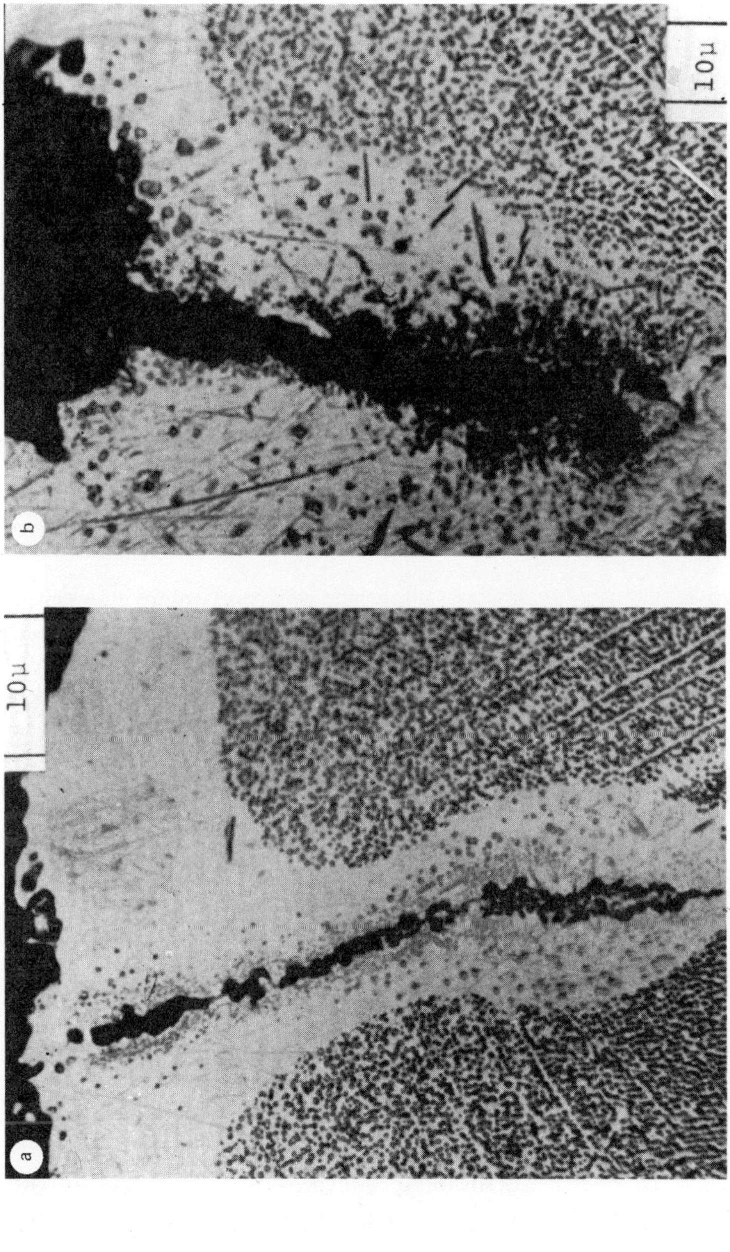

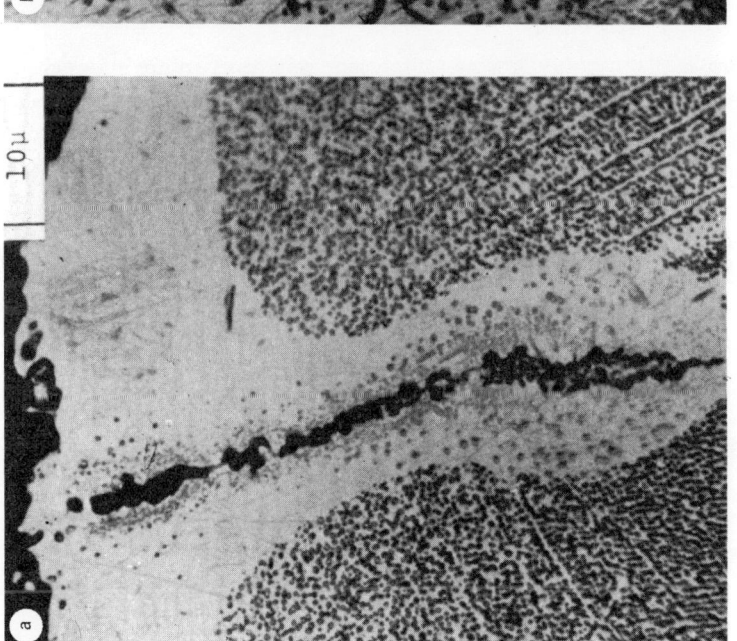

Fig. 14. Micrographs of air-crept and stress-ruptured superalloy at 982°C, 15.6 ksi, and 315 hr, showing (a) oxidized and denuded surface along grain boundary, and (b) region at tip of blunted intergranular crack. (After Aning.[14])

creep-tested samples as opposed to samples tested in less oxidizing environments. Whether or not this is a result of increased grain boundary sliding in oxidizing environments can only be confirmed by direct comparative observation of such sliding in different environments.

It has also been repeatedly suggested[18,45,55] that such environmentally affected void formation and crack growth determine not only tertiary creep and stress rupture behavior, but are also responsible for environmental effects on steady-state creep rates. Although this hypothesis seems reasonable, metallurgical investigations have only recently been focused on the problem of hot crack growth in creep situations,[11] and existing steady-state creep theories have not treated to any extent contributions of cracking and void growth to steady-state creep.[1–6]

Conclusive analysis of when and where this surface and internal cracking is important in corrosion creep must await more in-depth and systematic analysis of all aspects of the creep and stress rupture. Until then, by again borrowing information from areas other than stress rupture and by noting observed differences in post-stress rupture microstructures in corrosion creep tests, we can only speculate as to environmental influences on hot creep cracking.

The impact of environments on hot fracture, or stress rupture in our case, can perhaps be most clearly discussed in terms of crack initiation and then growth. As a general rule one cannot presume that crack initiation is always enhanced by the corrosion-produced heterogeneities. Although premature initiation in oxidizing gases is often observed (e.g., Refs. 18–21 and 173), oxidizing environments also appear capable of delaying surface cracking (e.g., Refs. 25, 29, 61). This beneficial effect apparently occurs when corrosion products can smooth over potential surface inclusions that are more potent stress raisers than the corrosion products themselves. The potency of the general corrosion product phases in initiating cracks depends on the brittleness of the phases,[116] the precipitation stresses resulting from the products,[102] and the morphology of the product phases.[140] The morphological aspect is especially important when grain boundaries are preferentially oxidized, resulting in long wedgelike oxides.[181–21,103]

Corrosion products can also play a dualistic role in crack growth, depending on many factors. A dispersion of oxide particles can initiate cracks ahead of a propagating crack tip,[5] but can also serve to blunt the propagating crack.[25–29,152,155] Corrosion products formed internally[22] or formed at crack tips[27,29,155] can produce similar phenomena. Possibly, the stress concentration ahead of the crack can also be reduced by the corrosion product.[27] Such inhibition of crack growth is believed to be responsible for increasing the rupture time, t_r, when the creep rate, $\dot{\varepsilon}_s$, remains unaffected.[152] The meeting of oxides growing from opposing crack surfaces, resulting in the healing of cracks, is believed to be responsible for the sometimes observed pseudo fourth- and fifth-stage creep behavior, which can be viewed as a repetition of steady-state and tertiary creep.[22,30] In fact, this dynamic internal oxidation due to repeated cracking and oxide healing, if it leads to a strengthening network of oxides, can potentially render an alloy more creep-resistant.[27-29,157]

As in the case of grain boundary decohesion, one must also consider the diminutive effect of adsorbed gases with minimal formation of corrosion products, such as at low temperatures or in more inert atmospheres, to lower the alloy surface energy. This in turn could modify the relation for crack growth and facilitate such a phenomenon.[25,26,57,112,157,174-176]

On the other hand, in cases of faster crack growth due to oxidation, it has been suggested[18-21,173,177] that the stimulative effect of oxidation on surface cracking and crack propagation is similar to certain models for stress-corrosion cracking (SCC), such as the oxide wedging mechanism[102] and SCC mechanisms of surface film rupture and repassivation.[101,178-182] The stimulative aspects of both these mechanisms is based on stress-accelerated corrosion by way of repetitive film rupture followed by more corrosion. Indeed, accelerated creep rates observed in sodium chloride environments have been attributed to either this type of stimulative effect[183] or to the overall corrosion caused by the NaCl corrosion,[40] or to both processes occurring simultaneously.[184] In either case one would expect reductions of ductility, which is observed.[40]

Although obvious, we point out that early fracture can certainly occur when the load-carrying capacity of the structure is greatly reduced because of large-scale replacement of alloy cross section with that of brittle, and at times porous, oxides or other corrosion products. This usually occurs at very high temperatures and in the more aggressive environments where corrosion is rampant, and progresses deep into the substrate if not through the cross sections. In these cases, such oxide healing effects as bridging[29,30,103] are, of course, ineffectual because of the lack of neighboring metallic regions to be bridged. Examples of this severe behavior are plentiful,[40,103,185] and its occurrence in practice is not courted intentionally.

The preceding discussions on creep and stress rupture in different environments emphatically warn of the necessity to incorporate environmental factors into engineering design. Particularly in the more recent high-temperature alloy environments such as found in the obviously pugnacious coal gasification units, or in the unexpectedly deleterious "inert" environments of helium-cooled nuclear reactors, or even deep-space exploration apparatus, one simply cannot design creep strength according to data obtained in air tests.

This becomes dramatically evident if one considers the perhaps most widely utilized creep rupture life extrapolation, namely, the Larson–Miller relationship,[12] which states that

$$\log \sigma_A = f\left[T(C + \log t_r)\right] \qquad (20)$$

where C is constant, 20. Recently, this parameter C has been related to the creep parameters determined in matrix studies[186]:

$$C = \log\left\{\left[A \exp\left(-Q_c^*/RT\right)\right]^{\beta}/B\right\} \qquad (21)$$

For the 300-μm-grain-sized Udimet-700 tested in air at 760°C and 982°C, the derived values of C are 23 and 18, respectively, which are reasonably close to the expected value of 20.[14] In vacuum (10^{-5} torr) tests on the same alloy, the Larson–Miller parameters are 41 and 33, respectively. Since environmental conditions do affect creep behavior, crack initiation, and growth differently, there is no reason, *a priori*, to expect the parameter in these stress rupture empirical relations to be constant under all conditions, and obviously they are not.

CONCLUDING REMARKS

In this chapter we have reviewed the present state-of-the-art understanding of corrosion creep and stress rupture. A major portion of this generalized understanding is based on extrapolations from numerous corrosion creep studies lacking in parametric analysis, and even from related fields such as ambient temperature tensile behavior of plated metals or corrosion fatigue. Predominant among these extrapolations are the strengthening effects of surface oxide scales and the degradation of creep and rupture properties in hot corrosive environments due to surface scale disruption and chemical attack of the underlying alloy itself. Although internal oxidation effects also seem to universally strengthen alloys, the effects of corrosion on grain boundaries and their sliding still remain inconclusive. Corrosion effects on stress rupture and hot crack growth in oxidative environments are even less well understood at present, and can only be understood in terms of competing phenomena such as oxide weding, crack blunting, and gaseous adsorption effects. Indeed, grain size, temperature, and stress transitions in corrosion creep behavior suggest that this entire hyphenated field can only be understood in terms of competing phenomena, such as outlined in Table 5.

The need for complete parametric studies in corrosion creep and stress rupture aimed at determining the roles of these competing effects has been emphatically pronounced. In modern metallurgy, the gathering of essential microstructural data invariably lags behind the accumulation of mechanical test data. This chapter certainly indicates that corrosion creep and stress rupture might be regarded as an exception to the rule. Indeed, high-temperature corrosion has been amply studied under the microscope to discern where and how it occurs in alloys. What is needed now is a correlation of these environmental factors to creep and stress rupture properties in more systematic studies, such as those that have been suggested in this chapter.

ACKNOWLEDGEMENT

The authors are grateful to the National Science Foundation for supporting portions of this work through grants NSF-DMR75-09878 and NSF-DMR77-11281.

REFERENCES

1. J. K. Tien, M. K. Malu, and S. Purushothaman, in *Alloy and Microstructural Design* (J. K. Tien and G. S. Ansell, eds.), Academic Press, New York (1976).
2. D. McLean, *Mechanical Properties of Metals*, New York (1965).
3. F. Garofalo, *Fundamentals of Creep and Creep Rupture in Metals*, Macmillan, New York (1965).
4. J. C. Freche, in *Alloy and Microstructural Design* (J. K. Tien and G. S. Ansell, eds.), Academic Press, New York (1976).
5. J. K. Tien, *Alloy Design with Oxide Dispersoids and Precipitates*, in Proceedings of the Battelle Colloquium on Fundamental Aspects of Structural Alloy Design, Paper III-2, Harrison Hot Springs, Wash, and Vancouver, B.C., September, 1975 (R. I. Jaffee and E. A. Wilcox, eds.), Plenum Press, New York (1975).
6. *Creep and Recovery*, ASM, Cleveland, Ohio (1957).
7. J. M. Davidson and J. K. Tien, in *Properties of High Temperature Alloys* (Z. A. Foroulis and F. S. Pettit, eds.), The Electrochemical Society, Princeton, N.J. (1977).
8. A. R. C. Westwood and N. S. Stoloff, eds., *Environment Sensitive Mechanical Behavior*, Gordon and Breach, New York (1966).
9. *Stress Corrosion Cracking of Metals—A State of the Art*, ASTM STP 518, ASTM, Philadelphia (1972).
10. C. H. Wells and C. P. Sullivan, in *Fatigue at High Temperature*, ASTM STP 459, pp. 59–74, ASTM, Philadelphia (1969).
11. S. Purushothaman and J. K. Tien, A Theory for Creep Crack Growth, *Scripta Met.* **10**, 663–666 (1976).

12. F. R. Larson and J. Miller, A Time Temperature Relationship for Rupture and Creep Stresses, *Trans. ASME* **74**, 765 775 (1952).
13. S. Purushothaman and J. K. Tien, Role of Back Stress in the Creep Behavior of Particle Strengthened Alloys, *Acta Met.* **26**, 519 528 (1978).
14. K. Aning, Engineering Science Doctoral thesis, Columbia University, New York (1976).
15. M. J. Donachie, Jr., and R. G. Shepheard, Creep-Rupture Behavior of Hastelloy-X in Vacuum, *Mater. Res. Stand.* **1964**, 495–498.
16. M. J. Donachie, Jr., and R. G. Shepheard, Creep-Rupture Behavior of Hastelloy-X, *Proc. ASTM* **62**, 981–987 (1962).
17. D. J. Duquette, The Effect of High Vacuum on the Creep Properties of a High Strength Nickel Alloy Single Crystal, *Scripta Met.* **4**, 633–636 (1970).
18. P. N. Chaku and C. J. McMahon, Jr., The Effect of an Air Environment on the Creep and Rupture Behavior of a Nickel-Base High Temperature Alloy, *Met. Trans.* **5**, 441–450 (1974).
19. P. N. Chaku, The Effect of Air Environment on the Creep and Stress Rupture Properties of Udimet-700, Master's thesis, University of Pennysylvania, Philadelphia (1972).
20. M. L. Sessions, The Effect of an Air Environment on the Stress Rupture Behavior of Cast Udimet-700, Master's thesis, University of Pennsylvania, Philadelphia (1974).
21. M. L. Sessions, M. J. McMahon, Jr., and J. L. Walker, Further Observations on the Effect of Environment on the Creep/Rupture Behavior of a Nickel-Base High Temperature Alloy: Grain Size Effects, *Mater. Sci. Eng.* **27**, 17–24 (1977).
22. R. Widmer and N. J. Grant, The Role of Atmosphere in the Creep-Rupture Properties of 80 Ni–20 Cr Alloys, *Trans. ASME* **D82**, 882–886 (1960).
23. P. Shahinian and M. R. Achter, in *High Temperature Materials* (R. F. Hehemann and G. M. Ault, eds.), pp. 448–465, Wiley, New York (1957).
24. P. Shahinian, Effect of Environment on Creep-Rupture Properties of Some Commercial Alloys, *Trans. ASM* **49**, 862–882 (1957).
25. P. Shahinian and M. R. Achter, The Effect of Atmosphere on Creep-Rupture Properties of a Nickel–Chromium–Aluminum Alloy, *Proc. ASTM* **58**, 761–772 (1958).
26. P. Shahinian and M. R. Achter, A Comparison of the Creep-Rupture Properties of Nickel in Air and in Vacuum, *Trans. AIME-TMS* **215**, 37–41 (1959).
27. P. Shahinian and M. R. Achter, Temperature and Stress Dependence of the Atmosphere's Effect on a Nickel–Chromium Alloy, *Trans. ASM* **51**, 244–255 (1959).
28. P. Shahinian, Creep-Rupture Behavior of Unnotched and Notched Nickel-Base Alloys in Air and in Vacuum, *Trans. ASME* **D87**, 344–350 (1965).
29. T. C. Reuther, P. Shahinian, and M. R. Achter, Anomolous Fracture in the Creep of Nickel, *Proc. ASTM* **61**, 956–969 (1961).
30. T. R. Cass and M. R. Achter, Oxide Bonding and the Creep-Rupture Strength of Nickel, *Trans. AIME-TMS* **224**, 1115–1119 (1962).
31. H. G. Nelson and D. P. Williams, The Effect of Vacuum on Various Mechanical Properties of Magnesium, *Proc. Soc. Aer. Proc. Eng.* **11**, 291–297 (1967).
32. C. E. Price, Some Oxygen Effects in Silver, *Acta Met* **11**, 160 (1963).
33. C. E. Price, On the Creep Behavior of Silver, *Acta Met.* **14**, 1781–1799 (1966).
34. H. E. McCoy, The Influence of Various Gaseous Environments on the Creep Rupture Properties of Nuclear Materials Selected for High Temperature Service, *Corros. React. Mater.* **1**, (Proc. IAEA Conf., Salzburg, 1962), 263–292 (1962).
35. H. E. McCoy, W. R. Martin, and J. R. Weir, Effect of Environment on the Mechanical Properties of Metals, *Proc. Inst. Environ. Sci.* **1961**, 163–176.
36. C. Tyzack, D. S. Wood, B. J. L. Darlaston, H. C. Gowen, J. Rands, and N. Grandison, The Behavior of Construction Materials in Mark III Helium Cooled Reactors, Paper 15, in Component Design in High Temperature Reactors Using Helium as a Coolant, pp. 1–29, Inst. Mech. Eng. Symposium, London (May 1972).

37. D. S. Wood, M. Farrow, and W. T. Burke, A Preliminary Study of the Effect of Helium Environment on the Creep and Rupture Behavior of Type 316 Stainless Steel and Incoloy 800 (Proc. Int. Conf. Corros., BNES), pp. 213–228. Clowers, London (1971).

38. J. Board, The Effect of a Helium Environment on the High Temperature Properties of Structural Materials, BNES **9**, 101–112 (1970).

39. R. W. Buckman, Jr., Operation of Ultra High Vacuum Creep Testing Laboratory, in *Transactions Vacuum Metallurgy Conference 1966* (M. A. Orehoski and R. F. Bunshah, eds.), pp. 25–37, American Vacuum Society, Boston (1966).

40. N. E. Paton, W. M. Robertson, and F. Mansfeld, High Temperature Behavior of Superalloys Exposed to Sodium Chloride: 1. Mechanical Properties, *Met. Trans.* **4**, 317–320 (1973).

41. G. B. Wilkes, Jr., in *Symposium on Corrosion of Materials at Elevated Temperatures*, ASTM STP 108, pp. 11–25, ASTM, Philadelphia (1950).

42. C. T. Evans, Jr., in *Symposium on Corrosion of Materials at Elevated Temperatures*, ASTM STP 108, pp. 59–113, ASTM, Philadelphia (1950).

43. D. Sunamoto and T. Nishida, The Effect of Corrosive Environment due to High Temperature on the Stress Rupture Strength and Fatigue Strength of Heat Resisting Alloy Inconel 700, *J. Mater. Sci. Japan* **20**, 381–386 (1971).

44. A. M. Talbot and E. N. Skinner, in *Symposium on Corrosion of Materials at Elevated Temperatures*, ASTM STP 108, pp. 42–49, ASTM, Philadelphia (1950).

45. P. Rodriguez, Effect of Oxygen Environment on the Creep Behavior of 304 Stainless Steel, *Trans. Indian Inst. Met.* **20**, 213–219 (1967).

46. H. T. McHenry and H. B. Probst, in *High Temperature Materials* (R. F. Hehemann and G. M. Ault, eds.), pp. 466–485, Wiley, New York (1957).

47. R. H. Atkinson and D. E. Furman, Creep Characteristics of Some Platinum Metals at 1382°F, *Trans. AIME-TMS* **194**, 530–531 (1952).

48. O. C. Shepard and W. Schalliol, in *Symposium on Corrosion of Materials at Elevated Temperatures*, ASTM STP 108, pp. 34–41, ASTM, Philadelphia (1950).

49. M. R. Pickus and E. R. Parker, Creep Behavior of Zinc Modified by Copper in the Surface Layer, *Trans. AIME-TMS* **191**, 792–796 (1951).

50. J. R. Cuthill and W. N. Harrison, Effect of Ceramic Coatings on the Creep Rate of Metallic Single-Crystal and Polycrystalline Specimens, *WADC Tech. Rept. Nat. Bur. Stand.* (April 1956).

51. M. Metzger and T. A. Read, The Effect of Surface Films on the Creep of Cadmium Crystals, *Trans. AIME-TMS* **212**(2), 236–243 (1958).

52. O. Parr, J. Moreau, and J. Manenc, Quelques Observations Relatives à l'Influence de l'Oxydation sur le Fluage d'un Acier Doux, *Mem. Sci. Rev. Met.* **59**(10), 629–642 (1962).

53. B. W. Lee, Effect of Helium on High Temperature Properties of $2\frac{1}{4}$Cr–1Mo and Type 316 Stainless Steels, *Trans. Iron Steel Inst. Japan.* **10**, 325–331 (1970).

54. L. May, K. J. Truss, and P. S. Sethi, The Effects of Some Gaseous Environments on the Creep of a Stainless Steel, *Trans. ASME* **D91**, 575–580 (1969).

55. E. C. Scaife and P. L. James, Influence of Environment on Intergranular Cavitation, *Met. Sci. J.* **2**, 217–220 (1968).

56. L. H. Kirschler and R. C. Andrews, A Limited Comparison of the Mechanical Strength of Austenitic Steel in 1200°F Sodium, Air, and Helium, *Trans. ASME* **D91**, 785–791 (1969).

57. R. L. Stegman, P. Shahinian, and M. R. Achter, The Weakening Effect of Oxygen on Nickel in Creep Rupture, *Trans. AIME-TMS* **245**, 1759–1763 (1969).

58. Y. Hosoi and S. Abe, The Effect of Helium Environment on the Creep Rupture Properties of Inconel 617 at 1000°C, *Met. Trans.* **6A**, 1171–1178 (1975).

59. J. T. Agnew, G. A. Hawkins, and H. T. Solberg, Stress-Rupture Characteristics of Various Steels in Steam at 1200°F, *Trans. ASME* **68**, 309–314 (1946).

60. J. M. Francis and K. E. Hodgson, Interaction of High Temperature Oxidation and Creep Processes in an Austenitic Steel, *Mater. Sci. Eng.* **6**, 313–319 (1970).

61. R. Kossowsky, D. G. Miller, and E. S. Diaz, Tensile and Creep Strength of Hot-Pressed Si_3N_4, Westinghouse Research Laboratories Scientific Paper 74-904-Foram-P8 (August 7, 1974).

62. K. Natessan, O. K. Chopra, and T. F. Kassner, Effect of Sodium on the Creep-Rupture Behavior of Type 304 Stainless Steel, Int. Conf. Liq. Met. Tech., Energy Prod., Am. Nucl. Soc., Seven Springs (May 1976).

63. F. Garofalo, Effect of Environment on Creep and Creep-Rupture Behavior of Several Steels at Temperatures of 1000° to 1200°F, *Proc. ASTM* **59**, 973–984 (1958).

64. D. D. Lawthers and M. J. Manjoine, in *High Temperature Materials* (R. F. Hehemann and G. M. Ault, eds.), pp. 486–500, Wiley, New York (1957).

65. F. J. Wall, T. Hengstenberg, and H. B. Gayley, Creep-Rupture and Fatigue Properties of Three Nickel-Base Alloys in Helium Environment, *Proc. ASTM* **61**, 970–979 (1961).

66. E. O. Müller, The properties of Some Moly-Base Alloys under Helium High-Temperature Conditions, Paper 18, in *Component Design in High Temperature Reactors Using Helium as a Coolant*, pp. 55–62, Inst. Mech. Eng. Symposium, London (May 1972).

67. F. D. Richardson and J. H. E. Jeffes, The Thermodynamics of Substances of Interest in Iron and Steel Making from 0°C to 2400°C, *J. Iron Steel Inst.* **160**(3), 261–270 (1948).

68. G. C. Wood, High Temperature Oxidation of Alloys, *Oxid. Met.* **2**(1), 11–57 (1970).

69. I. A. Kvernes and P. Kofstad, The Oxidation Behavior of Some Ni–Cr–Al Alloys at High Temperatures, *Met. Trans.* **3**, 1511–1519 (1972).

70. F. S. Pettit, Oxidation Mechanisms for Nickel–Aluminum Alloys at Temperatures between 900° and 1300°C, *Trans. AIME–TMS* **239**, 1296–1305 (1967).

71. F. S. Pettit and J. K. Tien, Hot Gas Environment—Alloy Reactions and Their Relations to Fatigue, in *Proceedings International Conference on Corrosion Fatigue*, NACE, Houston (1972).

72. F. S. Pettit, C. S. Giggins, J. A. Goebel, and E. J. Felten, in *Alloy and Microstructural Design* (J. K. Tien and G. S. Ansell, eds.), Academic Press, New York (1976).

73. P. Kofstad, *High-Temperature Oxidation of Metals*, Wiley, New York (1966).

74. J. M. Davidson, M. K. Malu, J. K. Tien, J. Tabock, and N. A. Gjostein, Oxidation and Heat Resistant Alloys for Emission Control Exhaust Systems, in Proceedings of the Sixth National Technical Conference of SAMPE, Dayton (1974).

75. B. Chattopadhyay and G. C. Wood, The Transient Oxidation of Alloys, *Oxid. Met.* **2**(4), 373–399 (1970).

76. R. A. Rapp, Kinetics, Microstructures, and Mechanism of Internal Oxidation—its Effect and Prevention in High Temperature Alloy Oxidation, *Corrosion* **21**, 382–401 (1965).

77. J. H. Swisher, in *Oxidation of Metals and Alloys*, pp. 235–267, ASM, Cleveland, Ohio (1971).

78. R. Hales and A. C. Hill, The Oxidation of Nimonic-80A in Oxygen at Reduced Pressure, *Corros. Sci.* **14**, 553–562 (1974).

79. G. C. Fryburg, F. J. Kohl, and C. A. Stearns, Enhancement of Oxidative Vaporization of Chromium(III) Oxide and Chromium by Oxygen Atoms, NASA TN D-7629 (1974).

80. S. A. Jansson and E. A. Gulbransen, in *High Temperature Gas–Metal Reactions in Mixed Environments* (S. A. Jansson and Z. A. Foroulis, eds.), pp. 2–32, AIME-TMS, New York (1973).

81. A. M. Beltran and D. A. Shores, in *The Superalloys* (C. T. Sims and W. C. Hagel, eds.), pp. 317–340, Wiley, New York (1972).

82. J. A. Goebel and F. S. Pettit, Na_2SO_4-Induced Accelerated Oxidation (Hot Corrosion) of Nickel, *Met. Trans.* **1**, 1943–1954 (1970).

83. J. A. Goebel and F. S. Pettit, The Influence of Sulfides on the Oxidation Behavior of Nickel-Base Alloys, *Met. Trans.* **1**, 3421–3429 (1970).

84. E. L. Simons, G. V. Browning, and H. A. Liebhafsky, Sodium Sulfate in Gas Turbines, *Corrosion* **11**, 505t–514t (1955).
85. G. Lucas, M. Weddle, and A. Pierce, The Liquidus of Metal-Oxide/V_2O_5 Systems, *J. Iron Steel Inst.* **179**, 342–347 (1955).
86. G. W. Cunningham and A. des Brasunas, The Effects of Contamination by Vanadium and Sodium Compounds on the Air-Corrosion of Stainless Steel, *Corrosion* **12**, 389t–405t (1956).
87. R. C. Hurst, J. B. Johnson, M. Davies, and P. Hancock, in *Deposition and Corrosion in Gas Turbines* (A. B. Hart and A. J. B. Cutler, eds.), pp. 143–157, Wiley, New York (1973).
88. N. S. Bornstein and M. A. DeCrescente, The Relationship between Compounds of Sodium and Sulfur and Sulfidation, *Trans. AIME-TMS* **245**, 1947–1952 (1969).
89. A. Moskowitz and L. Redmerski, Corrosion of Superalloys at High Temperatures in the Presence of Contaminating Salts, *Corrosion* **17**, 131–138 (1961).
90. A. V. Seybolt, Contribution to the Study of Hot Corrosion, *Trans. AIME-TMS* **242**, 1955–1961 (1968).
91. A. V. Seybolt and A. Beltran, in *Hot Corrosion Problems Associated with Gas Turbines*, ASTM STP 421, pp. 21–37, ASTM, Philadelphia (1966).
92. V. I. Nikitin and T. N. Grigorieva, Effect of Ash Deposits on the Creep Strength of the Material of the Working Blades of Gas Turbines, *Fiz. Khim. Mekh. Meter.* **8**(5), 19–26 (1972).
93. C. Sims, ASME Preprint 70-GT-24 (May 1970).
94. H. J. Grabke, in *High Temperature Gas–Metal Reactions in Mixed Environments* (S. A. Jansson and Z. A. Foroulis, eds.), pp. 130–142, AIME-TMS, New York (1973).
95. K. Natesan, Corrosion–Erosion Behavior of Materials in a Coal-Gasification Environment, *Corrosion* **32**(9), 364–369 (1976).
96. I. Kozman, in *High Temperature Gas–Metal Reactions in Mixed Environments* (S. A. Jansson and Z. A. Foroulis, eds.), pp. 155–167, AIME-TMS, New York (1973).
97. R. F. Hochman, in *Properties of High Temperature Alloys* (Z. A. Foroulis and F. S. Pettit, eds.), The Electrochemical Society, Princeton, N.J. (1977).
98. G. Hörz, in *Properties of High Temperature Alloys* (Z. A. Foroulis and F. S. Pettit, eds.), The Electrochemical Society, Princeton, N.J. (1977).
99. E. B. Evans, N. Isangarkis, H. B. Probst, and N. J. Garibotti, in *High Temperature Gas–Metal Reactions in Mixed Environments* (S. A. Jansson and Z. A. Foroulis, eds.), AIME-TMS, New York (1973).
100. G. Wand, B. S. Hockenhull, and P. Hancock, The Effect of Cyclic Stressing on the Oxidation of a Low Carbon Steel, *Met. Trans.* **5**, 1451–1455 (1974).
101. D. A. Vermilyea, A Theory for the Propagation of Stress Corrosion Cracks in Metals, *J. Electrochem. Soc.* **119**(4), 405–407 (1972).
102. L. J. Weirick and W. L. Larsen, The Effect of Stress on the Low-Temperature Oxidation of Niobium, *J. Electrochem. Soc.* **119**(4), 465–476 (1972).
103. M. Kaufman, Examination of the Influence of Coatings on Thin Superalloy Sections, General Electric Company Aircraft Engine Group Report. 121115 (August 1972).
104. R. A. U. Huddle, in *Deposition and Corrosion in Gas Turbines* (A. B. Hart and A. J. B. Cutler, eds.), pp. 158–177, Wiley, New York (1973).
105. V. V. Gerasimov, V. A. Shuvalov, and Z. I. Emel'yantseva, Effect of Stresses on the Electrochemical Behavior of Stainless Steels, *Prot. Met.* **7**(2), 150–152 (1970); trans. from *Zashch. Met.* **7**(2), 178–181 (1971).
106. C. F. Knight and R. Perkins, The Effect of Applied Tensile Stress on the Corrosion Behavior of Zircaloy-2 in Steam and Oxygen, *J. Nucl. Mater.* **36**, 180–188 (1970).
107. M. L. Sessions and C. J. McMahon, Jr., in *Grain Boundaries in Engineering Materials—Proceedings of the 4th Bolton Landing Conference* (J. L. Walter, ed.), pp. 477–489, Claitor's, Baton Rouge, La. (1975).

108. N. N. Koslova, E. N. Fufaeva, and I. V. Paizov, Oxidation of Heat-Resistant Steels and Alloys under Stress, *Met. Sci. Heat Treat.* **15**(5), 457–460 (1973). Translated from *Metalloved. Term. Obrab. Met.* **1973**(6), 11–14.

109. K. H. Wiedemann, The Effect of Mechanical Tensile Stressing on the Suppression of Corrosion of Zircaloy-2, *J. Nucl. Mater.* **36**, 340–342 (1970).

110. C. S. Giggins and F. S. Pettit, Oxide Scale Adherence Mechanisms, Pratt & Whitney Aircraft M.E.R.L. Quarterly Report (February 28, 1975).

111. J. K. Tien and F. S. Pettit, Mechanism of Oxide Adherence on Fe-25Cr-4Al (Y, Sc), *Met. Trans.* **3**, 1587–1599 (1972).

112. W. H. Chang, Tensile Embrittlement of Turbine Blade Alloys after High-Temperature Exposure, in *Superalloys—Processing, Proceedings of the Second International Conference on Superalloys*, pp. V1–V41, AIME-TMS, New York (1972).

113. I. R. Kramer and L. J. Demer, Effects of Environment on Mechanical Properties of Metals, *Prog. Mater. Sci.* **9**(3), 133–199 (1961).

114. E. S. Machlin, in *Strengthening Mechanisms in Solids*, pp. 375–404, ASM, Cleveland, Ohio (1962).

115. D. A. Douglas, in *High Temperature Materials* (R. F. Hehemann and G. M. Ault, eds.), pp. 429–447, Wiley, New York (1957).

116. A. R. C. Westwood, in *Environment Sensitive Mechanical Behavior* (A. R. C. Westwood and N. S. Stoloff, eds.), pp. 1–65, Gordon and Breach, New York (1966).

117. R. Roscoe, Strength of Metal Single Crystals, *Nature* **133**, 912 (1934).

118. M. A. Adams, The Plastic Behaviour of Copper Crystals Containing Zinc in the Surface Layer, *Acta Met.* **6**, 327–338 (1958).

119. F. D. Rosi, Surface Effects on Plastic Properties of Copper Crystals, *Acta Met.* **5**, 348–350 (1957).

120. S. Harper and A. H. Cottrell, Surface Effects and the Plasticity of Zinc Crystals, *Proc. Phys. Soc.* **B63**, 331–345 (1950).

121. J. Takamura, Effect of Anodic Surface Films on the Plastic Deformation of Aluminum Crystals, *Mem. Fac. Eng. Kyoto Univ.* **18**(3), 255–279 (1956).

122. J. J. Gilman, in *Symposium on Basic Effects of Environment on the Strength, Scaling, and Embrittlement of Metals at High Temperatures*, ASTM STP 171, pp. 3–13, ASTM, Philadelphia (1955).

123. J. J. Gilman and T. A. Read, Surface Effects in the Slip and Twinning of Metal Monocrystals, *Trans. AIME-TMS* **194**, 875–883 (1952).

124. L. C. Weiner and M. J. Gensemer, Effects of Solid Environments on the Brittle Fracture of Zinc Single Crystals, *J. Inst. Met.* **85**, 441–448 (1956–57).

125. S. Shapiro and T. A. Read, Effect of Oxide Films on the Internal Friction of Cadmium Single Crystals, *Phys. Rev.* **82**(2), 341 (1951).

126. C. S. Barrett, An Abnormal After-Effect in Metals, *Acta Met.* **1**, 1–7 (1953).

127. J. C. Grosskreutz and C. Q. Bowles, in *Environment Sensitive Mechanical Behavior* (A. R. C. Westwood and N. S. Stoloff, eds.), pp. 67–106, Gordon and Breach, New York (1966).

128. E. N. Da C. Andrade and R. F. Y. Randall, Surface Effects with Single Crystal Wires of Cadmium, *Nature* **162**, 890–891 (1948).

129. J. C. Fisher, Discussion of Ref. 49, *Trans. AIME-TMS* **194**, 531–532 (1952).

130. H. G. F. Wilsdorf, The Influence of Surface Films on Mechanical Properties, Rept. MS-3572-101-75 (1975).

131. J. V. Cathcart, ed., *Stress Effects and the Oxidation of Metals*, AIME-TMS, New York (1975).

132. J. Stringer, Stress Generation and Relief in Growing Oxide Films, *Corr. Sci.* **10**, 513–543 (1970).

133. N. B. Pilling and R. E. Bedworth, The Oxidation of Metals at High Temperatures, *J. Inst. Met.* **29**, 529–591 (1923).

134. J. D. Noden, C. J. Knights, and M. W. Thomas, Growth of Austenitic Stainless Steels Oxidized in Carbon and Oxygen Bearing Gases, *Br. Corros. J.* **3**, 47–55 (1968).

135. J. K. Tien and J. M. Davidson, in *Stress Effects and the Oxidation of Metals* (J. V. Cathcart, ed.), pp. 200–220, AIME-TMS, New York (1975).

136. J. H. Westbrook, in *Environment Sensitive Mechanical Behavior* (A. R. C. Westwood and N. S. Stoloff, eds.), pp. 247–268, Gordon and Breach, New York (1966).

137. H. Forestier and A. Clauss, Influence des Gaz Adsorbés sur la Résistance Mécanique d'un Film Métallique, *Rev. Met.* **52**(12), 961–964 (1955).

138. J. Lunsford and N. J. Grant, Relative High Temperature Properties of the Hexagonal Close-Packed and Body Centered Cubic Structures in Iodide–Titanium, *Trans. ASM* **49**, 328–337 (1957).

139. D. L. Douglas, The Effect of Oxidation on the Mechanical Behavior of Nickel at 600 C, *Mater. Sci. Eng.* **3**, 255–263 (1968–69).

140. P. Hancock and R. Fletcher, The Oxidation of Nickel in the Temperature Range 700–1100°C, Proc. Journées Int. d'Étude Sur l'Oxyd. des Métaux, pp. 70–78 (1965).

141. J. K. Tien and W. H. Rand, The Effect of Active Element Addition on Void Formation during Oxidation, *Scripta Met.* **6**, 5–57 (1972).

142. W. E. Pennell, Structural Materials Aspects of LMFBR Core Restraint System Design, *Nucl. Technol.* **16**, 332–353 (1972).

143. J. Gittus, *Creep, Viscoelasticity, and Creep Fracture in Solids*, pp. 204–273, Wiley, New York (1975).

144. S. Mahajan and L. Himmel, Strengthening Mechanisms in Internally Oxidized Silver-Based Alloy Single Crystals, *Acta Met.* **20**, 1319–1323 (1972).

145. O. Preston and N. J. Grant, Dispersion Strengthening of Copper by Internal Oxidation, *Trans. AIME-TMS* **221**, 164–172 (1961).

146. D. L. Wood, Internal Oxidation of Copper-Aluminum Alloys, *Trans. AIME-TMS* **215**, 925–932 (1959).

147. E. Gregory and G. C. Smith, The Effects of Internal Oxidation on the Tensile Properties of Some Silver Alloys at Room and Elevated Temperatures, *J. Inst. Met.* **85**, 81–87 (1956–57).

148. G. S. Ansell, T. D. Cooper, and F. V. Lenel, eds., *Oxide Dispersion Strengthening*, Gordon and Breach, New York (1968).

149. J. L. Meijering and M. J. Druyvesteyn, Hardening of Metals by Internal Oxidation, *Philips Res. Rep.* **2**(2), 81–102 (1947).

150. R. A. U. Huddle, *The Influence of HTR Helium on the Behavior of Metals in High Temperature Reactors* (Proc. Int. Conf. Corros., BNES), pp. 203–212, Clowers, London (1971).

151. W. R. Martin and H. E. McCoy, Effect of CO_2 on the Strength and Ductility of Type 304 Stainless Steel at Elevated Temperatures, *Corrosion* **19**, 157–168 (1963).

152. R. W. Swindeman and D. A. Douglas, Improvement of the High-Temprature Strength Properties of Reactor Materials after Fabrication, *J. Nucl. Mater.* **1**, 49–57 (1959).

153. H. H. Bleakney, The Ductility of Metals in Creep-Rupture Tests, *Can. J. Technol.* **30**, 340–351 (1952).

154. H. H. Bleakney, The Creep Embrittlement of Copper, *Can. Met. Quart.* **4**(1), 13–29 (1965).

155. E. J. Dulis and G. V. Smith, Formation of Nitrides from Atmospheric Exposure during Creep Rupture of 18 pct Cr–9 pct Ni Steel, *Trans. AIME-TMS* **194**, 1083–1084 (1952).

156. G. V. Smith, E. J. Dulis, and E. G. Houston, Creep and Rupture of Several Chromium–Nickel Austenitic Stainless Steels, *Trans. ASM* **42**, 935–979 (1950).

157. P. Shahinian and M. R. Achter, Influence of Environment on Crack Propagation at High Temperatures, *Proc. Crack Prop. Symp. Cranfield* **1**, 29–75 (1961).

158. M. F. Ashby, in *Strengthening Methods in Crystals* (A. Kelly and R. B. Nicholson, eds.), pp. 71–128, Wiley, New York (1971).

159. J. W. Coombs, R. E. Allen, and F. H. Vitovec, Creep and Rupture Behavior of Low Alloy Steels in High Pressure Hydrogen Environment, *Trans. ASME* **D87**, 313–318 (1965).

160. G. D. Collins and E. L. Zebroski, U.S. AEC Report (AI-AEC 12721) (1969).

161. P. G. Shewmon, *Diffusion in Solids*, McGraw-Hill, New York (1963).

162. T. Kê, A Grain Boundary Model and the Mechanism of Viscous Intercrystalline Slip, *J. Appl. Phys.* **20**, 274–281 (1949).

163. J. H. Westbrook and D. L. Wood, A Source of Grain-Boundary Embrittlement in Intermetallics, *J. Inst. Met.* **91**, 174–182 (1962–63).

164. R. Resnick and L. Seigle, Nucleation of Voids in Metals during Diffusion and Creep, *Trans. AIME-TMS* **209**, 87–94 (1957).

165. J. Intrater and E. S. Machlin, Grain Boundary Sliding and Intercrystalline Cracking, *Acta Met.* **7**, 140–142 (1959).

166. N. V. Perstov, N. I. Ivanova, Yu. V. Goryunov, and E. D. Schukin, Effects of the Diffusion of Liquid Metals into Solid Metals on the Manifestation of the Adsorption-Induced Reduction in Strength, *Sov. Mater. Sci.* **6**(6), 726–728 (1973); translated from *Fiz. Khim Mec.* **6**(6), 79–82 (1973).

167. W. R. Johnson, C. R. Barrett, and W. D. Nix, The Effect of Environment and Grain Size on the Creep Behavior of a Ni–6 pct W Solid Solution, *Met. Trans.* **3**, 695–698 (1972).

168. R. J. Sherman and M. R. Achter, Crack Propagation in Air and in Vacuum for Nickel and a Nickel–Chromium Aluminum Alloy, *Trans. AIME-TMS* **224**, 144–147 (1962).

169. *High-Temperature, High-Strength, Nickel-Based Alloys*, p. 22, International Nickel Company, Inc., New York (1977).

170. R. C. Gifkins, Grain-Boundary Sliding and Its Accommodation during Creep and Superplasticity, *Met. Trans.* **7A**, 1225–1232 (1976).

171. D. K. Matlock and W. D. Nix, The Effect of Sample Size on the Steady State Creep Characteristics of Ni–6 pct W, *Met. Trans.* **5**, 1401–1412 (1974).

172. J. M. Davidson, Engineering Science Doctoral Thesis, Columbia University, New York (1978).

173. C. J. McMahon, Jr., On the Mechanism of Premature In-Service Failure of Nickel-Base Superalloy Gas Turbine Blades, *Mater. Sci. Eng.* **13**, 295–297 (1974).

174. H. H. Smith, P. Shahinian, and M. R. Achter, Fatigue Crack Growth Rates in Type 316 Stainless Steel at Elevated Temperature as a Function of Oxygen Pressure, *Trans. AIME-TMS* **245**, 947–953 (1969).

175. D. J. Duquette and M. Gell, The Effect of Environment on the Mechanism of Stage I Fatigue Fracture, *Met. Trans.* **2**, 943–951 (1971).

176. G. J. Danek, Jr., H. H. Smith, and M. R. Achter, High-Temperature Fatigue and Bending Strain Measurements in Controlled Environments, *Proc. ASTM* **61**, 775–788 (1961).

177. C. J. McMahon, Jr., and L. F. Coffin, Jr., Mechanism of Damage and Fracture in High-Temperature, Low-Cycle Fatigue of a Cast Nickel-Based Superalloy, *Met. Trans.* **1**, 3443–3449 (1970).

178. A. J. McEvily, Jr., in *Proceedings of Conference on Fundamental Aspects of Stress Corrosion Cracking* (R. W. Staehle and D. Van Rooyen, eds.), pp. 72–81, NACE, Houston (1969).

179. A. J. McEvily and A. P. Bond, in *Environment Sensitive Mechanical Behavior* (A. R. C. Westwood and N. S. Stoloff, eds.), pp. 421–453, Gordon and Breach, New York (1966).

180. A. J. McEvily and A. P. Bond, On the Initiation and Growth of Stress Corrosion Cracks in Tarnished Brass, *J. Electrochem. Soc.* **112**(2), 131–138 (1965).

181. A. J. Forty and P. Humble, in *Environment Sensitive Mechanical Behavior* (A. R. C. Westwood and N. S. Stoloff, eds.), pp. 403–420, Gordon and Breach, New York (1966).

182. A. J. Forty and P. Humble, The Influence of Stress Tarnish on the Stress Corrosion of α-Brass, *Phil. Mag.* **8**(8), 247–264 (1963).

183. H. A. Johnson, T. J. Boswork, and R. T. Jorgenson, Evaluation of Metallic Materials for Space Shuttle Booster Thermal Protection Systems, *SAMPE Space Shuttle Mater.* **3**, 347–357 (1971).

184. V. E. Belyakov and V. V. Romanov, Influence of Anions on the Long-Term Strength of MA2-1 Magnesium Alloy, *Sov. Mater. Sci.* **7**(4), 400–402 (1971). Translated from *Fiz. Khim. Mekh. Mater.* **7**(4), 27–30 (1971).

185. J. A. Harris, Jr., J. F. Schratt, and M. C. Van Wanderham, Creep-Rupture Properties of Materials in High Pressure Gaseous Hydrogen at Elevated Temperature, *SAMPE Space Shuttle Mater.* **3**(2), 1–8 (1971).

186. F. T. Furillo, S. Purushothaman, and J. K. Tien, Understanding the Larson–Miller Parameter, *Scripta Met.* **11**, 493–496 (1977).

187. D. L. Douglas, in *Oxidation of Metals and Alloys*, pp. 137–156, ASM, Metals Park, Ohio (1970).

188. G. V. Samsonov, ed., *The Oxide Handbook* (trans. by C. Nigel Turton and Tatiana I. Turton), p. 125, IFI Plenum Press, New York (1973).

THE ROLE OF METALLURGICAL VARIABLES IN HYDROGEN-ASSISTED ENVIRONMENTAL FRACTURE

Anthony W. Thompson

Science Center, Rockwell International
Thousand Oaks, California

and

I. M. Bernstein

Department of Metallurgy and Materials Science
Carnegie-Mellon University
Pittsburgh, Pennsylvania

INTRODUCTION

This chapter focuses on two kinds of environmental fracture, stress corrosion cracking (SCC) and hydrogen embrittlement, and discusses the role in such failures of a number of metallurgical variables. These include chemical composition; microstructural components such as precipitate type and structure, and grain size and shape; crystallographic texture; heat treatment and its effect on the foregoing variables; and processing, particularly the thermomechanical treatments (TMT), which are attracting increased attention for property optimization. These variables are expected to be of great importance in the development of new engineering materials to meet demanding service conditions.

In recent years, understanding of the role of metallurgical variables in hydrogen embrittlement has improved significantly. One purpose of the present review is to summarize this progress. Since hydrogen is believed to be broadly involved in many environmental cracking phenomena, it is a second purpose of the review to extend that improved understanding to at least some cases of SCC. It might be expected that similar role(s) of metallurgical variables would be observed for both hydrogen embrittlement and hydrogen-assisted SCC, particularly when hydrogen plays a dominant role.

The usual definition of SCC is that both a corrosive environment and a tensile stress (residual or applied) act together to cause cracking, which has a macroscopically brittle appearance. It is implicit in this definition that SCC is a phenomenon, not a mechanism, and it is so treated in this chapter. Hydrogen embrittlement is also a phenomenon, which may or may not occur during SCC. Hydrogen as a gas or as a species evolved from chemical or electrochemical reactions could be regarded as an SCC corrodent. But since classical hydrogen embrittlement studies have been conducted with hydrogen dissolved within the metal, which corrodents usually are not, such studies have been cited in the past as evidence against a link between SCC and hydrogen embrittlement. This review concludes, as have others, that such a conclusion cannot be general. There are now known to be a number of cases in which hydrogen participates in SCC fracture, and the accepted balance between hydrogen cracking and, say, anodic dissolution as components of SCC appears to need reappraisal or revision. It is a purpose of this chapter to identify such areas.

It should be strongly emphasized that we do not accept the proposition that there must be a *single* causative process for SCC. There are indeed instances of purely dissolution processes, or purely hydrogen embrittlement processes, giving rise to what is generally described as SCC. But in many cases there appear to be contributions from both types of processes. This chapter emphasizes the hydrogen-related behavior in order to make maximum use of the new understanding of hydrogen embrittlement, but that emphasis should not be construed as being a mechanistic emphasis.

The combined literature record on SCC and hydrogen embrittlement is very extensive (above 10,000 papers in one estimate), and it is not our intention to review it here in detail. We have relied on earlier reviews where appropriate, but have emphasized recent work, with a focus on a specific point: the extent to which SCC and hydrogen embrittlement can be controlled through manipulation of metallurgical variables. That evidence also helps clarify the extent of hydrogen participation in SCC. Our emphasis on metallurgical (internal) variables should not be interpreted as slighting the importance of environmental (external) variables. Instead, this emphasis is intended to highlight the ways in which alloy design improvements can contribute to improved environmental resistance.

Phenomenology

Hydrogen embrittlement is relatively simply manifested through changes in mechanical properties; as the name implies, the most pronounced changes

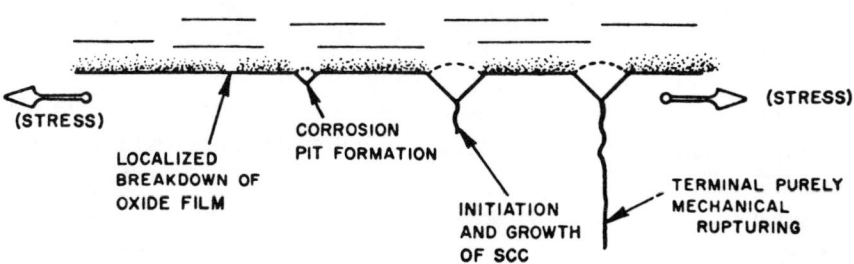

Fig. 1. Schematic sequence of events (left to right) which could occur in SCC testing of a smooth specimen. (From Brown.[1])

are often in the ductility parameters. In SCC, however, there is by definition a reaction with the environment, and this can greatly complicate the phenomenon. There are now known to be a great many different environment–material combinations which result in SCC. Emphasis in this chapter is on those environments which (under at least some conditions) can generate hydrogen, so that we can apply our understanding of hydrogen behavior to them; this includes most of the more common types (e.g., chloride-containing aqueous solutions).*

To clarify the comments on testing methods (next section), it is useful to present a general view of the SCC process. Figure 1 is such a view. The initiation process can be summarized as on the left side of Fig. 1, and it is evident without going into detail that chemical and electrochemical factors will often be predominant at this stage. Toward the right, however, the fracture process is both electrochemical and mechanical in character. There may be an interplay between these processes, especially in a ductile material which tends to resist crack growth by a blunting process, and in these cases localized electrochemical dissolution, or pitting, can then resharpen the crack tip and cause another increment of crack growth. We emphasize that a cooperative series of steps such as these, required to operate in sequence, may occur in many instances of SCC. In some cases, such as titanium alloys, a sharp precrack may have to be introduced by fatigue in order to observe any SCC at all. At the opposite extreme of very low toughness materials, fast fracture can be caused by the pit alone, acting as a critical-size flaw.

* As described below, such a focus incorporates situations in which local hydrogen production occurs despite general anodic conditions.

Testing Methods

Hydrogen testing is usually designed to investigate one of two behavior types. One is the short-time or "instantaneous" behavior, where little or no entry of hydrogen into the metal occurs by diffusion. Such behavior is modeled by a tensile or fracture mechanics test in low- or high-pressure gas. The second is the behavior of the material when hydrogen has been introduced into the lattice, as would occur by accumulation during prolonged exposure to hydrogen in service. This behavior can be simulated by tests on specimens which have been supersaturated by gas phase or electrolytic charging. In either case, the technique may comprise tensile, stress rupture, fatigue crack growth, or sustained load crack growth tests.

There are many specimens and test methods for SCC evaluation; in the past many consisted in essence of a smooth specimen, for example, a thin strip bent into a U-shape, which was exposed to the environment, and the environmental susceptibility was (hopefully) quantified as time-to-failure (TTF). Unfortunately, as several authors have discussed,[1-3] TTF testing of smooth specimens must be interpreted with caution, because crack initiation, crack propagation, and unstable or fast fracture (see Fig. 1) all occur during the test. The characteristics of the first two processes can depend differently on the material–environment interaction, while all three processes can be different for different materials.[1] For these reasons, TTF data for smooth specimens must be used with care and after thorough evaluation, if they are not to be highly misleading.

An alternative (now widely recognized) is to measure TTF for precracked specimens or to use a method which provides quantitative data on the crack propagation part of the test. This is realistic from the viewpoint of structural materials design, since few real structures can be considered to be without the scratches, crevices, inclusions, or other flaws which accelerate the initiation of SCC. Similar comments also apply in many cases of hydrogen embrittlement testing. The results of the crack-growth type of test are usually presented as in Fig. 2, showing crack velocity V as a function of stress intensity K. The existence of three stages or regions in this curve was first noted by Wiederhorn.[4] The critical value of K for fast fracture, which is called K_Q, K_{1x}, or in some circumstances K_{1c}, may intervene to prevent observation of the full three-stage curve in constant load tests, but parts of this curve are observed for many materials.

When crack growth tests are conducted under conditions of decreasing K

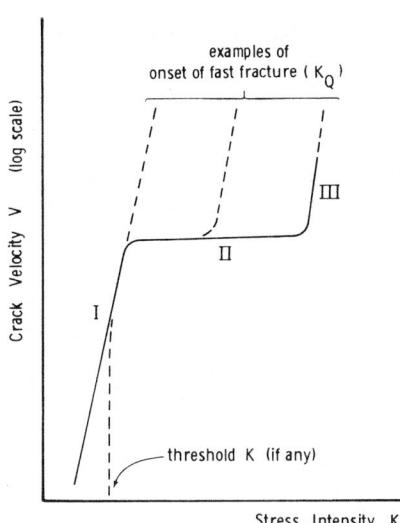

Fig. 2. Generalized depiction of the dependence of crack velocity on stress intensity. The three "regions" or stages of crack behavior were first described by Wiederhorn.[4] In cases where the region I data attain a vertical slope, the corresponding K value is called the threshold K, K_{Ienv}, or K_{Iscc}: not all alloys exhibit such behavior. Under some circumstances, the K_Q of fast fracture is called K_{Ic}.

(e.g., in a bolt-loaded, precracked specimen), crack "arrest" can occur. The K value at arrest, if one exists, is called the threshold K for crack growth (designated K_{th}). For environmental tests, K_{th} is often called K_{Iscc} or K_{Ienv}. This type of test is usually also found to be able to define clearly the K-independent or region II behavior. Region II behavior has been used[2] as the "maximum" V in SCC, since region III is usually associated with environment-independent fracture. Development of this testing method has been presented in some detail,[2,3,5] and it is to be hoped that future studies will incorporate this kind of testing. Crack-growth-rate data and TTF data, properly interpreted, can be complementary,[6] and both should be measured.

Some alloy classes exhibit little cracking under sustained load or deflection conditions, as just described, but do exhibit losses of ductility in tensile or fatigue tests. It is common practice with these alloys to use the reduction of area (RA) as a measure of ductility, and to compare the RA in the environment of interest with that in a neutral environment. The percent RA loss is then defined as 100 [1 − (environmental RA/neutral RA)]. Such alloys, in general, are less subject to environmental damage than those which crack readily. Characterization of their behavior is nevertheless of importance, because large losses in fracture ductility can affect the failure resistance of a material in service.

The several alloy families discussed below are metallurgically very diverse. Each is presented and analyzed in a separate section to minimize complications and cross references, and the important factors for each family are then compared and discussed in the concluding sections. The order of presentation is not intended to have great significance, but an approximate order of commercial-use magnitude can be detected, with the overall sequence being steels, aluminum alloys, titanium alloys, and nickel alloys.

FERRITIC AND MARTENSITIC STEELS

This section concentrates on steels which have in common the body-centered cubic (bcc) crystal structure. These materials range from nearly pure iron to highly alloyed stainless steels; compositions of all steels referred to are shown in Table 1, together with other typical alloys. Although very different microstructures are found in the various steels of Table 1, and different specific behaviors have been observed, we group them together for convenience because of their bcc structure and because, as we demonstrate, their environmental response contains many common elements of behavior. Later we discuss the important variable of microstructure at some length.

Composition

It is difficult to discuss composition alone, because composition changes tend to result in changes in microstructure and other variables as well. In what follows, we address patterns in compositional variables, keeping in mind that interactions do occur, and that synergistic effects can be expected. Those complications are discussed at the end of this section.

Purity

It is broadly true that most mechanical properties of a steel are improved by elimination, or control, of residuals. This tends to be the case for environmental embrittlement as well. For example, vacuum remelting improved both the resistance to hydrogen cracking in a 410 martensitic

Table 1. Typical Compositions of Ferritic and Martensitic Steels (wt %)

Steel	Nominal yield strength (MN/m²)	C	Ni	Cr	Mo	Mn	V	Si	Co	Ti	Al
Ferritic											
Armco iron[a]	50	0.01	—	—	—	—	—	—	—	—	—
Fe–Ti	60	0.005	—	—	—	—	—	—	—	0.15	—
1020	200	0.20	—	—	—	0.45	—	0.1	—	—	—
1080	400	0.80	—	—	—	0.80	—	0.1	—	—	—
1330	500–900	0.30	—	—	—	1.75	—	0.25	—	—	—
Martensitic (low alloy)											
4130	1100–1600	0.30	—	1.0	0.2	0.5	—	0.3	—	—	—
4340	1000–1900	0.40	1.9	0.8	0.25	0.7	—	0.6	—	—	—
D6aC	1200–1700	0.45	0.6	1.2	1.0	0.8	0.05	0.25	—	—	—
HY-130	900–1000	0.12	5.0	0.6	—	0.9	—	0.2	—	—	—
H-11	1200–1650	0.40	—	5.0	1.3	0.3	0.5	0.9	—	—	—
20 CND 10	600–1130	0.20	0.8	2.7	0.3	0.6	—	0.2	—	—	—
Martensitic (high alloy)											
410 stainless	700–1000	0.15	—	12.0	—	0.5	0.4	0.4	—	—	—
9–4–25	1100–1400	0.25	8.5	0.5	0.5	0.1	0.1	0.1	4.0	—	—
9–4–45	1300–1700	0.45	8.5	0.3	0.2	0.1	0.1	0.1	4.0	—	—
Maraging											
18 Ni (200)	1400–1600	0.03	13.0	—	3.0	—	—	—	8.0	0.2	0.1
18 Ni (300)	1800–2100	0.03	13.0	—	5.0	—	—	—	9.0	0.6	0.1
PH stainless											
AFC 77[a]	1100–2000	0.16	0.2	14.0	5.0	0.2	—	—	13.0	—	—
17–7 PH[a]	1000–1400	0.07	7.0	17.0	—	0.5	—	0.3	—	—	1.0
PH 13–8 Mo[a]	1000–1500	0.04	8.0	13.0	2.0	0.1	—	0.05	—	—	1.0

[a] Trade names of Armco Steel Corp.

stainless steel,[7] and the static corrosion fatigue life of a 30% Cr steel.[8] Sulfur and phosphorus appear particularly detrimental,[9,10] and this may be related to the close connection between hydrogen embrittlement and temper embrittlement.[11,12]

Substitutional Elements

The common substitutional additions to steels are chromium, manganese, silicon, nickel, and molybdenum. Elements of lesser importance are vanadium, aluminum, titanium, and cobalt. We begin by summarizing the effects of these elements on environmental embrittlement, and then turn to more detailed consideration.

There appears to be agreement that chromium,[10,13] particularly at low concentrations, and manganese[7,14,15] increase susceptibility to environmental embrittlement, while silicon[9,15-17] decreases embrittlement, particularly in higher-strength steels. Titanium also decreases embrittlement[10,18] except in maraging steels, which show a greatly enhanced susceptibility as the Ti content is increased. The behavior of Mo is ill defined; results over a range of strength levels, microstructures, and environments show variable and contradictory behavior.[10,19,21] The information for nickel is even more sketchy and inconsistent than for Mo.

The means by which these elements impart or diminish susceptibility have been the subject of speculation.[20] In the case of chromium, for example, it has been suggested that its deleterious effect is due to microstructural changes[10] or to an enhanced corrosion rate,[20] possibly by the establishment of local concentration cells at chromium carbides. We postpone consideration of such issues until later.

Turning to detailed discussion of alloying elements, there seems to be general agreement for Mn. Sandoz showed[21] that an increase from 0.1 to 2.7% Mn in a 4340-type steel resulted in a decrease in K_{Iscc} in salt water of as much as 60% for anodic, cathodic, or open-circuit conditions, dependent on the Mn content. Some of these results are shown in Fig. 3 for steels of two strength levels. The same pattern of behavior was seen with hydrogen charging and with an external gaseous hydrogen environment.[15,22] It has been suggested[15,23] that behavior in aqueous systems stems from the ability of manganese to render the crack tip increasingly cathodic, but hydrogen results[15,22,24] with a similar pattern lead us to suggest instead that the fracture

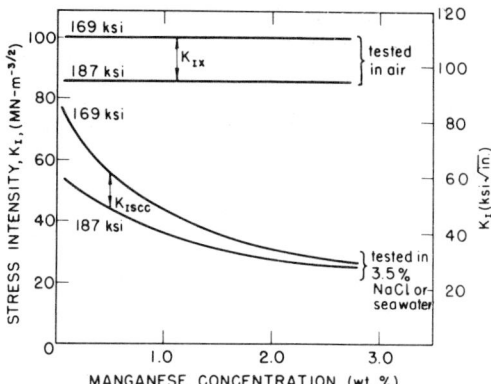

Fig. 3. Effect of manganese on the SCC resistance of 4340-type steel heat treated to two strength levels; 145 psi = 1 MN/m². K_{IX} refers to unstable fast fracture in air. (From Sandoz.[21])

behavior itself may be changed by Mn additions. Such a conclusion is supported by the large losses in ductility reported for hydrogen tests on a series of Fe–Mn alloys,[24] extending to more than 10% Mn content. It is important to emphasize that the results of Fig. 3, which are for relatively strong steels, may well apply also to lower-strength steel. For example, an increase from 1.35% to 1.60% Mn in linepipe steels (X52 and X70 grades) resulted in significantly increased sensitivity to SCC in boiling 20% ammonium nitrate solution, under cathodic polarization.[25] Systematic studies of crack growth rate as a function of Mn content would be appropriate as a supplement to the K_{Iscc} data[15,21,22] and tensile data[24] just discussed, but it would appear that Mn may be undesirable in terms of environmental performance.[7,14,15] The addition of increasing amounts of manganese to strengthen construction-grade steels, particularly linepipe steel and oil and gas well tubing, must be viewed with caution when such steels are used in hostile environments.

The effectiveness of silicon is reasonably well documented.[9,17] Carter[17] examined silicon additions in a 4340-type steel, and for a variety of strength levels, found general improvements up to silicon contents in excess of 2%. At high strength levels [i.e., 2000 MN/m² (~290 ksi) tensile strength], this improvement was attributed to a decreasing crack velocity, as shown in Fig. 4; here it appears that a minimum of 1% Si must be added. At medium-strength levels, improvement came instead from an increase in K_{Ienv} at 0.5 to 1.0% Si, which was related to changes in tempering behavior. These improvements are reflected in a commercial modification of 4340, called 300M, which contains 1.5 to 1.8% Si. Mechanistically, it has been suggested that silicon's role is

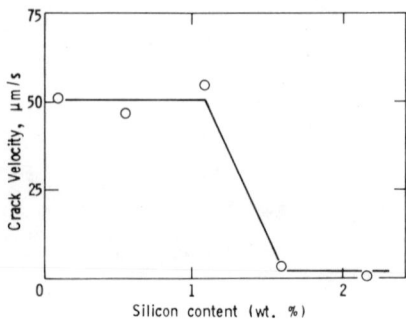

Fig. 4. Effect of silicon content on crack velocity in 4340-type steels of about 2000 MN/m^2 tensile strength (290 ksi). Applied stress intensity was 60 MN/m$^{3/2}$; environment was 3.5% NACl solution. (Data are those of Carter.[17])

related to its ability to prevent epsilon carbide from acting as a cathodic site for hydrogen discharge[9,17]; that the carbide improves crack resistance by acting as a hydrogen trap[26]; and that silicon reduces the diffusivity of the damaging species, e.g., hydrogen.[15,16] Silicon is thus an element whose role is unclear and may be complex, but the beneficial effects are well documented, at least in high-strength steels. The environmental resistance conferred by Si stands in contrast to the poor performance caused by Mn, and silicon could well be the solute of choice to achieve strengthening and hardenability in steels for use in hostile environments. Silicon additions, however, can adversely affect fabricability and weldability of steels, so that use of high Si contents would necessitate careful alloy development.

The behavior of titanium is complex and probably depends on the specific microstructure. In maraging steels, Ti is present as the intermetallic Ni$_3$Ti. In those steels, whose hardening behavior differs from most steels in this section, it promotes hydrogen embrittlement,[27,28] even when a probable variation in yield strength with increasing titanium content is taken into account. In other ferritic and martensitic steels, however, Ti has usually been found extremely beneficial in both H$_2$ and H$_2$S environments for a wide range of Ti concentrations, stress levels, and microstructures.[10,19,20,28,29] This positive behavior was once attributed to titanium's ability to limit the amount of retained austenite, thereby reducing the dangers of subsequent martensite formation.[28,30] Recent results suggest instead that its primary role is to act as a preferential trap for dissolved hydrogen, even at room temperature, when it is present as a substitutional solute or combined as a fine, uniformly distributed carbide, and as a surface inhibitor which reduces the amount of adsorbed hydrogen able to go into solid solution.[31] Support for this proposed affinity between hydrogen and titanium is an observed reduction in the frequency of cracking of cathodically charged iron–titanium alloys.[31]

The elements Cr, Ni, and Mo are essential for attainment of high hardenability, strength, and toughness in steels. The evidence for behavior of these elements is less complete than for those just discussed, and there are conflicting claims as to effects.[10,14,19,27] For example, small additions of Cr appear deleterious to SCC performance,[10] but since the threshold stress in both hydrogen and salt water is unaffected by up to 2 % Cr,[21,22] the effect must be upon the crack growth rate. At lower tensile strength levels of 700 MN/m^2 ($\sim$ 100 ksi), however, addition of 3.3 % Cr and 0.4 % Mo reduces losses in ductility of hydrogen-charged specimens, compared to a plain carbon steel[32] (Fig. 5). More work would be desirable here.

Nickel tends to be somewhat deleterious in the amounts usually found in steels. However, large nickel additions (e.g., above 8 %) can be beneficial in a variety of environments.[33] The evidence thus suggests a minimum-type behavior, with Ni additions decreasing environmental resistance up to some critical amount, and increasing resistance above that amount. The critical amount depends on the environment and perhaps also on strength level. For

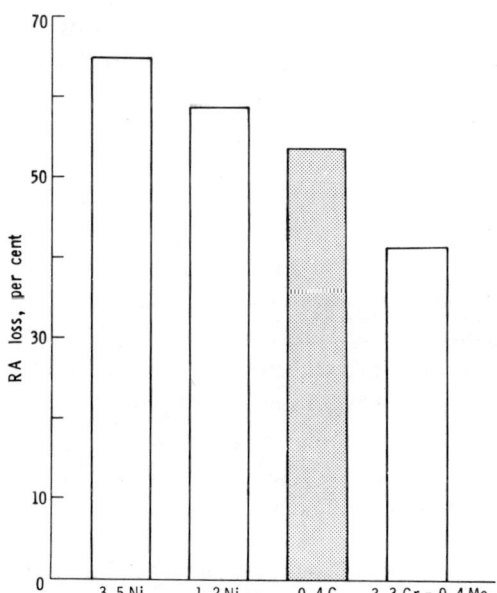

Fig. 5. Ductility (RA) loss for steels charged with 8 wt ppm H and tested in air. Reference steel (shaded) was 0.4 % C: others were 1.2 % and 3.5 % nickel steels, and a 3.3 % Cr–0.4 % Mo steel. All steels had an ultimate strength of 700 MN/m^2 ($\sim$ 100 ksi). (Data of Hobson and Sykes.[32])

example, the evidence suggests that at a constant high strength, K_{Ienv} decreases above about 4% Ni,[21] but the effect is small. At lower strengths, Fig. 5 shows that additions of 1.2% Ni and 3.5% Ni are deleterious in hydrogen tests, although there is not much difference from a plain carbon steel.[32] In nitrate environments,* there is a critical amount of about 2% Ni, above and below which performance is better.[34] In both boiling 42% $MgCl_2$ and in boiling 33% NaOH containing chloride, the critical amount appears to lie around 8%, although the composition of minimum resistance was not precisely identified and no microstructural or strength details were given.[33]

Additions of Mo appear detrimental in Fe–Ni alloys,[19] possibly due to the formation of intermetallic Ni_3Mo, analogously to the formation of Ni_3Ti in maraging steels described above. The Mo content was found to be either quite damaging[21] in salt water, or beneficial[34] in nitrate solutions, in Fe–C–Mo alloys, but variations in Mo had little effect in more highly alloyed steels of the 4340 type in salt water.[21,22] Additional study, especially of crack growth rate effects to complement the existing work oriented toward K_{Iscc}, is indicated for Mo and Ni, as was the case for Cr.

Interstitial Elements

The important interstitial elements in steel are carbon and nitrogen, and their behavior seems generally predictable.[20] Sandoz,[21–23] in his comprehensive studies on the role of alloying elements in environmental embrittlement of high-strength steels, found that increasing carbon from 0.15 to 0.53%, in a 4340 series, significantly decreased K_{Ienv}, but only under open-circuit conditions; he found no effect on K_{Ienv} for cathodic or anodic polarization.[22] The open-circuit data are shown in Fig. 6; there is some uncertainty[21] about the improvement beyond 0.4%C. The suggestion to explain Fig. 6 was that the crack tip changed from an anodic to a cathodic condition with increasing carbon content.[15,23] This deleterious effect of carbon (and nitrogen) has also been found by others,[19,34,35] although there is evidence that carbon may improve the cracking behavior of 18 Ni maraging steel.[13]

In lower-strength Fe–C alloys, when hydrogen is introduced by cathodic charging, an increasing carbon content increases the extent of hydrogen

---

* Nitrate environments are strongly oxidizing, but as discussed below, local hydrogen production is usually possible nevertheless.

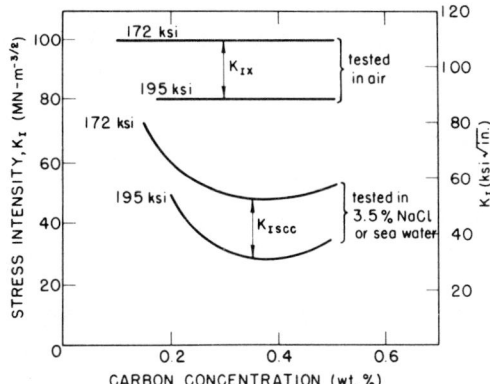

Fig. 6. Effect of carbon on the SCC resistance of 4340-type steel heat-treated to two strength levels; 145 psi = 1 MN/m². K_{IX} refers to unstable fast fracture in air. (From Sandoz.[21])

cracking,[36] or reduces the time to failure for both low-strength iron[19] and iron–nickel alloys.[19] In SCC studies performed in nitrate solutions,[34,35] similar results were obtained in water-quenched alloys; if the alloys were furnace-cooled, the behavior was more complex. Since these higher-carbon alloys had a martensitic structure when quenched and a ferritic–pearlitic one when slow-cooled, microstructural effects were probably dominant. There does appear, however, to be a generally deleterious effect of carbon under conditions in which cracking is the manifestation of hydrogen damage.

On the other hand, carbon does not appear to be harmful in those cases where hydrogen damage is manifested as a reduced reduction of area with no obvious change in fracture mode, as is often the case when lower-strength alloys are tested in high-pressure hydrogen,[37–39] pulled to failure after cathodic charging,[39–41] or tested in bending.[19] For example, Rauls and Hoffman[37] found no changes in the reduction of area, or in the relative degree of embrittlement, for a series of iron carbon alloys tested in pressurized hydrogen, for carbon contents ranging from 0.1 to 0.5 wt %. Although the variation in strength level may have affected their results,[37] Jewett et al.[38] found a similar result comparing Armco iron to 1020 steel of comparable strength levels. More work is needed to generalize the results, but considering the omnipresence of water vapor in service environments, and the well-documented deleterious effects of carbon on toughness and weldability, carbon content clearly has an important role to play in hostile-environment service. It should be noted again that nitrogen is expected to follow a behavior similar to carbon.

As an indication of how the effects of the substitutional and interstitial elements can interact, Fig. 7 compares the failure behavior of five high-strength steels in distilled water.[42] All were tempered to similar strengths: a yield strength of about 1485 MN/m² (215 ksi) and an ultimate strength of about 1650 MN/m² (239 ksi). The order-of-magnitude variations in failure times of these precracked specimens of different steels suggest an effect of composition.[15,42] Table 1 shows that careful analysis would be needed to isolate the critical elements; but note the increase in Cr and Si and decrease in Ni and Mn in the series 4340, D6aC, H-11. For the high-Ni, low-Cr 9Ni–4Co steels, note the (apparent) effects of C and Mo. It must be emphasized, however, that the microstructures of these steels were not described,[42] so that part of the spread in Fig. 7 could result from microstructural variation. There was also a spread in tempering temperatures,[42] which, as we will see, is another important variable in environmental susceptibility.

Table 2 summarizes our understanding of the role of alloying additions in terms of the two major fracture mechanics parameters, the ease of producing a crack as manifested by K_{Ienv}, and the ease of moving a crack, as manifested by the crack growth rate. Much of this correlation is styled after that of

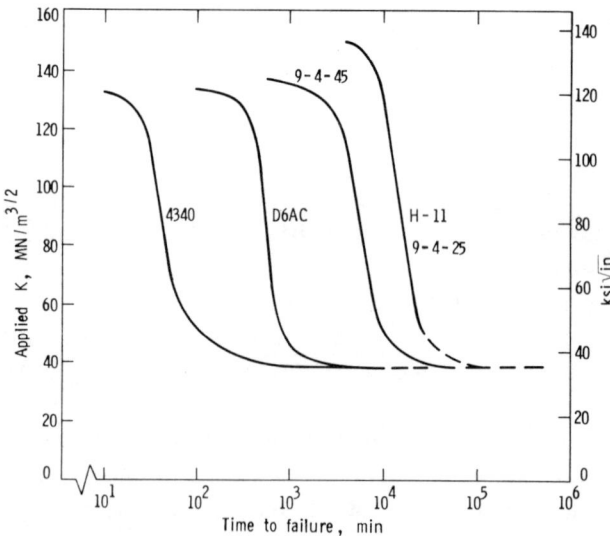

Fig. 7. Failure time as a function of applied stress intensity for precracked specimens of five steels of constant strength (see the text). Environment was distilled water. Ultimate tensile strength, 1650 MN/m². (From Steigerwald and Benjamin.[42])

Table 2. Effect of Solutes on Susceptibility of Steel to Hydrogen Embrittlement

Element	General susceptibility	K_{Ienv}	Crack growth rate
Manganese	Increases	Reduces	Not known
Sulfur and phosphorus	Appears to increase	Little effect	Should increase
Carbon (in carbon martensites)	Appears to increase	Reduces	Not known
Chromium	Increases	Little effect	Should increase
Titanium (in general)	Reduces	Not known	Not known
Titanium (in maraging steels)	Increases	Not known	Not known
Silicon	Reduces	Reduces	Reduces
Molybdenum	No consistent behavior	Little effect	Not known
Nickel	Not known	Little effect	Not known

Gerberich,[15] who concluded from his work, and that of others,[21] [23] that chromium, molybdenum, sulfur, phosphorus, cobalt, and silicon have a minimal effect on K_{Ienv}, while carbon and manganese reduce it significantly. Silicon was also found to reduce the crack growth rate. Table 2 has interleaved these results with others,[20] and reinterpreted some of the early data, so that although some of the conclusions are similar, a number of others are distinctly different. Even a glance at Table 2 makes it obvious that much research remains to be done on composition effects, and it is to be hoped that future studies will attempt to measure not only K_{Iscc} or K_{Ienv}, but also the crack-growth-rate behavior. Only in this way will insight be possible into the complex interactions among alloying elements. Finally, as we discuss in a later section, the development of models for the environmental fracture process cannot proceed without detailed data of the kind implied by Table 2.

Microstructure and Thermal Treatment

There is a close correlation between microstructure and alloying elements, making it difficult to separate these variables. In the previous section we have tacitly ignored microstructures. In this section we will do the same for alloying elements. This is not as arbitrary as it may appear, since the data to be presented show that a number of microstructures exhibit fairly fixed behavior, even with wide variations in alloy content.

There is general agreement that untempered martensite promotes environmental embrittlement.[10,27,43] This appears to be in large part due to the brittle nature of the martensite plates.[10] Specifically, it has been suggested that the high elastic stresses associated with plate formation are the major factor in embrittlement, since it was found that high residual stresses promote hydrogen-induced cracking even in the absence of martensite.[44] Such a conclusion would be in agreement with results on TRIP steels in hydrogen.[45,46] Diffusion does not seem important since hydrogen diffuses more slowly in untempered than in tempered martensite.[14]

At high strength levels, it appears that the best environmental resistance, at least against hydrogen, occurs in well-tempered martensite or bainite microstructures, which have been ausworked* to produce refined plate sizes and a uniform dispersion of fine carbides.[47,48] Medium and low strength levels are somewhat more difficult to assess, but such assessment is important because of the widespread use of steels in this strength range. A specific difficulty is that grain size and composition effects can modify the ranking of microstructures.[49] With these constraints in mind, however, we can list the microstructures from best to poorest environmental performance. The order would be as follows: quenched and tempered bainite or martensite is best,[14,50] followed by a spheroidized structure of uniformly dispersed carbides, with fine carbides being preferred to coarse ones[10,16,23]; a normalized structure is poorest,[10] although exceptions can be found,[20,51] particularly for spheroidal and pearlitic structures. These rankings will now be discussed in more detail.

The susceptibility difference between a normalized pearlite and a ferrite–spheroidal carbide microstructure is important because of the interest in using such structures in medium-strength applications. While the evidence is somewhat mixed,[20] especially when results are compared for cathodic charging and for nitrate or caustic solutions, most investigators consider spheroidal structures to be superior.[10,16,23] However, one study[51] has shown pearlite to be superior. Hobson and Hewitt,[51] for the same strength level of

---

* Ausworking or ausforming is the process of heavily deforming an austenitized steel at bainite bay temperatures prior to transforming to martensite.

550 MN/m² ($\sim$80 ksi), found a spheroidized carbide structure to be three times more susceptible to hydrogen embrittlement than a ferrite–pearlite mixture. Part of the discrepancy could lie with fracture mode changes or possibly grain size effects. Henthorne and Parkins[49] have shown an inversion effect; at equal strength levels, a coarse-grained spheroidal structure is more resistant to environmental cracking than pearlite, but this is reversed for fine grain sizes. They account for this by the interesting observation that the initiation stress for cracking is grain-size-dependent for pearlitic steels, but not for spheroidal ones.

Quenched and tempered bainite or martensite is superior to a pearlitic microstructure,[14,50] but there is no clear consensus about spheroidized steels.[14,23,51] To indicate the degree of complexity, the previously cited result of Hobson and Hewitt[51] can be compared with other results of the same experimental series. At 640 MN/m² ($\sim$93 ksi) strength, spheroidized steel is only slightly poorer than tempered bainite, which, in turn, at strengths of 865 and 1240 MN/m² (125 and 180 ksi), shows an equivalent resistance to hydrogen embrittlement as does tempered martensite. This is an area that will require careful, systematic study under controlled conditions.

The Hobson and Sykes data[32] in Fig. 8 illustrate the interrelation between microstructure and strength level for a Cr–Mo steel. As the tempering temperature was raised, strength was reduced, as was RA loss due to hydrogen. But when tempering temperatures reached 700°C, spheroidization began, and

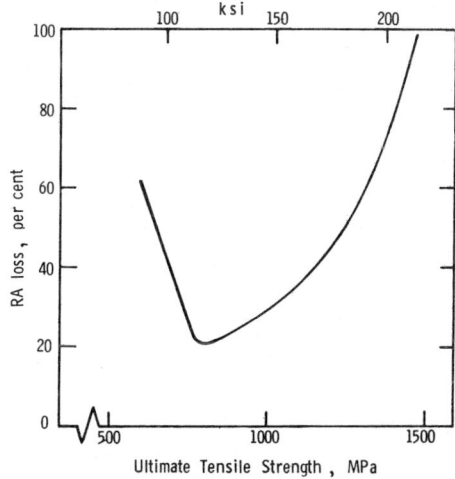

Fig. 8. Ductility (RA) loss as a function of strength level, in a 3.3 % Cr–0.4 % Mo–0.26 % C steel containing 3.6 wt ppm H. Changes in strength obtained by tempering to produce different microstructures. (Data of Hobson and Sykes.[32])

both the strength and hydrogen performance trends were reversed. The lowest-strength material in Fig. 8 was significantly spheroidized.[32] The problem is therefore one of careful consideration of both the microstructure and strength-level variables in any given case, to determine which one is of predominant importance. In addition, as mentioned, the fracture mode (brittle or ductile) may modify microstructural rankings.

A major difficulty in accurately establishing relative microstructural rankings is that most investigators have neither controlled their tempering conditions nor established the relation between microstructure and crack path. An obvious manifestation of this is the relation between environmentally induced and temper-induced embrittlement, for a variety of microstructures. This has been demonstrated in a 4340 steel, with a quenched and tempered microstructure,[52] in a 20 CND 10 (French) bainitic steel,[53] in martensitic stainless steels,[54] in Ni–Cr steels,[11,41] and in HY 130 steels.[12] It has been shown that these mutual embrittlement effects can occur when the steel has been temper embrittled at either 535 K or 810 K (500°F or 1000°F). This connected susceptibility can cause a change in environmental crack mode from transgranular to intergranular.[12,53] We believe this to be an extremely important observation, and one that may be of greatest importance in the development of effective microstructural control for environmental failure. It involves the general area of the effect of crack path on environmental susceptibility, as we now discuss.

Any analysis of crack path behavior must recognize two points. First, many reported studies on environmental fracture include little or no fractography, so not all data are equally comparable. Second, crack paths depend strongly on variables such as stress intensity and crack velocity; thus comparisons must be made at sensibly equal values of these parameters. This point is illustrated in Fig. 9, which shows the transition from ductile fracture at high K, through so-called (ductile) quasi-cleavage, to intergranular fracture at low K. This exact sequence is not always observed,[12] but it emphasizes the necessity of making comparisons with care. With these problems in mind, crack path effects in environmental fracture can be discussed.

It has often been stated (see also Fig. 8) that environmental cracking resistance improves with increasing tempering temperature,[9,15,23,27] with the provision that temperature ranges for temper embrittlement are avoided.[7,27,52] It has been further suggested that this may result from a change in hydrogen diffusivity,[15] from an increased ease of intergranular cracking,[9] or from the occurrence of mixed mode cracking.[54] However, there is little direct

Fig. 9. Schematic depiction of fracture surfaces for two 4340-type steels (0.15% and 0.28% C), under hydrogen-assisted (HAC) and SCC conditions. Crack propagation direction left to right. Wedge-loaded specimens exhibited "pop-in" (P-I) as K decreased to K_c: subcritical crack growth then occurred until crack arrest at K_{Iscc}. Fracture modes were ductile (D), "quasi-cleavage" (QC), and intergranular (IG); fast fracture (FF) subsequent to test also shown for specimens not fractured in environment. (From Beachem.[55])

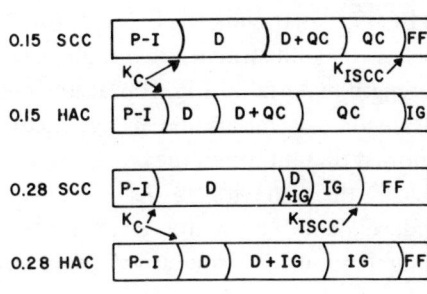

evidence to support any of these. More important, the generality of such a temperature–cracking relationship must be questioned, since similar behavior is not found in Cr martensitic stainless steels.[54,56]

It appears that variations in tempering temperature can modify the subsequent crack path[36,41]; the critical point is that the crack path which occurs must be compatible with the microstructure. For example, it would be of limited value to produce an extremely fine structure within the grains, if the environmentally induced crack path were intergranular. Such a contradiction might, for example, explain the observation that ausforming significantly improved the SCC resistance of D6aC steel, but had little effect on H-11, while improving the toughness of each[48]; the steels had been tempered at widely different temperatures. Compare this behavior to Fig. 7.

Cooling rate after tempering has also been found to affect susceptibility to hydrogen-induced failure, with air cooling better than water quenching.[19] This effect may be related to the relationships between cooling rate, crack path, and interstitial partitioning and distribution found in purified iron.[20] It is important to establish the generality of this observation for steels of various strength levels.

Grain Size and Texture

The role of grain size in environmental cracking of steels seems well documented. For a wide range of polarization conditions and environments, the general effect is that grain refinement improves the resistance to cracking.[16,18] This has been found in such varied iron-base alloys as 4340,[13] AFC 77,[23] maraging steel,[27,57] purified iron,[20,50] and Fe–Ti alloys.[20,58] This behavior is illustrated for a high-strength steel in Fig. 10a, and for a low-strength material in Fig. 10b. In the high-strength steel (Fig. 10a), where the behavior was monitored using fracture mechanics procedures, the results showed that although the crack growth rate decreased with decreasing grain size,[13] the behavior of K_{Icnv} was mixed, with it either increasing[23] or independent of austenitic grain size.[13] Again, it must be cautioned that to be unambiguous, the concurrent grain size/strength relationship must also be considered. The strong dependence of strength on grain size makes this a difficult variable to control.

The relation between texture developed through rolling, and environmental embrittlement, has been studied by several investigators. In both 4130 and 4340 steels, the transverse, and particularly short transverse grain directions were found to be more susceptible to stress corrosion cracking than the longitudinal direction,[59,60] and this has also been found to be the case for other steels.[26] Uhlig and co-workers[19,61] found that cold working improves the resistance of the material to cathodically induced cracking when it was stressed parallel to the rolling direction, but decreased it when stressed

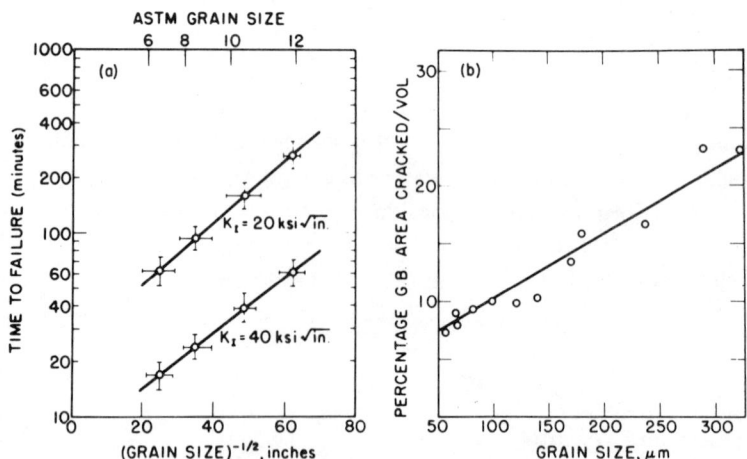

Fig. 10. Resistance to hydrogen cracking as a function of grain size: (a) 4340 steels of tensile strength 1880 MN/m² (273 ksi) at two applied K levels (0.9 ksi in.$^{1/2}$ = 1 MN/m$^{3/2}$) (from Proctor and Paxton[13]); (b) an iron–titanium alloy (Fe–0.15 % Ti) of low strength (from Rath and Bernstein[18]).

perpendicular to the rolling direction. Uhlig's contention[61] that SCC failures (claimed to be distinct from hydrogen failures) are independent of rolling direction requires considerably more support to be convincing.

While there is no indication of the operating crack path in the above cases, the results are compatible with a study of the relation of cathodic cracking to grain and grain boundary orientation.[58] Specifically, it was found that while low-angle boundaries are immune to cracking, other specific orientations exhibit extensive grain boundary cracking. Grain reorientation during rolling generally leads to texture formation, so that longitudinal and transverse stressing reflects the behavior of low- and high-angle boundaries, respectively. Suitable combinations of texture and crack path control would thus seem promising as a useful technique of inhibiting cracking.

An alternative interpretation of the above results would be that the effect is not one of crystallographic texture, as in the iron study,[58] but is one of mechanical texture or fibering. Hydrogen accumulation at grain boundaries or at inclusion-matrix interfaces should aid cracking,[62] and transverse directions of stress place larger areas of such boundaries under normal stresses. We shall return to this topic in more detail in discussing aluminum alloys, but the results of Ohnishi[63] support such an interpretation.

Synergistic Effects

Throughout this examination of the behavior of steels, we have, in general, assumed that the observed trends and effects are independent variables, and their total behavior could thus be arrived at by superposition of individual behaviors.[20] It is obvious that this cannot exactly be the case, and that considerable interaction between variables must exist.[10,21,22] However, this does not negate the validity of the basic premise that variable control is an effective technique to control failure.[64] For example, while grain size variations may affect the specific ranking of microstructures, this is secondary to the dominant and generally predictable role that microstructure plays. Nevertheless, such interactions are important and must be considered for alloy ranking and selection. Synergistic effects in relation to problems of alloy design have been discussed elsewhere at greater length.[64] The results also suggest that metallurgical variables play a similar role in bcc iron alloys of varying strength levels and when tested in many different types of environment.

AUSTENITIC STAINLESS STEELS

The austenitic stainless steels are generally different in character from the ferritic and martensitic steels just discussed. They have the face-centered cubic (fcc) structure, and aside from their resistance to general corrosion, they are as a class relatively resistant to hydrogen embrittlement and to some kinds of SCC phenomena. As we show and as others have demonstrated, however, their characteristics differ considerably from alloy to alloy, so that broad generalizations about these steels as a class are inappropriate. Again, the effects of composition and microstructure are important in these alloys.

Composition

Major Elements

Austenitic stainless steels are essentially iron–chromium–nickel ternary alloys, with the major alloying elements being 15–25 % Cr and 7–25 % Ni (Table 3). Ni and Cr act in concert; although Ni is normally regarded as an austenite stabilizer and Cr as a ferrite stabilizer, Cr tends to prevent martensite formation in the presence of Ni and therefore increases the range of

Table 3. Typical Compositions of Austenitic Stainless Steels (wt %)

Alloy	Cr	Ni	Mn	C	N	Si	Mo	Other
Standard grades								
304	19.0	10.0	1.5	0.06	—	0.5	—	
304L	19.0	10.0	1.5	0.03	—	0.5	—	
309	23.0	14.0	1.5	0.20	—	0.5	—	
309S	23.0	14.0	1.5	0.06	—	0.5	—	
310	25.0	20.0	1.5	0.20	—	1.0	—	
316	17.0	12.0	1.5	0.08	—	0.5	3.0	
321	18.0	11.0	1.5	0.06	—	0.5	—	Ti: 5XC
347	18.0	11.0	1.5	0.06	—	0.5	—	Nb: 10XC
Special grades								
18Cr–15Mn	18.0	—	15.0	0.04	0.5	0.5	—	
21Cr–6Ni–9Mn	20.0	7.0	9.0	0.04	0.3	0.5	—	
22Cr–13Ni–5Mn	22.0	12.5	5.0	0.05	0.3	0.5	2.0	

metastable austenite. When the Cr–Ni ratio becomes large, however, there is a tendency to retain δ, the high-temperature ferrite phase. Figure 11 shows the approximate extent of the various phases at room temperature[65]; it should be noted that this is a metastable rather than an equilibrium diagram.

The effect on SCC of increasing concentrations of both Cr and Ni (usually for tests in boiling 42% $MgCl_2$) shows different patterns. As Ni content is increased above about 8%, SCC resistance steadily increases.[33,66–69] Cr is beneficial up to about 12% in a 10% Ni matrix[66]; at higher nickel contents, Cr contents of 25% or more are most beneficial, while a content of about 10% continues to be beneficial.[70] The interesting feature, then, is a minimum in Cr benefit around 15–18%.[66–70] There is a similar trend of behavior for Ni and Cr in hydrogen embrittlement studies, as has been noted[68,71,72]; increased Ni improves hydrogen behavior, as do Cr contents larger or smaller than 18%.

A discussion of mechanisms is postponed until a later section, but it is important here to point out a correlation. The results for Cr and Ni just discussed appear to be correlated with the stacking fault energy (SFE) of the austenite. Neff *et al.*[73] have collected numerous results on the composition dependence of SFE, and their summary plot, augmented with a few later data

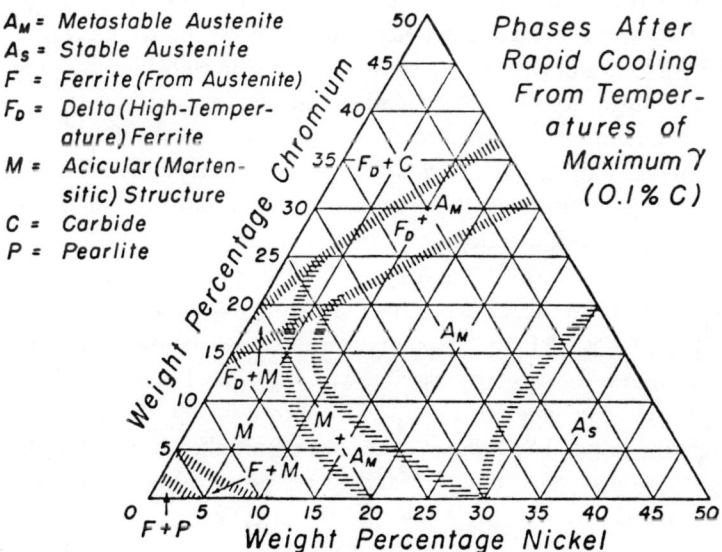

Fig. 11. Approximate diagram of phases present in Fe–Ni–Cr alloys after quenching; note that this is not an equilibrium diagram. (From the 1948 *Metals Handbook*.[65])

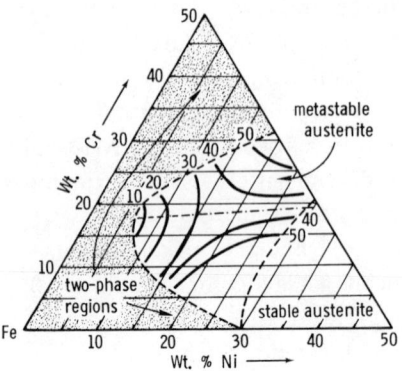

Fig. 12. Approximate iso-SFE contours in the iron-rich corner of the Fe–Ni–Cr diagram, based on data of a number of authors. Units of SFE are mJ/m^2 (ergs/cm^2). Note the minimum or "trough," indicated by a dash–dot line, at about 18% Cr. (From Rhodes and Thompson,[76] after Neff et al.[73]).

(e.g., Ref. 74), is shown in Fig. 12.* A minimum or "trough" is SFE at about the 18% Cr level is evident. Many studies of the effects of alloying additions have been conducted in this trough, thereby minimizing SFE throughout the alloy series, and leading to reduced generality of the results. The importance of SFE probably is that it is one of the factors which determine ease of cross-slip, or extent of slip planarity, during deformation.[68]

We believe that slip mode, which in these alloys can be controlled by SFE, is one of the important factors in determining how Ni and Cr affect resistance to both hydrogen embrittlement and SCC (at least in chloride environments). While other suggestions have been made to rationalize the effects of Ni and Cr content, for example the concept that Ni alters the crack tip electrochemistry by changing the local cathodic reaction rate,[77] the universality of such ideas is difficult to maintain in view of the parallels between some SCC data and tests in hydrogen gas. To illustrate this point, Fig. 13 shows data for three typical alloys in hydrogen tests,[72,74] and Fig. 14 shows these and other data as a function of SFE from Fig. 12. Note that the ductility loss appears to be a single-valued function of SFE (filled points are discussed below). SCC data for Fe–Ni–Cr alloys[70] are also correlatable with Fig. 12, as Fig. 15 shows; data for 18% Cr alloys of varying Ni content[78] are shown in Fig. 16 as a function of SFE. Here the original SFE data[78] have been multiplied by a factor of 2.3 to conform to modern SFE theory[79] and to the data of Fig. 12. Figure 16 is strikingly similar to Fig. 14. Thus slip planarity, expressed through SFE,

---

* There have recently been efforts[75] to supplement the experimental data of Fig. 12 with calculated values, but the failure of that technique to predict correctly the observed value[76] of SFE for AISI 310 casts doubt upon its accuracy. We therefore rely on the experimental determinations as summarized in Fig. 12.

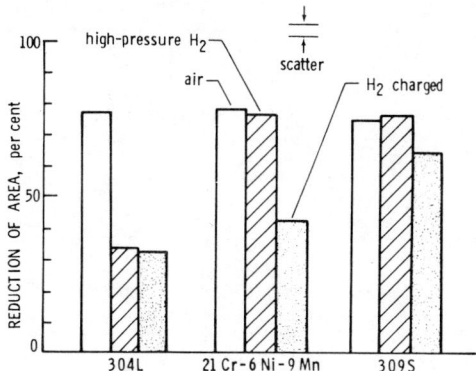

Fig. 13. Ductility (RA) in hydrogen for austenitic stainless steels, for tests in air, in 69-MPa hydrogen gas, and after hydrogen charging to supersaturation.[72,74]

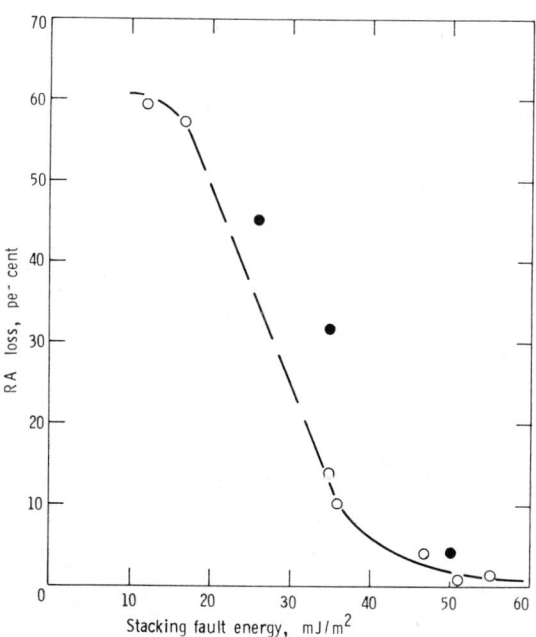

Fig. 14. Correlation of RA loss with SFE, using data from several sources.[39,72,74,83] Filled circles are for nitrogen-containing steels, discussed in the text.

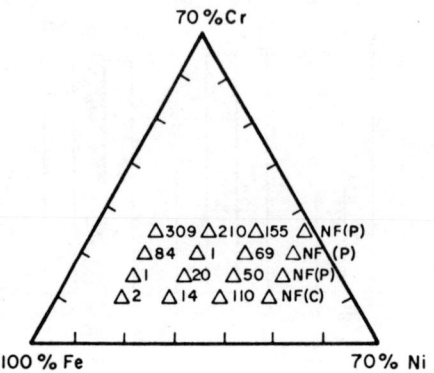

Fig. 15. SCC data for austenitic steel ternary alloys, specimens loaded at 140 MN/m^2 in 3% NaCl solution, under a pressure of 14 MPa oxygen. Numbers are times to failure (hours); NF (P) = not failed but pitted, NF (c) = not failed but cracked. (From Truman and Perry.[70])

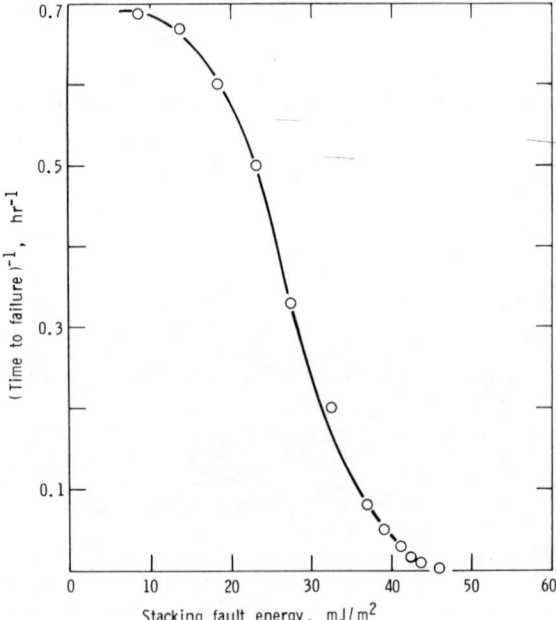

Fig. 16. SCC of austenitic stainless steels: time-to-failure (inverted) and SFE data (corrected by factor of 2.3) of Swann.[78]

strongly correlates with the environmental susceptibility in these alloys, whether it be hydrogen testing or SCC in chloride environments.

The similarity of Figs 14 and 16 cannot be simply interpreted as evidence that SCC and hydrogen embrittlement processes are equivalent in austenitic stainless steels; Fig. 14 refers to ductile failure of tensile specimens, while Fig. 16 refers to a fixed-load cracking process. (The 18% Cr content for Fig. 16 suggests roughly equivalent initiation times, so that the inverse TTF parameter essentially represents average crack velocity.) There is also evidence that slip planarity is not a sufficient condition for susceptibilty to SCC.[66,80] However, the similarity does suggest that the SCC process in these steels may have a hydrogen component as well as an anodic dissolution component, and also that metallurgical variables are important in both cases. Further pursuit of this point is postponed to the Discussion.

The other elements generally present in austenitic stainlesses are Mn (1–2%), C (0.03–0.25%), N (0.02–0.30%), and Si (1–3%); P is often present as a tramp element. Mn appears to be variable in its effect on SCC resistance; the "least ambiguous" experiments[66] showed no effect,[81] but beneficial as well as detrimental effects have been shown outside the 1–2% composition range.[66,68,69,82] There is some evidence, for aqueous test conditions, of diminished SCC resistance for Mn contents above ~3%.[83] In gaseous hydrogen environments, for higher concentration ranges, Mn is clearly deleterious.[39,84] These Mn additions, often as replacements for Ni, are made to increase nitrogen solubility and thus strengthening potential, so these effects may be due in part to slip planarity enhancement caused by the N, as discussed below; they may also occur because Mn increases SFE less than does Ni. Additional work on Mn and its effects on both SCC and hydrogen resistance is clearly needed.

As has been reviewed,[66, 68] amounts of C above 0.1% confer appreciable SCC resistance. On the other hand, increases in C in the range 0.01–0.05% appear damaging. It has been speculated[85] that the latter effect is due to interaction with other interstitials such as N rather than being intrinsic to C. There is in any case a carbon content of minimum resistance to SCC, at about 0.06%. It is well known that increasing C content accelerates sensitization in certain heat treatments, which in turn increases intergranular corrosion; but the occurrence of sensitization is not always correlated with either SCC[66] or hydrogen embrittlement.[68,74] Popular belief has it that SCC susceptibility is correlated with sensitization, and to the occurrence of intergranular corrosion in general; but as we shall show for other alloy systems in addition to stainless steels, such a correlation exhibits many exceptions and is therefore unreliable.

The behavior of N differs from that of C in that all additions of N accelerate SCC.[66–69,80,82,85–87] The acceleration is somewhat increased by low-temperature aging,[88] which may be due to interaction with C.[69,85] There is a similar adverse effect of N on hydrogen embrittlement. Figure 13 illustrates this for the alloys 309S and 21Cr–6Ni–9Mn. Both are stable austenites; that is, they form no strain-induced martensite on deformation, and they have very similar SFE values of about 35 mJ/m^2 (35 ergs/cm^2).[68] An example of the extended dislocation nodes present in 21–6–9 is shown in Fig. 17. But Fig. 13 shows that 21–6–9 performs more poorly than 309S when hydrogen charged, evidently because of the ~0.3% N added to 21–6–9 for strength purposes. Nitrogen effects of the same kind are evident in Fig. 14.

Swann showed that N produces very planar dislocation arrangements in stainless steels.[78] This effect, however, does not occur through a major change in SFE,[78] but seems to arise from the clustering of N atoms. The planar slip results from dislocation cutting of the clusters,[68.78] which makes subsequent slip on the same plane easier and inhibits subsequent cross slip. This interpretation is consistent with the aging effect,[88] which may develop the clusters. It is also consistent with a reduced effect of N in alloys of very easy cross slip,[80] and with increased RA loss due to increased planarity (Fig. 14). There is evidence[66,69,78,82,87] that phosphorus has similar effects on SCC as does N (both are group V elements). Slip planarity is increased by P with no change in SFE[78]; a synergistic interaction of P and N is suggested by the observation that P becomes less detrimental to chloride SCC as N content decreases.[89] The detrimental effects of P are also reduced by increased Si and

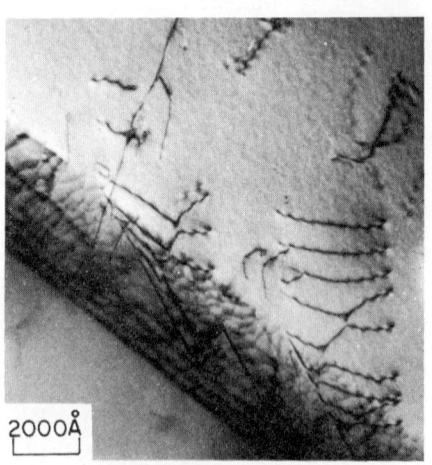

2000Å

Fig. 17. Transmission electron micrograph of 21Cr–6Ni–9Mn steel deformed about 1%. Note grain boundary dislocation sources and extended nodes at dislocation junctions.

decreased Ni.[69] P thus appears to be strongly affected by the presence of other elements.[83]

Silicon improves resistance to cracking and to ductility loss when present in sufficient amounts.[66,67,69,83,87,90] It is particularly effective above a concentration of 4%, with some evidence that the beneficial effect is for both crack initiation and propagation.[91] At such concentrations, however, Si stabilizes delta ferrite in the microstructure, so that the effect may be primarily microstructural, not compositional. As a solute in austenite, Si somewhat reduces the SFE.[77] This element may therefore serve as an example that a reduction in SFE does not necessarily lead to increased cracking or to other forms of damage, although the SFE reduction for small Si additions is slight and may be too small to be effective.[68] This latter conclusion is supported by the observation that Si contents in the range 0.8–1.5%, which have little effect on δ ferrite and are thus present in the austenite, do not alter SCC behavior.[69,82,92] In aqueous solutions, the role of silicon has been suggested to be electrochemical in nature,[66] although direct studies which investigate microstructure would be welcome; hydrogen gas studies would also be of interest.

Minor Elements

Some grades of austenitic steel contain 2–3% Mo, or Nb or Ti at concentrations controlled by the minimum C content (e.g., 5 times the minimum carbon). Mo is added to improve pitting resistance; for example, 316 stainless is essentially type 304 with 2.5% Mo. But studies of Mo additions have shown little effect[70,91] or detrimental effects,[66,69,81,82] evidently because a minimum in SCC resistance occurs at about 1.5% Mo.[66–68] Mo also shifts the carbon content of minimum SCC resistance from 0.06 to 0.30%.[66] It has been suggested[66] that the chemical similarity of Mo to Cr, both being group VIa elements, should mean that Mo would only be damaging if the (Mo + Cr) content lay in the range of poor performance due to Cr alone (i.e., 15–18%). There is evidence to support such a suggestion.[68,70] An alternative suggestion would be that Mo at low concentrations promotes planar slip through formation of clusters,[78] since such Mo clusters have been observed in stainless-like alloys.[93] At higher Mo concentrations, there is indirect evidence[75] that SFE is increased, thus accounting for the beneficial effect above 1.5% Mo. It should also be noted that Mo promotes intergranular SCC fracture in chloride environments,[92] contrary to the usual transgranular behavior.[66]

The elements Nb and Ti are added to obtain resistance to sensitization and thus to intergranular attack, since they tend to form carbides, but they decrease resistance to chloride SCC.[66,67,81,82,89] Both Nb and Ti decrease SFE of stainless steels when added in small amounts.[94] In the case of Ti, although small additions worsened resistance,[81,82,87] additions of 2% Ti appeared beneficial.[91] There may therefore be an intermediate Ti content which results in minimum SCC resistance. As in the case of Si, however, the benefits at large concentrations may arise from stabilization of δ ferrite. Benefits found for significant additions of Ti, Si, V, and Al have been explained[66,91] by assuming that volume fractions of δ ferrite as small as 5% can blunt cracks which are propagating in the austenite. We address this question below, but for the case of amounts which clearly remain in solution, Ti and Nb are detrimental.

Finally, there are several solute elements which have been studied in model alloys and are primarily of academic interest. These include V, which is somewhat beneficial,[91] especially when δ ferrite is stabilized,[66] and Al, which is variable in its effects.[67,82,91] The platinum elements (group VIII except for Fe, Co, and Ni) greatly accelerate SCC, evidently because of electrochemical effects.[66] They also promote a transition to intergranular fracture. A number of other elements have the behaviors indicated in Fig. 18, taken from Sedriks' review[67]; note that the present review has reassessed some of the results summarized in Fig. 18.

The group Vb elements in addition to N and P are evidently all detrimental: As, Sb, and Bi.[67] These latter elements are interesting since they are used as hydrogen recombination poisons in cathodic charging. Sulfur is also a poison and is known to be detrimental.[66,67] There is an increasing body of evidence, for several alloy classes, that solutes which apparently promote the stability of dissociated hydrogen in an alloy, also promote hydrogen embrittlement and cracking. However, the behavior of platinum group elements[66] indicates that the converse is not true. Two elements whose behavior should have a bearing on the viewpoint that SFE and environmental susceptibility are intimately related, are Cu and Co, since they affect SFE oppositely.[77,95] But as has been discussed,[68] neither the SFE changes nor changes in SCC resistance are more than modest,[77,82] so that little can be concluded from behavior of these elements.

As was mentioned for ferritic steels, multiple alloy additions are hard to assess as separate variables, particularly when an alloy contains six or more significant elements. The results cited for interactions of P with other elements are a graphic example. Detailed experiments to optimize alloys based on the

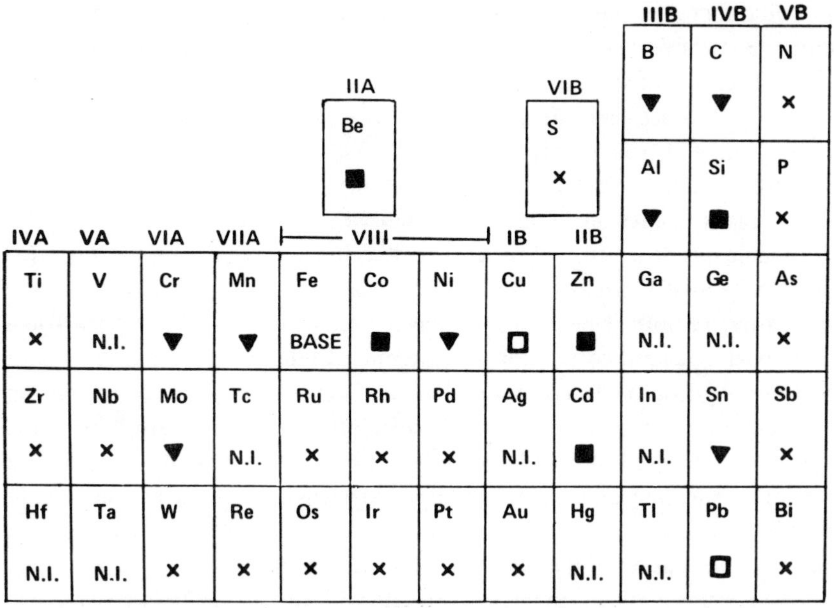

									IIIB	IVB	VB
									B	C	N
									▼	▼	×
			IIA			VIB			Al	Si	P
			Be			S			▼	■	×
			■			×					
IVA	VA	VIA	VIIA	⊢——— VIII ———⊣			IB	IIB			
Ti	V	Cr	Mn	Fe	Co	Ni	Cu	Zn	Ga	Ge	As
×	N.I.	▼	▼	BASE	■	▼	□	■	N.I.	N.I.	×
Zr	Nb	Mo	Tc	Ru	Rh	Pd	Ag	Cd	In	Sn	Sb
×	×	▼	N.I.	×	×	×	N.I.	■	N.I.	▼	×
Hf	Ta	W	Re	Os	Ir	Pt	Au	Hg	Tl	Pb	Bi
N.I.	N.I.	×	×	×	×	×	×	N.I.	N.I.	□	×

Fig. 18. Portion of periodic table indicating the effect of various elements on the resistance of stainless steels to SCC in chloride solutions: ■, beneficial; ▼, variable; □, no effect; ×, detrimental; N.I., not investigated. Recent data have altered certain of these conclusions, as discussed in the text. (From Sedriks.[67])

above assessments would therefore be important, and might contradict or at least modify in some respects the conclusions reached.

Microstructure and Thermal Treatment

The austenitic steels in general have a single-phase microstructure. The major exceptions are the presence of δ ferrite when sufficient ferrite-stabilizing elements, such as Cr, Si, or Ti, are present, and the formation of strain-induced martensite when some of these steels are deformed. The martensite may be bcc α' or hexagonal close-packed (hcp) ε, or both, depending on the steel. There is evidence that the presence of δ improves SCC resistance,[66,91,96] although the evidence would be more clear-cut if ferrite-free alloys had also been tested.[66,91] In hydrogen tests, in which the effects are manifested as a loss in RA, the presence of δ ferrite alters fracture morphology, with cracking occurring in austenite–δ interfaces.[97] These morphology changes resulted in no additional loss of RA relative to ferrite-free material in both 304L and 309S steels.[72,97,98]

It would therefore appear that δ ferrite can play a role in crack propagation, either as a less "crackable" phase, or as a phase in which crack tip electrochemical sharpening, as discussed at greater length below, occurs less readily.[60,64] Since such effects do not occur in hydrogen tests, it seems that δ plays a different role in ductile-type RA losses.

Formation of α' and ε martensite has often been suggested as a determinant in environmental susceptibility of austenitic stainless steels. But as Staehle has reviewed,[99] there are good arguments against a critical role for martensite in SCC behavior. Similar proposals for a causative martensite role in hydrogen embrittlement have also been discounted.[39,72,74,84,100,101] There does exist a general correlation of austenite stability with resistance to both SCC and hydrogen effects, although many exceptions are known. Thompson pointed out that this approximate correlation evidently arises from the correlation between SFE and austenite stability,[100] which in turn arises from the role of stacking faults in formation of ε martensite[102] as an intermediary in the formation of α' martensite.[103] Note, however, that hydrogen does not assist the formation of martensite,[104] and that both the SCC and hydrogen results discussed above tended to correlate with slip planarity in general, not with SFE alone. The pivotal example is the behavior of nitrogen, which increases environmental susceptibility without changing SFE. Thus, since it appears that martensite formation is neither necessary nor sufficient for either SCC or hydrogen embrittlement, it should be regarded as playing only a secondary or exacerbative role, not a causative role.

Another microstructural feature of significance is that of carbide precipitation at grain boundaries, or "sensitization." Cooling rates from the solution temperature (1250–1400 K) can have an effect when they produce carbide (usually $M_{23}C_6$) precipitation. This sensitized condition leads to accelerated intergranular corrosion, but has a mixed effect on chloride SCC,[66] including a sensitivity to the oxidizing potential in the environment. Cracking modes tend to be intergranular rather than transgranular, and may or may not be more rapid. Hydrogen gas results are generally parallel to these aqueous results.[68,74] It would be concluded that sensitization can alter the cracking process in SCC, either because of electrochemical changes[66,99] or structural behavior changes[74] in chromium-depleted regions near grain boundaries. As in the case of δ ferrite, such effects may well differ for cracking under electrochemical conditions and in hydrogen.

Deformation is another way in which the microstructure can be changed (independent of any martensite formation). Amounts of cold work up to about 10 % strain tend to accelerate SCC,[66] while larger strains reverse the trend and reduce SCC. For hydrogen embrittlement there is also evidence for initial worsening and then improvement at higher strain.[72,84] More pronounced changes are found with warm working, which permits partial retention of the worked structure. The effect of one method of such working, high-energy-rate forging (HERF), is shown in Fig. 19. The reason for the improved hydrogen performance appears to be the development of a dislocation structure characteristic of easy cross slip at the deformation temperature, although the alloy may deform by planar slip at room temperature.[84,101] Thus, the 304L in Fig. 19 exhibits greater HERFed improvement than 21–6–9 because its slip mode at room temperature is much more planar than its high-temprature mode, while 21–6–9 exhibits appreciable cross slip at both temperatures.

Grain Size and Texture

Austenitic stainless steels, when moderately deformed or annealed, are at best weakly textured, so that no large effects due to texture are expected. Grain size[116] can play a role, with grain size reduction offering a moderate decrease in cracking under dynamic loading[105] or statistically at a given percentage of the respective yield stresses.[101,106] There are preliminary indications of the same effect in hydrogen embrittlement of 304L.[107]

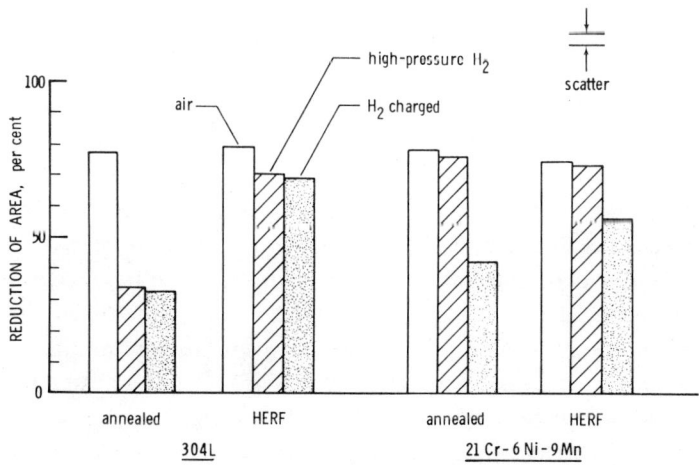

Fig. 19. Ductility of 304L and 21Cr–6Ni–9Mn alloys tested in air, in 69-MPa hydrogen, or after hydrogen charging to supersaturation. Effect of thermo-mechanical history (HERF) shown. (From Thompson.[72])

OTHER IRON-BASE AUSTENITES

There are a variety of iron-base alloys which are neither conventional (bcc) steels nor stainless (i.e., Cr-containing) austenitic (fcc) alloys. The information on some of these alloys provides a fuller picture of the behavior of steels, and is summarized briefly, although relatively little work has been reported.

Single-Phase Alloys

Binary alloys of iron and nickel are austenitic above about 30% Ni. As Smith has reviewed,[108] the ductility loss due to hydrogen charging decreases rapidly as Ni content increases above 30%, and at 50% Ni the loss is nil. Tests of an Fe–38% Ni alloy in a chloride-containing caustic solution and of an Fe–42% Ni alloy in boiling $MgCl_2$ both showed no cracking in 7–14 days,[33] which was a considerable improvement over Fe–8% Ni alloys (discussed under Steels). Alloys of 36% Ni and 51% Ni which were hydrogen-charged showed no loss in ductility.[109] Such nickel-rich alloys are stable austenites and exhibit easy cross-slip, although the benefit of Ni apparently extends to bcc alloys also. The Fe–Ni–Cr alloys of the Incoloy family will be discussed with the nickel alloys.

Significant effort has been devoted to Fe–Mn alloys, starting with the work of Uhlig.[24] Above about 14% Mn, these alloys tend to be austenitic at room temperature,[110,111] and as Mn content increases above 14%, SFE increases.[110,112] Microstructural observations in a series of alloys above 14% Mn[110] are parallel to observations on stainless steels of similarly increasing SFE. A broad correlation of resistance to hydrogen-induced ductility loss with increasing Mn content (i.e., planar slip expressed through SFE) would therefore be expected by analogy with behavior of stainless steels.

Studies of hydrogen behavior have uniformly shown that increasing Mn content above about 14% leads to increasing ductility in hydrogen.[24,113 115] Some of the experiments,[113,115] however, were performed in ~1-atm hydrogen gas, which is not a very severe test environment for austenitic alloys in general. More severe conditions, such as hydrogen charging or testing in 69-MPa (10,000-psi) hydrogen gas, showed greater ductility losses than the 1-atm tests, and were also in good agreement (e.g., the RA measured for 20% Mn alloys was 12% in hydrogen[24,114]). Addition of carbon improves performance of these alloys,[24] as does addition of Cr,[39,115] and both elements tend to stabilize the austenite and thus raise SFE. The behavior therefore seems generally parallel to that of stainless steels. The overall pattern of behavior, however, is that Fe–Mn alloys are rather distinctly embrittled by hydrogen, both in fracture mode[114] and as expressed by RA loss.[24,114] Thus, unless thermomechanical treatments prove successful,[113] these alloys do not seem promising for hydrogen service. More work would be appropriate, both with more complex alloys and with HERF or other processing.

Alloys used for glass–metal and ceramic–metal seals are typically austenitic Fe–Ni–Co ternary alloys, whose thermal expansivity matches that of the glass or ceramic. Two of these have been studied using hydrogen charging and 69-MPa gas tests[117]: Fe–29 % Ni–17 % Co (trade name Kovar) and Fe–27 % Ni–25 % Co (trade name Ceramvar), which have annealed yield strengths of about 320 MN/m². Results for one of these alloys are shown in Fig. 20; both alloys showed nil ductility loss in hydrogen tests.[117] These alloys are fairly stable austenites, and thus should exhibit nonplanar slip, a point which should be investigated as part of the general correlation between slip mode and hydrogen performance.

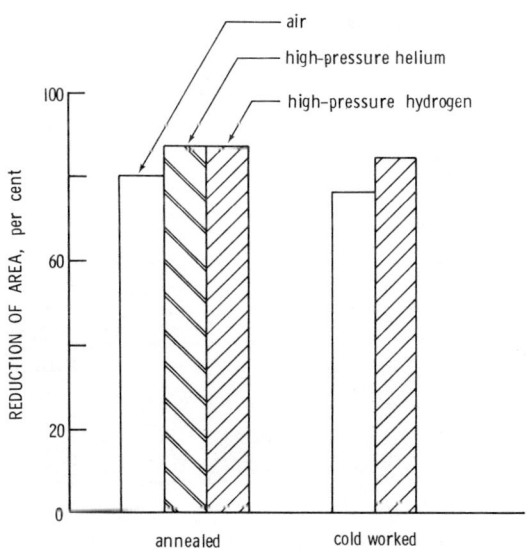

Fig. 20. Ductility (RA) of Fe–27Ni–25Co alloy (Ceramvar) for tests in air or 69-MPa helium and hydrogen. (From Thompson and Posey.[117])

Precipitation-Strengthened Alloys

In order to discuss precipitation-strengthened austenites, it is necessary to distinguish such alloys from the "precipitation-hardenable" or PH steels (see Table 1). The PH steels are heat-treated to produce a microstructure of precipitates (e.g., Fe−Ni−Al or Ni−Nb compounds) in a martensitic matrix; they are thus highly alloyed, high-strength, corrosion-resistant steels in the heat-treated condition. Their strength arises both from the precipitates and from the martensitic matix, so that the PH label is a slight misnomer. They are also rather susceptible to hydrogen embrittlement.[100,118,119]

This section discusses alloys of austenitic matrix which are non-martensitic and strengthened primarily by precipitation. The usual precipitate is the ordered γ' phase familiar in nickel-base superalloys, of composition $Ni_3(Al, Ti)$. For example, the alloy A-286 is a 15% Cr–25% Ni stainless steel, with additions of 2.25% Ti and 0.2% Al to form γ'. In the commercial A-286 alloy, both SCC[66,120] and hydrogen embrittlement[72,118,120,121] have been observed, as well as crack growth under sustained load in high-pressure hydrogen.[122] Figure 21 illustrates the degree of ductility loss, about 50%, and the fact that HERF working is not beneficial (cf. Fig. 19) when the material is subsequently aged to form γ'.

The presence of γ' therefore appears to be critical to the hydrogen-induced losses in RA in A-286. The presence of small, ordered, coherent precipitates of γ' should result in planar slip, since dislocations cut or shear

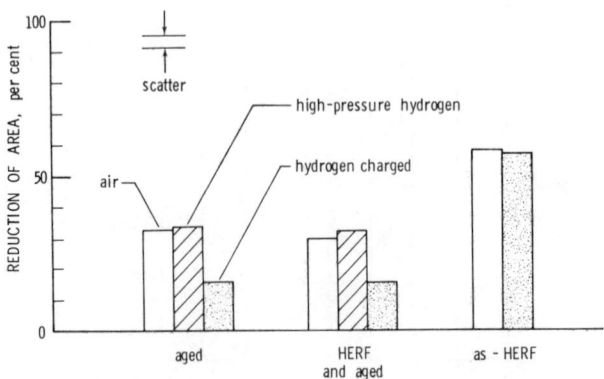

Fig. 21. Behavior of commercial A-286 with various histories, for tests after hydrogen charging to supersaturation from the gas phase. (From Thompson.[72])

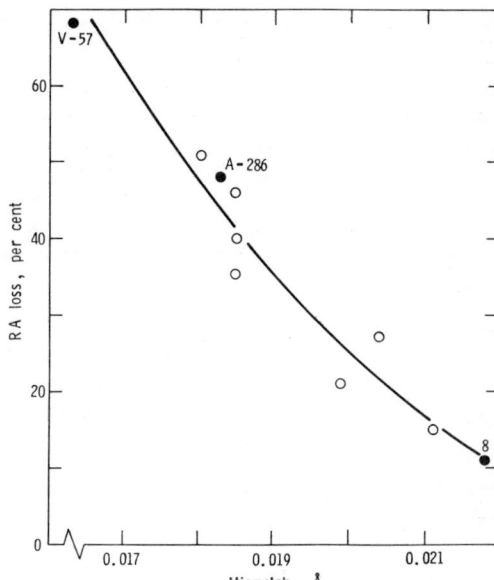

Fig. 22. Relation between ductility loss and matrix-precipitate mismatch for alloys similar to A-286. The commercial alloys are shown for comparison; alloy 8 is listed in Table 4. (From Thompson.[125])

such precipitates in the initial stages of deformation.[123] Since it is the first few percent strain which seem to be critical for hydrogen ductility losses,[39,72,84,100] the factors which influence slip planarity have been studied,[124] and it was observed that the crystallographic mismatch or misfit between γ' and the austenitic matrix is an important element in the extent of slip planarity.[124][126] When the mismatch δ is significant, coherency is lost at some plastic strain ε_c which is approximately given by[124,126]

$$\delta + \varepsilon_c \geqslant \mathbf{b}/r$$

where $\mathbf{b}$ is the Burgers vector and r the particle radius. Thus, for a given r, increasing δ corresponds to decreasing ε_c and to decreasing hydrogen accumulation at particles. A series of alloys of different composition, similar to A-286, were found to exhibit such a correlation, as shown in Fig. 22. Figure 22 is striking because it illustrates the consequences, in hydrogen tests, of slip planarity in general, whether caused by low SFE, by dislocation cutting of precipitates or clusters, or whatever cause. The trends in chloride SCC of A-286 with aging[66] also appear compatible with this picture. This conclusion has important implications for both iron-base and nickel-base alloys strengthened by γ', as we shall discuss.

Table 4. Composition of Precipitation-Strengthened Steels (wt %)

Alloy	Cr	Ni	Ti	Al	Mo	Si	Mn	C	B
A-286	15.0	25.0	2.25	0.2	1.2	0.6	1.5	0.05	0.005
Alloy 8[a]	14.5	30.5	2.1	0.25	1.2	0.15	0.1	0.02	0.001
V-57	14.8	26.0	3.0	0.25	1.25	0.55	0.25	0.05	0.008
CG-27	13.0	38.0	2.5	1.6	6.0	—	—	0.05	0.010

[a] Modification[124] of A-286 shown in Fig. 22.

The alloys V-57 and CG-27 are of the same type as A-286 (see Table 4). Both show ductility losses in hydrogen, with the loss for V-57 (Fig. 22) being similar to that for A-286.[125] The CG-27, with 6% Mo, exhibited very low ductility in hydrogen tests,[125] as did a modified A-286 composition with 5% Mo.[124] The Mo is evidently deleterious because it forms a compound which is very brittle, even without hydrogen.[124,125]

ALUMINUM ALLOYS

The majority of published work on the fcc matrix aluminum alloys has been on SCC behavior rather than on hydrogen embrittlement, although that situation is changing. We have discussed elsewhere[68] some of the current ideas on a hydrogen role in aluminum alloy SCC, and summarize that material below after reviewing the evidence on SCC. Aluminum alloys differ from nearly all the alloys discussed above in that they are almost exclusively strengthened by precipitation of second phases. The discussion accordingly is presented somewhat differently.

Composition

Composition of aluminum alloys can play a modest but significant role in SCC behavior. It will be necessary in discussing composition to refer to the physical metallurgy of the important alloy classes. The metallurgy of the binary alloys on which the commercial alloys are based has been reviewed in detail,[123,126,127] and the ternary and quaternary alloy behavior has also been summarized.[2,3,123,127-130] We shall largely restrict ourselves to description of precipitation sequences and morphologies, as these apply to SCC behavior.

The commercially important alloys of aluminum fall into four classes, by their Aluminum Association numbers: the 5000 series, essentially Al–Mg binary alloys; the 6000 series, essentially Al–Mg–Si ternaries; the 2000 series, Al–Cu or Al–Cu–Mg alloys; and the 7000 series, Al–Zn–Mg alloys which usually contain Cu. The latter are, in general, the strongest alloys, and also are most susceptible to SCC, while the 5000 and 6000 series are generally rather resistant to SCC but are only moderately strong. Compositions of typical alloys are shown in Table 5. The major elements in these alloys are sufficiently different that only a rather broad generalization on the role of composition seems appropriate: increasing the total amount of solute, that is, the amount which *can* be put in supersaturated solid solution, increases the susceptibility to SCC.[2,3,130] In 7000 alloys, there is some evidence[131] that a relatively low Mg/Zn ratio favors SCC resistance.

There are a variety of minor elements in aluminum alloys and these essentially have two kinds of effects. One is the tendency of many of them to form insoluble intermetallic particles, which pin grain boundaries and thereby stabilize the wrought grain shape, as shown in Fig. 23. They therefore prevent

Table 5. Typical Composition of Aluminum Alloys (wt%)

Alloy[a]	Cu	Zn	Mg	Si	Fe	Mn	Cr	Zr	Other
2014	4.4	0.2	0.4	0.8	0.6	0.8	0.1	—	—
2024	4.5	0.2	1.5	0.5	0.5	0.6	0.1	—	—
2219	6.2	0.1	0.02	0.2	0.3	0.3	—	0.15	0.1 V
2618	2.4	—	1.5	0.2	1.0	—	—	—	1.0 Ni
5083	0.1	0.2	4.5	0.4	0.4	0.7	0.15	—	—
5456	0.15	0.2	5.0	0.2	0.2	0.7	0.15	—	—
CS19	—	0.1	8.25	0.05	0.1	0.4	0.1	—	—
6061	0.25	0.2	1.0	0.6	0.6	0.1	0.25	—	—
7049	1.5	7.6	2.5	0.2	0.3	—	0.15	—	—
7050	2.3	6.3	2.3	0.1	0.1	—	—	0.1	—
7075	1.5	5.5	2.5	0.3	0.5	0.3	0.3	—	—
7079	0.6	4.3	3.3	0.3	0.3	0.2	0.2	—	—
7175	1.5	5.5	2.5	0.1	0.15	—	0.25	—	—
7178	2.0	6.8	2.7	0.4	0.5	0.3	0.3	—	—
MA52	1.2	6.7	2.5	0.1	0.1	0.1	—	0.15	—

[a] These alloys typically also contain 0.1–0.2 Ti, and up to 0.15 other elements.

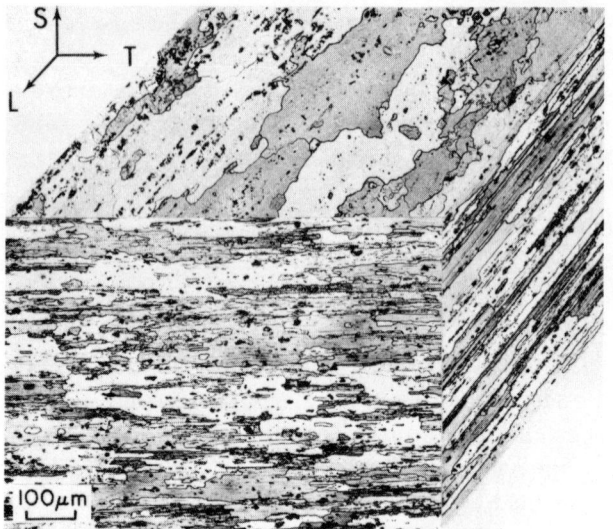

Fig. 23. Light micrograph of grain structure in 7075-T6 plate. Directions are as follows: L, longitudinal or rolling; T, transverse or width; S, short transverse or thickness.

development of an equiaxed structure. The elements of this class are Mn, Zr, and Cr, which affect grain shape in all four alloy classes. Grain shape is quite important in SCC of aluminum alloys, as we show below, and the many studies which have used model alloys of equiaxed structure must be regarded with caution. Such alloys can be used to separate the effects of individual element additions, but they do not model a commercial alloy, which is more complex chemically and microstructurally. It can thus be expected that at least some and perhaps many of the conclusions reached with model alloys will not apply to complex commercial alloys of wrought grain shape.

Minor elements can also affect precipitation of the strengthening precipitate, altering the kinetics and sometimes the morphology. This effect is particularly significant in 5000 alloys. In these alloys, the sequence seems to be[123]

$$\beta' \rightarrow \beta$$

where β is Mg_5Al_8; there is no clear evidence for GP zones preceding β' formation. The alloys can readily be prepared as metastable Al–Mg solid solutions, particularly at lower Mg contents, as in 5083 and 5456, since precipitation of the equilibrium β phase is quite sluggish. The β' nucleates heterogeneously, particularly at grain boundaries,[123] and as the β slowly develops, it tends to form a continuous layer. Evidently because they are strongly anodic to the matrix,[128] such β layers can cause considerable intergranular corrosion (not necessarily SCC). It is true in these alloys,[2] as we pointed out for other systems, that SCC susceptibility is sometimes but not always correlated with intergranular corrosion. Increasing Mg content, then, increases the alloy instability (i.e., the tendency to form β in service), and a variety of treatments and alloy additions have been developed to stabilize the 5000 alloys against formation of grain boundary β. For example, cold work followed by an anneal high in the $\alpha + \beta$ field stimulates nucleation of β' on dislocations throughout the matrix and reduces subsequent instability.[132]

Addition of Mn to binary Al–Mg alloys is beneficial in increasing the precipitation rate of β'; both Mn and Cr additions stabilize the wrought grain structure[133] and increase strength.[134] Additions of 0.2–0.4 % Bi also seem to help stabilize the alloy through formation of Bi_2Mg_3 particles;[135] additions of Cu and Zr have also been shown to improve SCC resistance.[136] When properly stabilized, 5000 alloys can perform fairly well in seawater environments,[2] although there is some evidence that strength increases at high-Mg levels are paralleled by slight decreases in SCC resistance.[134] New alloys, such as CS19, which are tough and weldable in addition to having yield strengths in excess of 200 MN/m² (30 ksi), exhibit maximum SCC sensitivity in the short transverse or S direction[134] (see Fig. 23); we shall see that many aluminum alloys behave similarly.

The precipitation sequence in 6000 alloys, such as 6061 and others, proceeds as follows[123]:

$$\text{GP zones} \rightarrow \text{ordered zones} \rightarrow \beta' \rightarrow \beta$$

where β is Mg_2Si. All these precipitates occur as needles. In commercial alloys, excess Si above the stoichiometric Mg/Si ratio causes precipitation of elemental Si, particularly at grain boundaries. Addition of about 0.2% excess Si increases strength, while lowering SCC resistance somewhat,[128] although these alloys in all conditions are rather resistant to SCC. Excess Si at grain boundaries accelerates intergranular corrosion,[2] but as we have noted before, the occurrence of such corrosion correlates poorly with SCC resistance. As in the 5000 alloys, additions of Mn and Cr increase strength and improve SCC resistance, partly through their effect on grain structure,[68] and partly through

Table 6. Mechanical and SCC Properties of Typical Aluminum Alloys[2,141,149]

Alloy, temper[a]	Product form	Minimum 0.2 % yield strength (MN/m²)	Fracture toughness K_{1c}[b] (MN/m³/²)	Stress-corrosion threshold stress[b] (MN/m²)
5083-H321	Plate	215	27–33	>175
6061-T651	Plate	240	30–35	>250
2014-T6	Forging	370	18–21	<60
7075-T6	Forging	435	17–22	<60
7075-T73	Forging	365	20–24	>295
7049-T7X	Forging	450	22–27	~150[c]
7050-T7X	Forging	450	25–33	~190[c]
MA52-T7X	Forging	415	~30	~170[c]

[a] Tempers are as follows: H refers to processing sequence at mill (not strictly heat treatment); T6, solution-heat-treated and aged: for 2014, typically aged 18 hr at 435 K (325°F), and for 7075, typically aged 24 hr at 395 K (250°F); T651, stress-relieved by cold stretching (~2%) before aging; T73 and T7x, solution-heat-treated and overaged: for 7075, this would typically be T6 aging followed by 24 hr at 435 K (325°F).
[b] In short transverse direction.
[c] Stress for 50% survival (mean critical stress) rather than threshold stress (see Ref. 149 for details).

reduction of slip planarity.[137] Cu additions also increase strength, but in amounts above 0.5% decrease SCC resistance.[2,138]

The two-alloy series just discussed, the 5000 and 6000 alloys, are only of moderate strength (see Table 6 for typical values). The two series to which we now turn are appreciably stronger and also distinctly more susceptible to SCC. The 2000 series alloys are either essentially Al–Cu binaries, such as 2219, or Al–Cu–Mg alloys such as 2014 and 2024. The precipitation sequence depends on the Cu/Mg ratio. When the ratio is very large, as in 2219, one observes the well-known sequence,[123]

$$\text{GP zones} \to \theta'' \to \theta' \to \theta$$

where θ is $CuAl_2$, in the form of a disc or plate. For the large Cu contents of most commercial alloys (i.e., $\geq 4\%$), maximum strength occurs[139] with the maximum amount of θ''. Mg additions accelerate the precipitation process, and at intermediate Cu/Mg ratios such as 7:1, give rise to a duplex precipitation sequence (i.e., the θ sequence above) together with[140]

$$\text{GPB zones} \to S' \to S$$

where S is Al_2CuMg. GPB zones are spherical like GP zones but are rich in Mg as well as Cu; S', like θ', is platelike. As the Cu/Mg ratio decreases to about 2:1, the S sequence alone is observed,[140] as in 2024. Maximum strength corresponds to mixed GPB zones and S'.

In 2000 alloys of low Cu/Mg ratio, additions of Cr and Mn increase strength[2]; Co and Mo additions have also been found to increase strength in developmental alloys.[142] Fe and Ni additions, as in 2618, stabilize grain size through formation of insoluble particles.[2] The British equivalent of 2618, R.R.58, which is extensively used on the Concorde supersonic transport, shows rather rapid region II crack growth (see Fig. 2), which may be due in part to the Fe and Ni additions.[2]

The strongest aluminum alloys in general are the 7000 series, and as Table 6 shows, they maintain adequate toughness at high strength. They are also rather subject to SCC under certain conditions, and considerable effort has been devoted to understanding and controlling SCC in order to extend the usefulness of the series. The precipitation sequence in Al–Zn–Mg alloys like 7075 and 7049 depends on the Mg content[123,127,130]; in general it is

$$GP \text{ zones} \rightarrow \eta' \rightarrow \eta \; (\rightarrow T)$$

where η is $MgZn_2$ and T, which is present for Mg contents above $\sim 3\%$, is $(AlZn)_{49}Mg_{32}$. At high Mg/Zn ratios, as in 7079, a T' phase is also observed. The peak strength occurs with predominantly GP zones, mixed with η'; η is usually present at grain boundaries. Figure 24 illustrates an overaged structure of predominantly η, which gives good resistance to SCC.

Additions of Cu are made to many 7000 alloys. Cu increases the volume fraction of η', and thereby increases both strength and SCC resistance[2,68,128,131,143] at the 1% level and above. Indeed, Cu is only omitted from 7000 alloys to obtain better weldability. Cr and Mn accelerate precipitation in these alloys; claims that this occurs partly because these elements precipitate during homogenization treatments and act as nuclei for later heterogeneous precipitation[130] have been disputed.[127] The effects of Cr and Mn on grain shape, referred to above, have been thoroughly studied in these alloys,[144] and Zr additions have a similar effect.[3,130] Of those three elements, however, Cr, followed by Zr, is most effective in decreasing SCC.[143,145] Very small additions of Ti (e.g., 0.04%) retard precipitation kinetics in model alloys[146] and improve SCC performance.[147] This

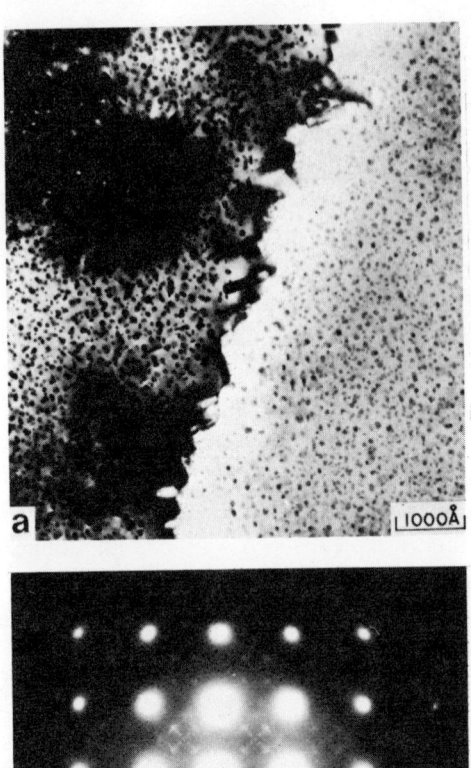

Fig. 24. Transmission electron microscopy of typical 7000-series alloy. (a)
Bright field of structure aged beyond T6
condition, becoming predominantly η
phase; large intermetallics at grain
boundary. (b) Diffraction of T73
condition of same alloy, matrix 100
zone, with characteristic precipitate
reflections. (Photographs courtesy of
N. E. Paton.)

composition level, however, is below the commercial maximum limit for grain-
refining Ti additions in all 7000 alloys[2,3] and is thus of limited interest. In the
same vein, additions of up to 0.6 % Li were found beneficial to SCC resistance
of Cu-free 7000 alloys,[148] but ingot casting problems have prevented use of Li
in commercial alloys.

A variety of other elemental additions has been studied,[2,68,142] as has
been reviewed in some detail by Mondolfo[130]; most such elements, however,
have little effect on SCC. Turning to new alloys, additions of Cr considerably

increase quench sensitivity (see below) in 7000 alloys. As Speidel and Hyatt[2] have discussed, much recent alloy development work has consequently been oriented to design of Cr-free 7000 alloys which would also resist SCC. Otherwise, these alloys cannot be used in section thicknesses above about 8 cm.

From this development work there have emerged four new alloys which have superior combinations of strength, toughness, and SCC resistance, compared to the widely used 7075.[2,131,149] Special processing and lower impurities in the 7075 composition led to the alloy 7175, while increased Zn and reduced Cr, Fe, and Si led to 7049. Somewhat lower in Zn than 7049 but higher than 7075 is 7050; it also has increased Cu, low Fe and Si and replaces Cr with Zr. Finally, the alloy MA52, not yet a standard alloy,[149] has low Fe and Si and replaces Cr with Zr. In the form of die forgings, all these alloys are superior to 7075,[149] particularly in toughness (see Table 6). The lowest quench sensitivity was shown by 7050 and MA52; MA52 was somewhat less resistant to SCC than the others, while 7050 seemed somewhat superior.[149]

An interesting subject in the history of the 7000 alloys has been the controversy over the value of Ag additions. Such additions might, by analogy, be expected to behave similarly to Cu additions. Considerable effort, especially in Europe, was devoted to development of Ag-containing alloys, and Polmear's review[150] suggested that improved SCC resistance was obtained. Subsequent work[2,131] has not confirmed those improvements, however, and further careful study by Staley and co-workers has explained the apparent contradiction. The interaction of Ag with vacancies refines precipitation[129,150] and leads to strength improvements after very fast quenches from solution temperature, or after fast heating to aging temperature[151]; such conditions are obtainable for small laboratory specimens. But these rates are not attainable commercially; indeed, at commercial rates, Ag degrades mechanical properties.[151] Resistance to SCC must be carefully evaluated, using silver-free microstructures which are fully comparable to those in the Ag-containing alloys.[151] When this is done, no improvement in SCC performance is found.[131,143] Further, Ag promotes intergranular corrosion and increases quench sensitivity.[130,131,143] In concert with the cost of silver, these findings have greatly reduced the interest in Ag additions to 7000 alloys.

Before leaving the subject of composition, we should note that the 2000 and 7000 alloys attain peak strength with a precipitate structure which is sheared rather than bypassed during deformation.[123,126] The same is probably true of 6000 alloys.[137] This, in turn, means that slip will be rather planar in the peak strength condition, becoming wavy as overaging progresses and precipitates become progressively incoherent.[123,126] Since SCC sensitivity tends to be high near peak strength and decline sharply with overaging, a correlation between slip planarity and SCC can be drawn in precipitation-strengthened aluminum alloys. The same may be true to a more limited extent in 5000 alloys, since Mg tends to increase difficulty of cross-slip.[152] We now turn to two other very important factors in aluminum alloy SCC: grain shape and thermal treatment.

Grain Shape

Wrought aluminum products exhibit to varying extents the same grain shape: a thin pancake considerably extended in the longitudinal and trans-verse directions and rather thin in the short transverse or thickness direction (see Fig. 23). Aluminum SCC is virtually always intergranular,[2,3,153] and for intergranular fracture the only easy or low-energy fracture path in the structure of Fig. 23 lies normal to the short transverse direction, or along the pancake faces. Thus SCC tests in the longitudinal and transverse directions, with essentially zero normal stress on the grain faces, show much lower cracking rates than do tests with the load in the short transverse direction[2,145,154]; indeed, SCC cracking rates have been shown to be a function of the resolved stress normal to the boundary.[2] The same grain orientation effect is familiar in smooth-bar TTF tests.[128,145]

This grain structure is the origin of the pervasive concern about "exposure of end grain" (i.e., alignment of the short transverse direction with the stress axis at a surface) in aluminum alloys, and it might be thought that a more nearly equiaxed structure would be preferable. Tests on equiaxed structures, however, have shown SCC susceptibility to be as great or greater than short transverse susceptibility.[2,3,145,153,155] Thus, the commercial grain structure not only restricts concern to one direction in the product, but improves SCC resistance over an equiaxed structure. We should repeat that studies on aluminum alloy SCC continue to be performed on model or "clean" alloys, in which the grain structure of Fig. 23 is not obtained. Cracking rates in such materials can be both higher and qualitatively different than in any commercial alloy,[2] and doubt is appropriate whether the results of such studies will have any application to commercial practice.

Thermal and Thermomechanical Treatment

By far the most common treatment of aluminum alloys to reduce SCC is overaging, particularly in the more susceptible alloys such as the 2000 and 7000 series. As the precipitates begin to lose coherency, and strength gradually declines, the SCC resistance often rises markedly. Thus only a modest penalty in strength may be paid in return for adequate resistance to SCC. In 7075, for example, overaging at 435 K (320°F) for 10 hr reduces yield strength only about 7%, although the smooth-specimen TTF "threshold stress" begins to increase sharply.[2] In fracture mechanics terms, Fig. 25 shows that the toughness K_{Ic} increases rapidly beyond 10 hr, while the plateau or region II maximum SCC crack velocity (see Fig. 2) in NaCl solution[2,3] falls by an order of magnitude. An additional order-of-magnitude reduction can be achieved with 10 hr further overaging, which also increases toughness. Behavior of 7178 is similar.

Selection of overaging treatments depends on the processing sequence and the consequent yield strength; thus, different product forms may require different aging times. One effort to meet this has been to monitor both yield strength level and electrical conductivity,[156] since the latter reflects the progress of the precipitation sequence. Good results have been obtained by such a surveillance technique for 7075-T73 products.[156] It is essential to the use of commercial high-strength aluminum alloys to be able to provide this kind of SCC resistance, since end grain is essentially impractical to avoid[2] and cannot be reliably protected by coating or painting.

An important aspect of thermal treatments for aluminum alloys is the quench rate from the solution treatment temperature. This can affect SCC performance of both 2000 and 7000 alloys. In naturally aged 2000 alloys, the rate of quench affects SCC resistance for rates less than about 550 K/s[2,128]; this has been attributed by Hunter et al.[157] to Cu-rich grain boundary precipitates formed during the slower cooling. Intergranular corrosion is also accelerated by lower quench rates.[128] Thus, thicknesses greater than about 6 mm in these alloys must be artificially aged,[2] since they cannot be quenched at a sufficiently high rate. (Artificial aging refers to aging at temperatures above room temperature.)

All 7000 alloys are artificially aged, and thus particular attention must be paid to the quenching rate, for two reasons. First, SCC performance can be affected; and second, tensile properties can be reduced. In relation to SCC, there is a critical temperature range from 675 down to 560 K (750–550°F), in which the cooling rate must be above 100 K/s.[127] Lower cooling rates reduce vacancy retention and produce coarse η' and η in both the grain boundaries and the matrix, and thereby accentuate susceptibility.[128,129] The short transverse SCC properties, however, are not improved by rapid cooling, although intergranular corrosion resistance is obtained.[128] Turning to the second problem, tensile properties, their decrease with decreasing cooling rate is called quench sensitivity. In 7000 alloys, Cr content causes considerable sensitivity,[2,3] but replacement of Cr with Zr or Mn greatly improves the situation.[2,130,131,143] Increased quench sensitivity results from additions of Ag[131] and Mo,[130] and to some extent from increased Zn and Mg content in 7000 alloys[2,143]; of these additions, only Zn improves SCC resistance.[131,143] For this reason, as discussed, new alloys like 7049, 7050, and MA52, which are intended for thick-section applications and thus require low quench sensitivity, are designed with increased Zn content, and Zr and sometimes Mn replacements for Cr[2,131,149] as grain shape stabilizers. These new alloys are thus improved in resistance to SCC.

A topic with major implications for future alloy processing and development is that of thermomechanical treatments (TMT). Until fairly recently, it had been thought that only modest improvements in properties were attainable.[2,3] For example, the strength of 2000 alloys can be increased somewhat by room-temperature deformation after quenching from the solution treatment temperature, and then normally aging (similar to a T3 temper). Efforts to follow the same procedure in 7000 alloys (as in the T8 temper) yielded little improvement in properties except in model alloys.[158]

The key to TMT of 7000 alloys seems to be warm work instead of cold work. Apparently the first workers to recognize this were Pavlov et al.,[159] who warm-worked their material to 20 % strain at 395 K (250°F) and subsequently also aged the material at 395 K. More recently it has been found that aging prior to working, to produce a precipitate structure roughly like that of the peak strength condition, is very beneficial. Warm work is then generally conducted at 445–470 K, followed by further aging. An impressive example of such work is that of Paton and Sommer,[160] who were able to obtain benefits in 7075, 7049, and also in 2024. In the case of 7075, they found that aging to the peak strength or T6 condition, followed by 15 % strain at 465 K and further aging 3 hr at 435 K gave essentially the T6 strength with the overaged or T73

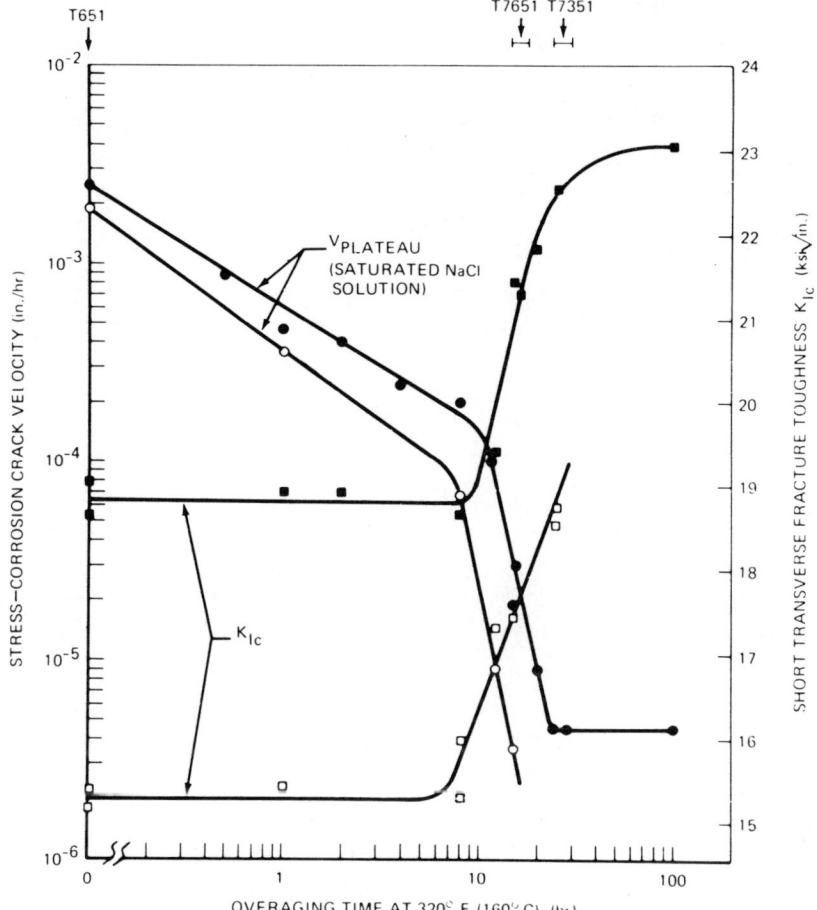

Fig. 25. Effect of aging time at 433 K (320°F) on the toughness and stress-corrosion crack velocity of 7075-T651 (●■) and 7178-T651 (○□); tempers indicated at top. 0.91 ksi in.$^{1/2}$ = 1 MN/m$^{3.2}$; 1420 in./hr = 1 cm/sec. Crack orientation: TL. (From Speidel and Hyatt.[2])

SCC resistance (cf. T7351 condition in Fig. 25). In each alloy, they were able to obtain "overaged" SCC properties at no sacrifice in strength or toughness from the peak strength condition.

    This use of warm work represents a means of avoiding the planar slip characteristic of room-temperature deformation. The worked structure is thus relatively homogeneous, and annealing it back to a lower strength (e.g., equivalent to the T6 condition where the process began) permits overaging the

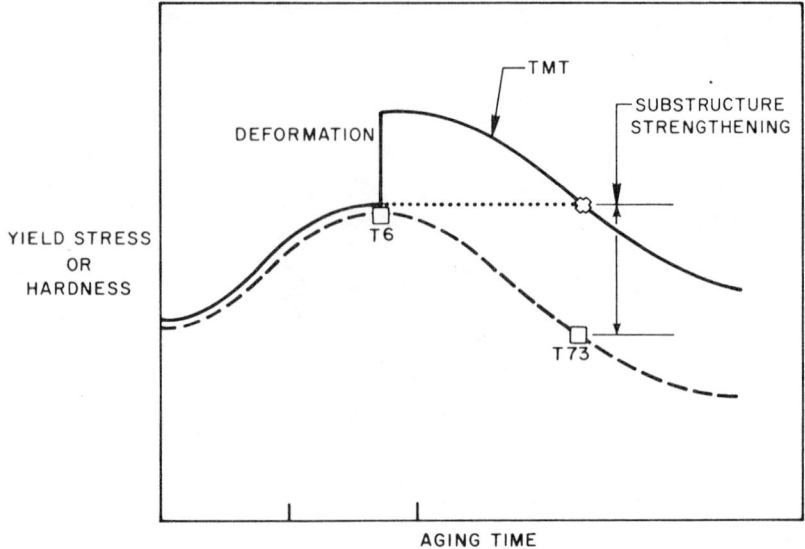

Fig. 26. Schematic comparison of strength as a function of aging time for conventional thermal treatment (dashed line) and for TMT (solid line). The TMT sequence was to age the material to the T6 condition, deform 15% at 465 K, and age further at 435 K. The resulting T73 SCC resistance was accompanied by T6 strength because of dislocation substructure strengthening. (From Paton and Sommer.[160])

precipitates at the same time. This process is shown schematically in Fig. 26. The attribution of the strength increase above the normal T73 condition to dislocation substructure is supported by transmission electron microscopy observations, as is the conclusion that the precipitates did become overaged.[160] It thus seems evident that TMT sequences deserve thorough investigation[160-162] as one means of manipulating microstructure to develop strong, tough, SCC-resistant aluminum alloys.

The Hydrogen Question

All the alloy systems treated in this chapter have long been known to suffer failures by the sole action of hydrogen, except for aluminum alloys. Yet the possibility of hydrogen involvement in aluminum alloy SCC has been discussed for some time,[163] and recently such discussions have begun to indicate that a hydrogen role in aluminum SCC is indeed a credible hypothesis. We shall present this work briefly here, and discuss it in greater detail in the concluding section.

There have been essentially three major arguments used in opposition to a hydrogen role in aluminum alloy SCC: first, that no embrittlement solely due to hydrogen has been demonstrated; second, that the evidence in hydrogen-containing environments is inconclusive; and third, that cathodic polarization of aluminum retards rather than accelerates SCC, as a naive view of the cathodic process would predict. All three have now been either weakened or disproved.

Regarding the first point, it is true that dry hydrogen gas environments, even at high pressure, have little effect on tensile properties[68,84,118] or on crack growth behavior[164-168] of aluminum alloys. But cathodic charging has now been shown to produce reversible embrittlement in aluminum,[169-171] which has the classical characteristics of hydrogen embrittlement as a function of strain rate and temperature.[170] It is therefore no longer tenable to say that hydrogen alone cannot cause embrittlement. Sufficient hydrogen fugacity to provide some minimum amount of hydrogen entry is apparently all that is required.

The requirement for higher fugacities leads one to reexamine the evidence related to hydrogen-containing environments. One example is air of varying humidity, in which SCC crack velocity V is less than in aqueous solution. V decreases with decreasing humidity in the manner shown in Fig. 27; as Speidel has discussed,[172] water should not be condensed at the crack tip below about 30% humidity, and the effect continues below the "knee" in the postulated adsorption isotherm for aluminum oxide, at about 5% humidity.[165] Thus, the easy hydration of aluminum oxide cannot account for Fig. 27. The reaction of water with aluminum to produce hydrogen at high fugacity is the only simple way to explain these data.[172] A similar explanation, advanced by Broom and Nicholson for fatigue,[164] has been accepted by a number of workers.[165-167]

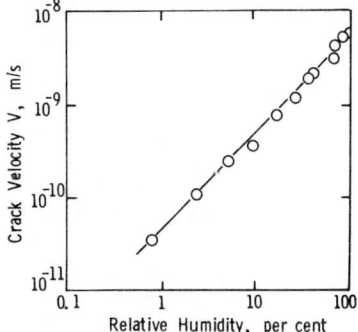

Fig. 27. Crack growth rate V in region II (see Fig. 2) as a function of relative humidity in air. Data are for 7075-T651 plate at $K = 19\,MN/m^{3/2}$. (From Speidel.[172])

Confirmation of this conclusion was obtained by Montgrain and Swann with the observation of hydrogen release during intergranular fracture,[173] and of effects of unstressed exposure to moisture.[173] Although no direct proof has been presented, the case for hydrogen involvement in such cases is increasingly strong.

The third argument, regarding cathodic polarization, has also been addressed in recent work. The critical observation[172,174] is that there is a minimum in cracking rate at intermediate potential, but that it increases again as the potential is made increasingly cathodic. This is illustrated in Fig. 28. This minimum in both cracking rate and permeability at intermediate potentials has also been observed in steels,[175] where there is little argument as to the role of hydrogen. It is thought that the minimum corresponds to a minimum in environmental fugacity of hydrogen, as shown by the

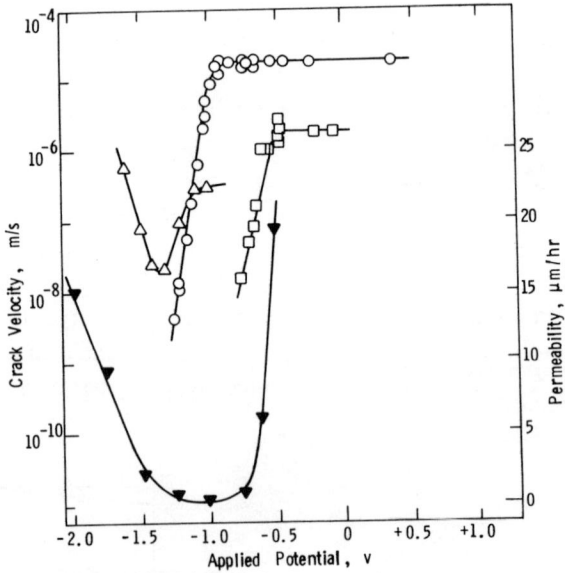

Fig. 28. Dependence of crack velocity and permeability on applied potential (volts vs. SCE) for several aluminum alloys, in the environments shown. Data on permeability for 7075-T651 (▼, 3% NaCl solution) are those of Gest and Troiano[169]; cracking data for 7075-T651 (□, 5 M iodide solution) and 7079-T651 (○, 5 M iodide solution) are from Speidel[153]; those for AlZnMg₃ (△, 1 M chloride solution), from Berggreen.[174] (The figure is taken from Speidel.[172])

permeability; increased permeation with very cathodic polarization is due to general hydrogen production, while increases with anodic polarization are due to acid generated by hydrolysis in pits or cracks,[2,172] leading to hydrogen generation locally even though the general potential is anodic. This process is now well established.[60,61,175-178] The electrochemical objection therefore seems to have been disposed of; as Speidel concluded,[172] and we concur, this leaves no strong objections to hydrogen participation in SCC of aluminum alloys.

It is also true that the results discussed above do not, of themselves, offer very positive proof that hydrogen is involved, only that its participation would not contradict the experimental facts. A very interesting effort to obtain more direct evidence was recently reported by Green et al.[179] They tested precracked specimens of 7075-T6 in torsion (i.e., Mode III) in an aqueous chloride chromate environment which causes rapid SCC. Mode III loading was chosen because there is no hydrostatic stress component which could cause hydrogen accumulation near the crack tip.[179-182] By comparison to tensile Mode I loading, then, at least part of any operative hydrogen behavior could be isolated. The results are shown in Fig. 29. It is evident that SCC

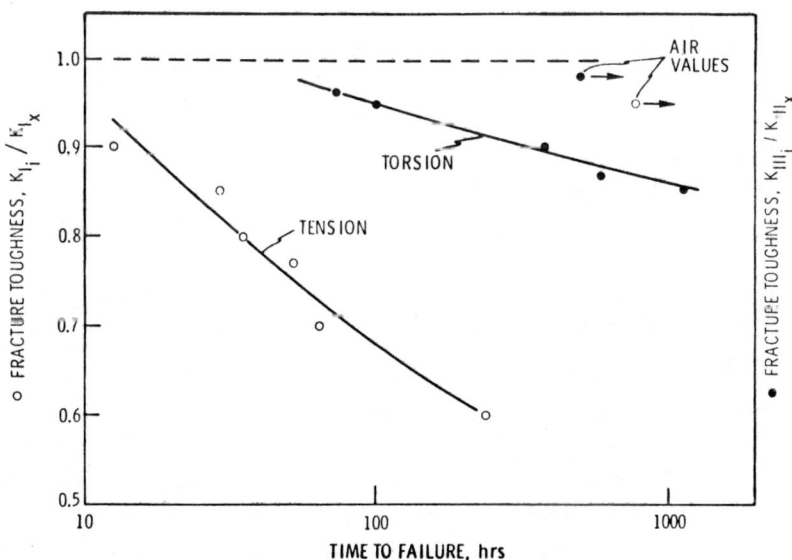

Fig. 29. Resistance to SCC of 7075-T6 as a function of loading mode (see the text), in an environment of 3.5 % NaCl and 3.0 % $K_2Cr_2O_7$ in water, with pH 3.2. (From Green et al.[179])

susceptibility is greatly reduced in torsion, but is not eliminated. This suggests that cooperative processes of hydrogen embrittlement and anodic dissolution are operating together; the same is shown by the behavior of arsenic poison in the environment.[179] This is consistent with our earlier statement that these two types of processes can occur simultaneously.

Clearly, much remains to be done on the question of hydrogen in aluminum alloys. But the recent work seems to show that the possibility of hydrogen participation in SCC is a viable one. We urge that experimental work continue[183] to be designed to consider hydrogen as one of the candidate processes in SCC.

TITANIUM ALLOYS

Titanium alloys are still undergoing intensive development work, and although considerable work has been done to understand their behavior, it can be expected that understanding will appreciably increase in the future. Thus, the work to be summarized below, complete as it appears in some areas, may well be modified or supplemented to a considerable extent as new work appears.

A problem in presenting SCC data for titanium alloys is the variety of environments which can cause SCC, ranging from distilled water and chloride solutions to alcohols and other organic liquids, hot solid and molten salts, nitrogen tetroxide, liquid metals, and others. Some selection of results from this range must necessarily be made; the emphasis, as elsewhere in this chapter, will be upon chloride-containing aqueous solutions, partly because the majority of the experiments have been made in that environment. Hydrogen gas data are similar in many respects to these SCC data. Where possible, corroborative data from other environments will be cited also.

Composition

As in the case of aluminum alloys, some discussion of the physical metallurgy of titanium alloys must be included in a presentation of the SCC and hydrogen data. For further details on this metallurgy, several thorough reviews[184–187] may be consulted. There are essentially three types of titanium alloys in commercial use: the α phase alloys, in which the structure consists of the hexagonal close-packed (hcp) α phase; the $\alpha + \beta$ alloys, whose equilibrium constitution consists of a mixture of α and the bcc β phase; and the metastable β alloys. These alloys are created by adding three types of elements to titanium,[183,185] as follows: (1) α-stabilizing, in which the α phase is increased in stability by alloying; (2) β-isomorphous, in which the β phase is stabilized by alloying; and (3) β-eutectoid, in which β is stabilized but a eutectoid is observed. The commercial alloys thus formed (alloy designations are the nominal composition in wt %) are typified by the all-α Ti–5Al–2.5Sn; by dispersions of β in α, as in Ti–6Al–4V; and by β matrices which contain α dispersions, as in the alloy Ti–11.5Mo–6Zr–4.5Sn, also called Beta III.

The α and β phases in these alloys can assume a wide variety of sizes, morphologies and volume fractions; titanium alloys can also contain martensitic phases, precipitates, and intermetallic compounds.[186] Commercial alloys are rarely used as other than $\alpha + \beta$ or $\beta + \alpha$ microstructures,[185] but as will be seen, such commercial microstructures can be quite complex.

The α-stabilizing elements Al, Sn and oxygen tend to worsen SCC resistance.[186,188-191] The similar alloying behavior of Ga, In, and Pb would suggest that they, too, would worsen SCC.[187,191] Replacement of Al with Sn or Ga was found to give no improvement in SCC behavior.[192] The effect of Zr is less clear, being inferred from behavior of ternary alloys, but effects on SCC performance appear variable.[186,190,192,193] The interstitial element oxygen is always added to Ti alloys in some concentrations to increase strength, but oxygen is also a ubiquitous contaminant in these alloys, rarely being present much below 0.1 wt % (1000 ppm). Its reduction can improve not only SCC resistance,[189,190] but fracture toughness as well.[186,189] There seems to be a critical value of Al content, 5–6 %, above which SCC properties are especially worsened.[186,188,191] Al and oxygen additions also increase the tendency for these alloys to exhibit planar slip. Such slip becomes marked at about 4–5 % Al, which is just the composition range in which SCC susceptibility begins to increase markedly.[186,188] The slip mode change, due to both Al and oxygen, apparently occurs because of short range order,[190,194,195] but there have also been claims that the SFE is lowered.[196] In any case, this is another example of the correlation between planar slip and environmental susceptibility, which we have identified several times in this chapter.

There are four β-isomorphous elements, Mo, V, Nb, and Ta. Only one of these, Mo, has been investigated in binary alloys,[186] but analysis of the behavior of multicomponent alloys indicates that all four elements reduce or

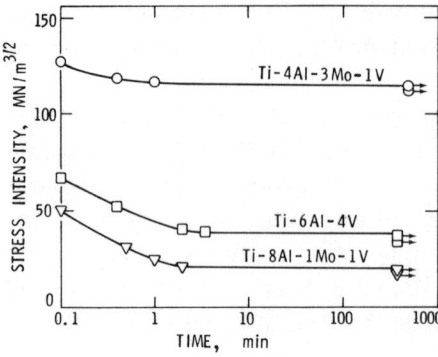

Fig. 30. Time to failure of precracked specimens of three commercial $\alpha + \beta$ alloys tested in 3.5 % NaCl solution. All materials were in the mill-annealed condition. (From Blackburn *et al.*[186])

eliminate susceptibility to SCC,[186] [188,192] with Mo and V being the best characterized of the four. But mere use of β-stabilizing elements is not sufficient.

At least some of the β-eutectoid elements are known to *increase* SCC, such as Fe, Cu, Mn, Co, Si, and possibly Cr.[186,188,190] This may occur because of the tendency of these elements to form compounds such as Ti_2Cu or $TiCr_2$ as equilibrium is approached.[186,190] It has been shown, for example, that formation of TiMn in Ti–8Mn induces SCC susceptibility,[186] although alloys such as Ti–4.5Al–4.5Mn show a ductile form of cracking in hydrogen[197]; no TiMn was reported in the optical microscopy of the latter study.[197] The compounds Ti_2Cu and Ti_5Si_3 have also been implicated in findings of poor SCC resistance.[190,191]

These results on α-stabilizing and β-isomorphous elements make it possible to understand the data of Fig. 30, in which three commercial alloys are compared. It is clear that decreasing amounts of Al, particularly below about 5 %, or increasing Mo + V, improve SCC resistance. It should be pointed out that care must be used in making such comparisons, since the oxygen content, amounts of α and β, and toughness level vary among these alloys. Nevertheless, the major compositional trends are clear, and are being used in current alloy development work.[198] We shall now turn to consideration of microstructural effects, which are important in titanium alloys.

Microstructure and Thermal Treatment

In predominantly α alloys, particularly those which are high in aluminum, the important microstructural element is the ordered phase α_2

(Ti$_3$Al). This appears at about 6% Al, the exact value depending on oxygen content; as first shown by Lane and Cavallaro,[199] α_2 considerably worsens SCC resistance. Later work has shown that increases in α_2 volume fraction, which are caused by lowered aging temperature or by increased Al, Sn, or oxygen content, worsen SCC behavior.[186,190,192] The presence of α_2 causes planar slip due to dislocation cutting of the precipitates, as shown in Fig. 31, and thus enhances the planar slip already present in the matrix; α_2 also increases slip coarseness.[194,200,201]

Heat treatments have been devised to cause dislocation bypassing of the α_2 particles, thus restoring slip homogeneity,[200] but slip remains planar. Overaging of α_2 might be expected to cause bypassing in the same way, and recovery of SCC resistance with overaging has been reported[202]; subsequent work, however, did not confirm that observation.[192] Thus, continued planarity, even with bypassing, means that SCC susceptibility is maintained. That conclusion is supported by the behavior of high-Al alloys which are quenched to suppress α_2 but have planar slip: they are susceptible to SCC.[191] When α_2 forms heterogeneously, as it does in binary Ti–Sn alloys as Ti$_3$Sn, it is less deleterious to SCC,[203] but Al + Sn alloys, in which Ti$_3$(Al, Sn) forms homogeneously,[190] perform poorly.[188] Heat treatment of other α-isomorphous alloys, such as those containing In, can apparently suppress α_2 and obtain improved SCC resistance.[192]

The most important alloy type commercially is the $\alpha + \beta$ alloy, particularly for high-strength applications. A wide variety of microstructures and mixtures of α and β can be produced in these alloys. For example, solution treatment in the $\alpha + \beta$ field followed by slow cooling results in an equiaxed

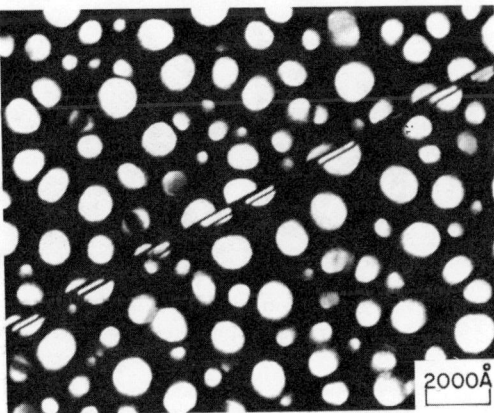

Fig. 31. Planar slip due to shearing of α_2 particles in Ti–10% Al alloy. Slip band lies along prism trace, (10$\bar{1}$0); particles are shown in dark field. (Photography courtesy of J. C. Williams.)

2000Å

structure of large primary α particles surrounding islands of β, as shown in Fig. 32(a). Cooling directly from the β field, on the other hand, results in an acicular structure, usually (because of the limited hardenability of most Ti alloys) of the Widmanstätten type shown in Fig. 32(b). These α plates are surrounded by thin but largely continuous sheets or strips of β, as depicted in Fig. 32(c).

The properties (other than strength) of the acicular structure tend to be superior, both in toughness and in SCC resistance, as indicated in Fig. 33. Here the "band" or separation between K_{Ic} and K_{Iscc} is about the same size in each

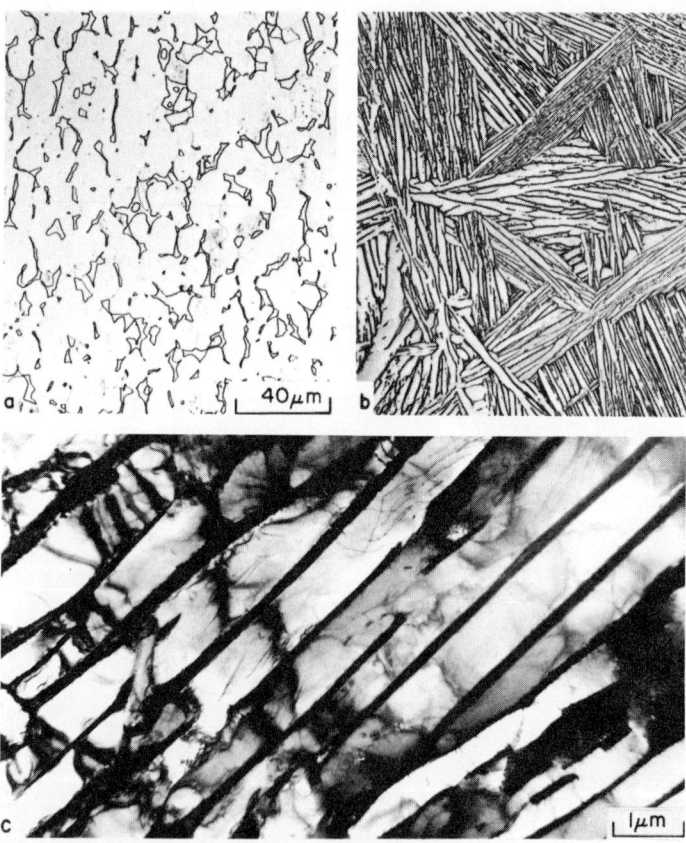

Fig. 32. Microstructure types in $\alpha + \beta$ alloys illustrated with Ti-6Al-4V. (a) Light micrograph of material slow cooled from 1200 K (1700°F) to produce equiaxed $\alpha + \beta$. (b) Light micrograph, same magnification as (a); air cooled from 1310 K (1900°F) to produce acicular (Widmanstätten) $\alpha + \beta$. (c) Transmission electron micrograph of acicular structure such as (b), showing virtually continuous β between α plates.

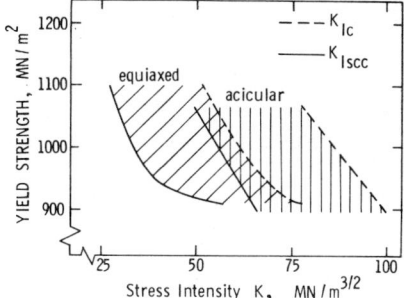

Fig. 33. Relation among yield strength and stress intensity for either unstable fast fracture or SCC in salt water, for Ti–6Al–4V. A variety of microstructures of each type was tested. (From Blackburn et al.[186])

case, but the acicular properties are displaced to higher values. This has been widely observed[186,188,191,192,204,205]; in particular, lower solution-treatment temperatures or precipitation of α_2 (as in Ti–8Al–1Mo–1V) appreciably increase SCC.[189,191] When martensitic rather than Widmanstätten acicular structures are produced by quenching, they, too, are resistant. Tempering of the martensite causes some precipitation of fine β particles but preserves the acicular morphology; SCC resistance is then intermediate between untempered martensite and the equiaxed structures.[204] Thus, acicular microstructures, whether Widmanstätten, or platelike or acicular martensite, are generally more resistant to SCC. Examples include Ti–6Al–4V[186] and Ti–4Al–3Mo–1V.[190,192]

The above discussion of SCC resistance is closely paralleled by the behavior of $\alpha + \beta$ alloys in hydrogen gas.[206–208] The equiaxed or continuous α structures differ from the acicular or continuous β structures, but the ranking depends on hydrogen pressure, as Fig. 34 illustrates. Also included in Fig. 34 are data (dashed lines) for saltwater SCC.[209] The apparent agreement of SCC data with hydrogen results at about 1 N/m^2 is supported by fractography observations, which showed that both test modes resulted in similar fracture surfaces.[209] This conclusion is at first surprising, since it suggests that if hydrogen is equally involved in each case, the effective fugacity produced from the salt solution is rather low. We return to this point below, but it should be recalled that Ti alloys are relatively unreactive in salt water,[186] and thus should produce little hydrogen by corrosion reactions.

Before leaving the subject of $\alpha + \beta$ alloys, it is important to point out that cracking is often observed to take place in the $\alpha-\beta$ interface, both in SCC[186] and in hydrogen.[209] In this connection, the recent observations of Rhodes and Williams[210] may be significant: they have described an "interface phase" which generally can be found along $\alpha-\beta$ interfaces in these alloys. The phase

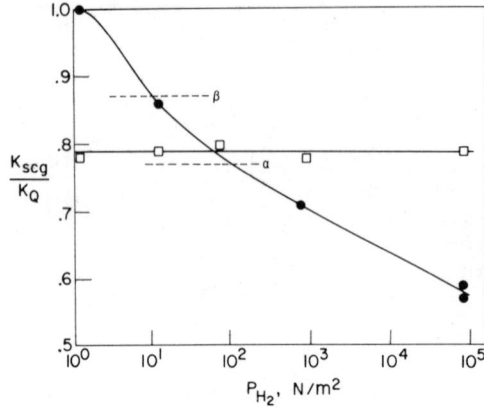

Fig. 34. Crack growth susceptibility as a function of hydrogen pressure for two microstructures (see Fig. 32) of Ti–6Al–4V. ●, Continuous β matrix; □, continuous α matrix. Susceptibility shown as ratio of stress intensities for subcritical crack growth (SCG) and unstable fast fracture (Q), tests at 297 K. Dashed lines for 3.5% NaCl solution (From Nelson.[208])

transfer and other processes may be significantly affected at α–β interfaces by the presence of this structure, and future work on microstructural effects in environmental fracture should examine carefully the role of non-Burgers α now often called "Williams/Rhodes α," after its discoverers) at interfaces.

The strongest Ti alloys are the metastable β alloys; they are also the most complex microstructurally. We shall discuss a variety of these structures below, but it should be emphasized that to date the β alloys are only being used in the solution-treated (metastable β) or aged ($\beta + \alpha$) conditions. The β can decompose[186,187] at low temperatures to produce, in addition to α, the ω phase, several compound phases, and a phase separation reaction into $\beta + \beta'$ (formerly called $\beta_1 + \beta_2$).[187]

The metastable β alloys unfortunately show varying behavior, and more work oriented toward microstructural variables is needed. We summarize the principal observations which are available. For the single-phase β condition (quenched from above the β transus), these alloys may either be immune or rather susceptible to SCC. For example, Ti–8Mn,[191] Ti–Mo alloys,[193] Ti–8Mo–8V–3Al–2Fe,[186] and Ti–11.5Mo–6Zr–4.5Sn or Beta III[211,212] were essentially immune, while Ti–16Mn[186] and Ti–13V–11Cr–3Al[186,204] were susceptible. When Ti–8Mn is aged to produce an $\alpha + \beta$ condition, it is quite susceptible to SCC;[186] as annealing temperatures are reduced, the volume fraction of β decreases, as does SCC susceptibility. That pattern might suggest that the β phase is susceptible. In several other alloys, however, evidence suggests that the α is susceptible, and that the size of the α dispersion controls behavior.[186,211] In support of such a view, the Ti–Mo alloys, which are

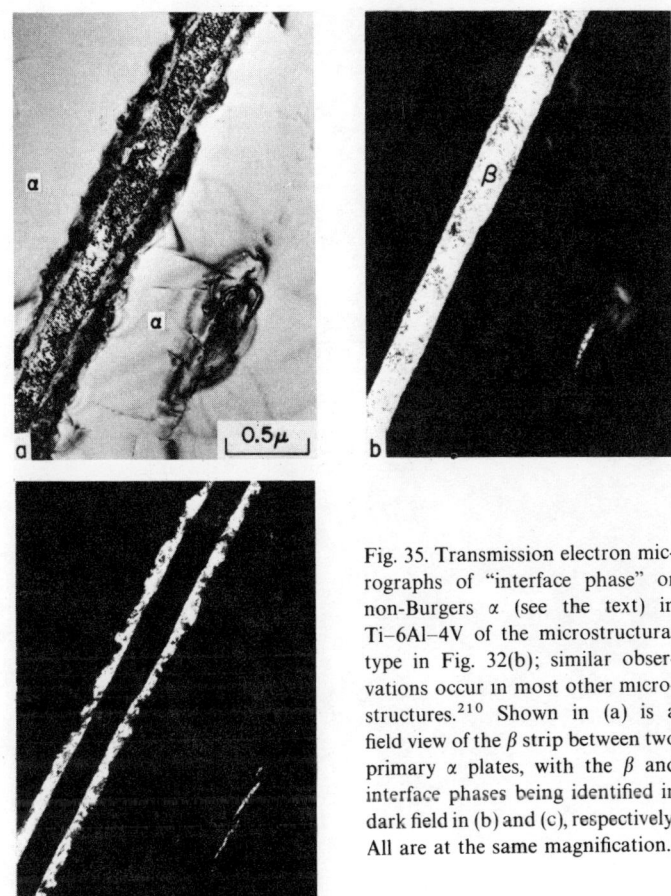

Fig. 35. Transmission electron micrographs of "interface phase" or non-Burgers α (see the text) in Ti–6Al–4V of the microstructural type in Fig. 32(b); similar observations occur in most other microstructures.[210] Shown in (a) is a field view of the β strip between two primary α plates, with the β and interface phases being identified in dark field in (b) and (c), respectively. All are at the same magnification.

immune when all-β, become susceptible when aged to form fine dispersions of α.[211] Similarly, the β-immune alloy Beta III is quite susceptible when aged below 810 K (1000°F) in the $\alpha + \beta$ field,[211] (see Fig. 36a); the same occurs in Ti–8Mo–8V–3Al–2Fe.[186] It is important to note that K_{Iscc} is essentially equal to K_Q in aged Beta III, but that crack growth rates are markedly accelerated in salt solution.[211] [213]

The alternative viewpoint on SCC of these alloys relies on the observation that when aged to form intragranular α, they also tend to form layers of α phase at the grain boundaries,[212,213] as shown in Fig. 36(b). These layers are less prominent at lower aging temperatures,[212] and their presence could account for at least some of the intergranular SCC fractures widely observed in

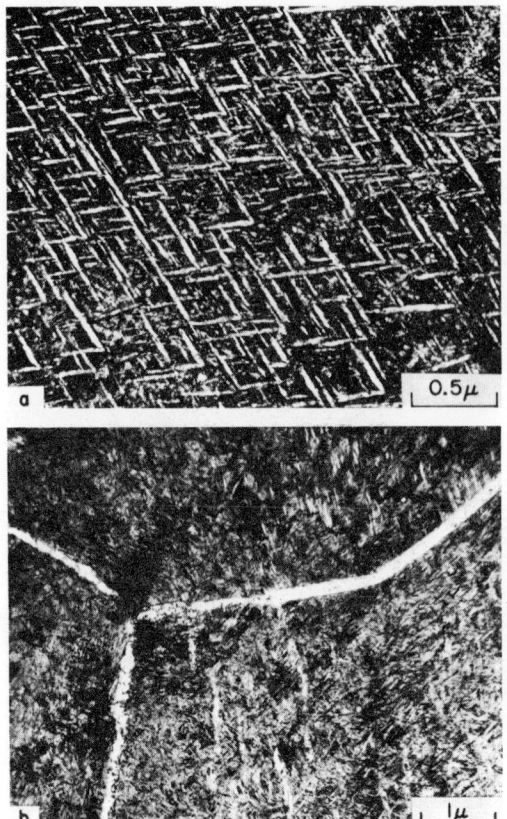

Fig. 36. Transmission electron micrographs of $\beta + \alpha$ microstructures in high-oxygen (0.28 %) Beta III. (a) Fine Widmanstätten α needles in β matrix after aging 12 hr at 700 K (800°F). (b) Virtually continuous grain boundary layer of α in material water quenched from 1045 K (1450°F) and aged 8 hr at 755 K (900°F). (Photographs courtesy of J. C. Williams.)

metastable β alloys, including Ti–8Mn, Ti–Mo alloys, Ti–8Mo–8V–3Al–2Fe, Beta III, Ti–3Al–8V–6Cr–4Mo–4Zr, and Ti–6Al–2Sn–4Zr–6Mo.[212,214,215] Presumably, the fracture takes place either by α–β interface cracking,[215] as in $\alpha + \beta$ alloys,[186] or by fracture in the α layer.[211] These layers could, in principle, be eliminated by aging techniques,[215] although that has not been demonstrated; and they may also be avoidable through TMT, as discussed below.

The other microstructures which have been studied in metastable β alloys include $\beta + \omega$, which in Ti–8Mn, Ti–11Mo, and Beta III seems essentially immune to SCC,[186,211,212] and the $\beta + \beta'$ structure in Ti–8Mo–8V–3Al–2Fe, which is likewise resistant.[186] It should be pointed out, however, that these microstructures are very low in fracture toughness and are therefore not very useful for applications.

Increases in Mo content in these alloys above about 10% tends to promote SCC failure,[186,211,212] and the same increases also promote increasing planarity of slip in the β matrix.[216] Vanadium has the same effect,[216] and the susceptible alloy Ti–13V–11Cr–3Al also deforms by planar slip.[212] These and other examples[186] demonstrate that there is a general correlation in metastable β alloys between slip planarity and SCC, just as there is in the α alloys.

Thermomechanical treatments (TMT) have been found effective in reducing SCC of several commercial alloys[190]; the most interesting example is that of heavy deformation applied to Beta III, followed by heat treating below the β transus to avoid recrystallization.[213] The α layers at boundaries seem also to be absent. The resulting structure consists of pancake grains, reminiscent of those found in wrought aluminum alloys (Fig. 23) and is only slightly susceptible to SCC.[213] Although such structures may be susceptible in the short transverse direction, by analogy with aluminum alloy behavior, the overall performance is improved relative to equiaxed structures.

Grain Size and Texture

It has been shown repeatedly in Ti alloys that refinements in microstructural size improve resistance to environmental fracture. For example, reduction of α grain size improved performance of Ti–5Al–2.5Sn,[190,191] Ti–7Al,[202] Ti–8Al,[217] and commercially pure Ti,[218] the latter in hydrogen. Refinement of primary α grain size in $\alpha + \beta$ alloys such as Ti–4Al–4Mo–2Sn–0.25Si[190] and in Ti–6Al–4V[186] also improves SCC resistance, as does refinement of prior β grain size to control subsequent martensite packet size.[186,190] The same is true for resistance to hydrogen gas cracking.[206] As was the case for each of the alloy systems described earlier in this chapter, therefore, we observe again that refinements in grain size or α plate size are beneficial.

For the α or $\alpha + \beta$ alloys, the anisotropy of the hcp α phase means that considerable crystallographic texture can be developed, and such texture significantly affects properties.[191] For the essentially α alloy Ti–8Al–1Mo–1V and the $\alpha + \beta$ alloy Ti–6Al–4V, the origins and effects of texture can be summarized in Fig. 37. The three groupings shown in Fig. 37 were described by Blackburn *et al.*[186] as follows:

(i) Processing in the β-phase field frequently leads to the development of the α-phase texture shown in Fig. [37a]. Such a texture gives rise to relatively isotropic properties and relatively high values of K_{Iscc}.

(ii) Processing through the $\alpha + \beta$ transus often causes development of the texture shown in Fig. [37b], which in turn causes both poor transverse SCC values, and also very anisotropic mechanical properties.

(iii) Processing at lower temperatures, 1085–1200 K (1085–1700°F), often results in development of the texture shown in Fig. [37c], which is typical of some sheet material. Such processing results in an equiaxed structure of $\alpha + \beta$, unless a β-solution treatment is used subsequently—thus the toughness and SCC resistance are usually lower than in material processed in the β field [Fig.37a].

This information suggests in a broad way that β processing and its consequent relatively random $(0002)_\alpha$ texture is the most desirable. The problem with hot working at such high temperatures is that banding of commonly oriented α plates can occur,[186] with pronounced effects on fracture behavior, which might extend to SCC. In addition, strength levels of β-processed material are typically reduced by 30–60 MN/m². Uniform structures of random texture, however, clearly seem desirable for resistance to SCC[186] or hydrogen gas embrittlement.[206]

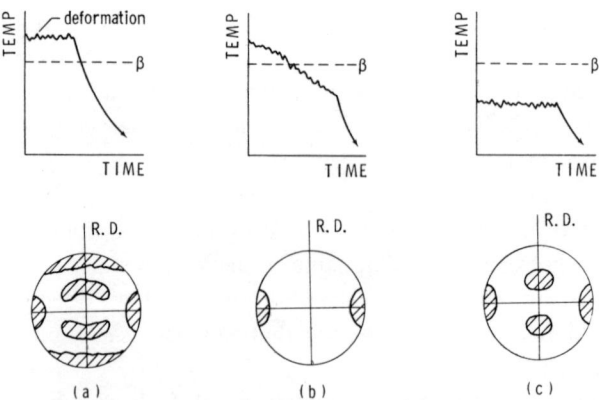

Fig. 37. Schematic representation of the influence of processing schedules (top) on resulting $(0002)_\alpha$ texture (bottom); the cracking behavior of these textures is described in the text. (From Blackburn *et al.*[186])

Texture is also important in its effect on fracture morphology. As discussed in the following section, transgranular failures in both the α and β phases frequently have crystallographic aspects and a cleavagelike appearance. In the α phase, the fracture plane tends to lie about 15° away from the basal plane,[219] while in the β, fracture usually occurs on {100}.[204] Thus, the texture of the material, in combination with the direction of stressing, can affect the ease with which a preferred cracking mode can occur.

Fracture Morphology and Hydrides

Fractures can assume a variety of transgranular and intergranular morphologies in SCC of Ti alloys[186,191,212]; cracking in methanolic solutions is particularly likely to be intergranular.[186,212] In $\alpha + \beta$ alloys, cracking in the α–β interface has been observed in both SCC[186] and in hydrogen gas,[206,209] while similar interface cracking has been reported in $\beta + \alpha$ alloys.[215] Transgranular cracking of the α is of particular interest, since it tends to exhibit an unusual crystallographic behavior. Because of its appearance, this fracture mode is usually referred to as "cleavage." Since cleavage on a high index plane is unusual, some authors have referred to the mode as "cleavagelike" or "nonclassical cleavage." This fracture mode is only observed at low K values; as K nears the critical value for unstable fast fracture, dimpled rupture becomes predominant.

The cleavage-like morphology is largely observed on planes 12–16° from the basal plane,[219] i.e., near planes of the type (10$\bar{1}$7) or (11$\bar{2}$12).[191,219,220] This is shown in Fig. 38, from Meyn[220]; note that other planes are also

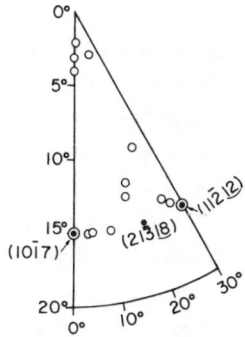

Fig. 38. Determination of crystallographic planes of cleavage facets in Ti–6Al–4V, measured by X-ray back reflection, and shown on the central portion of the hcp unit triangle, with (0001) at the apex. Filled points are the specific planes noted. (From Meyn.[220])

observed, some near the basal orientation. It has been conventional to use such results to rule out hydride participation in the fracture process,[186,220] since massive hydrides give rise to cracking[221] on $\{10\bar{1}0\}$, the predominant hydride habit plane of pure titanium,[222] rather than on basal or near-basal planes. Such a conclusion, however, does not appear justified, for two reasons.

First, there are complications in the massive-hydride result. Since environmental cracking in Ti–8Al–1Mo–1V tends to follow α–β boundaries,[221,223] and it is possible that $\{10\bar{1}0\}$ is involved in the $\beta \rightarrow \alpha$ transformation, the observation of hydride habit plane cracking may merely represent α–β interface cracking.[223] In that connection, it will be important to consider whether the structure of Fig. 35 is involved in the cracking behavior. Furthermore, the observation that applied stress affects both hydride habit plane and effective hydrogen solubility[224] could complicate this result[221]; the ability of hydrides to form by a strain-induced mechanism[225] at hydrogen contents well below the apparent solubility limit[224] considerably magnifies the experimental difficulties in this area.

The second reason to suspect the conclusion on hydride participation is the recent work by Paton and Spurling,[226] who have observed that as Al content increases, the $\{10\bar{1}0\}$ habit plane gives way to basal and near-basal habit planes (cf. Fig. 38). In particular, $(10\bar{1}7)$ hydrides were observed,[226] just as they are in zirconium alloys.[227] Figure 39 is an example. As hydrogen content increased, $\{10\bar{1}0\}$ hydrides were also observed, consistent with the earlier result.[221] But as Fig. 34 suggests, relatively low hydrogen contents are expected in SCC. These observations mean that hydride participation in SCC fractures cannot be ruled out on crystallographic grounds.

The important point here is the possible relation between what is called "slow strain rate hydrogen embrittlement"[186,224] and environmental cracking such as aqueous or methanolic SCC. As Paton and Williams pointed out in their review,[224] it has long been considered possible that strain-induced hydrides are responsible for slow rate embrittlement, although direct evidence has been lacking. Efforts to rule out hydrides in sustained-load cracking[220,228] are inconclusive, because of the hydrogen solubility and hydride habit plane problems described above. The same can be said regarding efforts to rule out hydrides in SCC.[186,229] These hydride observations also seem to reopen the question of whether the "cleavage" fracture mode represents matrix behavior or is caused by hydrides.

There have recently been efforts to go beyond the hydride questions just discussed, and address directly the issue of hydrogen involvement in SCC. In aqueous solutions, there is strong electrochemical evidence[230,231] that

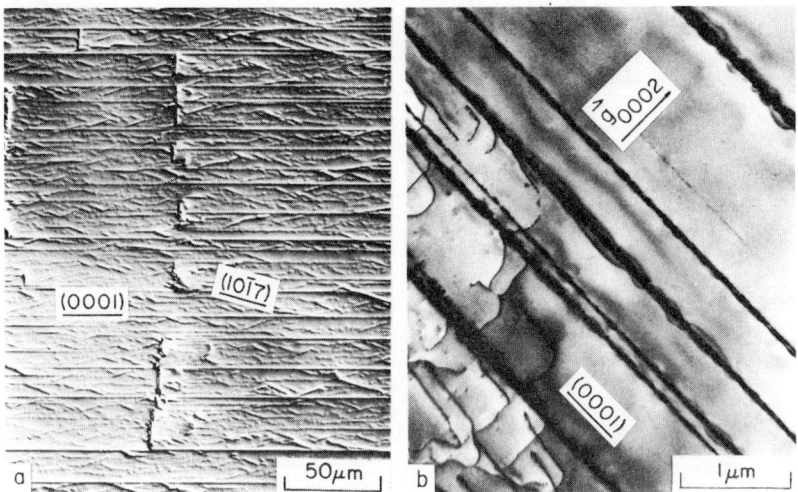

Fig. 39. Observation of (0001) and (1017) hydrides in Ti–Al alloys. (a) 1120 face of a Ti–2.9Al single crystal with 670 wt ppm H_2; plane traces shown. Discontinuous feature normal to (0001) is a subboundary. (b) Transmission electron micrograph of Ti–5Al with 780 wt ppm H_2, oriented with 1120 zone normal; operating vector and plane trace shown. (From Paton and Spurling.[226])

hydrogen is at least partly responsible for SCC; an important auxiliary role for chloride was also identified.[230] Similar conclusions have been reached for methanolic solutions.[231] In a saltwater environment, Mode III loading was found to cause no SCC, although Mode I did cause SCC.[179,182] These results are shown in Fig. 40, together with results on arsenic additions; as expected for a hydrogen process, these additions accelerate SCC in Mode I, but not in Mode III. Hot salt SCC is well established as a hydrogen-dominated process also.[232,233] On the other hand, anhydrous *molten* salt experiments appear to show no hydrogen effects[234]; the same is true of experiments in N_2O_4.[186] Although sustained-load cracking due to *internal* hydrogen[220] must be kept in mind for such cases, it would seem most general to regard all these experiments as expressions of a varying balance between two cooperative processes, dissolution and hydrogen embrittlement.[235]

We have attempted above to identify the patterns of behavior in reported work on environmental fracture of Ti alloys. We must add, however, that to date the understanding of microstructural effects in Ti alloys is incomplete, and few generalizations applicable across the whole range of these alloys appear possible. The correlations drawn here may be greatly changed by

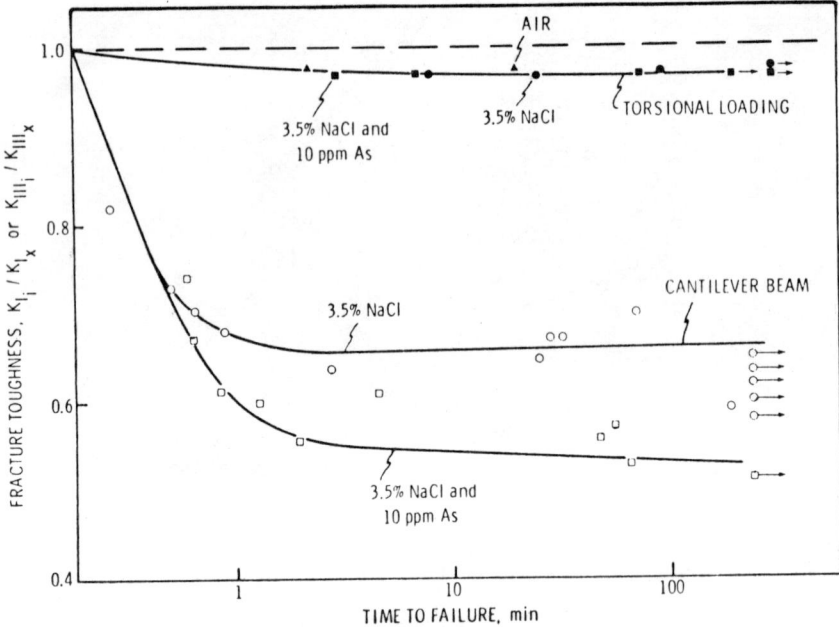

Fig. 40. Resistance to SCC of Ti–8Al–1Mo–1V as a function of loading mode (see the text), in 3.5 %
NaCl solution with and without arsenic additions to facilitate hydrogen entry, at −500 mV vs.
SCE. (From Green et al.[179])

thorough, systematic investigation. There may, in addition, be synergistic
effects among variables which are not yet evident, as we have discussed
elsewhere.[64]

NICKEL ALLOYS

The extent of research on nickel-base alloys under the kind of
environmental conditions we have been discussing is not as great as for some
of the other alloy systems described, and we shall thus present a relatively brief
account. The alloys referred to below are listed in Table 7. There are three
broad classes of nickel alloys, all with an fcc matrix: (1) the single-phase alloys,
such as Ni–30Cu, Ni–20Cr, and others; (2) the precipitation-strengthened
alloys, largely high-temperature superalloys which are aged to precipitate γ';
and (3) the dispersion-strengthened alloys, in which the strengthening second
phase is introduced in a manner other than by precipitation from solid

solution. Before discussing these classes individually, however, it is important to understand the behavior of nominally pure nickel.

The introduction of hydrogen into nickel polycrystals, for example by thermal charging from the gas phase, causes a change from the usual highly ductile rupture to brittle intergranular fracture.[108,236-239] But when single crystals are tested, rupture remains ductile and little ductility loss occurs.[108,240] It is therefore evident that hydrogen effects in nickel occur predominantly at the grain boundaries; intergranular SCC of Nickel 200 has also been reported.[241] Recent suggestions that solutes which are hydrogen recombination poisons are responsible for this behavior,[239] as is the case for temper-embrittled steels,[12] are of great interest and will therefore be evaluated.

It was argued[239] that Sn and Sb were responsible for the observed embrittlement, but the reported observation of 10 to 20 atomic layers substantially enriched in both solutes is puzzling. For a variety of solutes, particularly those known to be responsible for temper embrittlement of steel,[12] the characteristic distances[242] which have been reported indicate that

Table 7. Typical Compositions of Nickel Alloys (wt %)

Alloy[a]	Cr	Fe	Cu	Mn	C	Co	Mo	Al	Ti	Other
Nickel 200	—	0.1	0.1	0.1	0.08	—	—	—	—	—
Monel 400	—	1.2	31.5	1.0	0.15	—	—	—	—	—
Monel K-500	—	1.0	29.5	0.7	0.13	—	—	2.7	0.6	—
Inconel 600	15.5	8.0	—	0.5	0.08	—	—	—	—	—
Inconel 625	21.5	2.5	—	0.2	0.05	—	9.0	0.2	—	4 Nb
Ni-o-nel	20.0	32.0	1.85	0.9	0.05	—	3.25	—	1.0	—
Incoloy 800	21.0	46.0	0.4	0.7	0.05	—	—	0.3	0.3	—
Incoloy 804	29.5	25.0	—	0.7	0.05	—	—	0.3	0.6	—
Incoloy 825	21.5	30.0	2.25	0.5	0.03	—	3.0	0.1	0.9	—
Hastelloy X	22.0	18.5	—	0.5	0.10	1.5	9.0	—	—	
Hastelloy B	1.0	5.0	—	1.0	0.05	2.5	28.0	—	—	—
Hastelloy C	15.5	5.0	—	1.0	0.08	2.5	16.0	—	—	4 W
Inconel 718	18.6	18.5	—	0.2	0.04	—	3.1	0.4	0.9	5 Nb
René 41	19.0	—	—	—	0.09	11.0	10.0	1.5	3.1	—
IN-100	10.0	—	—	—	0.18	15.0	3.0	5.5	4.7	1 V
Waspaloy	19.5	—	—	—	0.08	13.5	4.3	1.3	3.0	—
Udimet 700	15.0	—	—	—	0.08	18.5	5.2	4.3	3.5	—
MAR-M 200	9.0	—	—	—	0.15	10.0	—	5.0	2.0	12.5 W

[a] Monel, Inconel, and Incoloy are registered trademarks of International Nickel Co.; Hastelloy is a registered trademark of Stellite Div., Cabot Corp.; René 41 is a registered trademark of Allvac-Teledyne; Udimet is a registered trademark of Special Metals Corp.; and MAR-M is a trademark of Martin Metals Co.

only one or two atomic layers are enriched.[242-245] The same has been reported for sulfur in nickel,[246] a solute which is well known for its effect on both Ni[246-249] and Ni-base alloys[250] in the *absence* of hydrogen. It would appear worthwhile to examine nickel samples of the type used[239] for second phases at grain boundaries, either the $Ni_{15}Sb$ superlattice or Ni_3Sn, since neither Sb nor Sn has high solubility in Ni; thin platelets of Ni_3S_2 have been detected in connection with very narrow enriched layers.[246]

It should be emphasized that the Sn + Sb report[239] refers to samples heat-treated at a high enough temperature that sulfur should not have been involved in the behavior.[246] Subsequent work,[239] however, which did not provide Auger spectroscopy of the fracture surfaces,[251] did employ heat treatments in the sulfur embrittlement range.[246] Sulfur is effective at levels as low as 1 ppm,[248] and care must be taken to separate the effects of hydrogen and sulfur.[246,252]

Single-Phase Alloys

Addition of 30 % or more of Cu to Ni has little effect on the ductility loss and intergranular fracture induced in pure Ni by hydrogen,[108] and thus the alloy family based on the 70Ni–30Cu binary composition (trade name Monel) would be expected to show, and does show, intergranular fracture in both SCC[241] and hydrogen embrittlement.[253] The precipitation-hardening Monel alloy K-500, although not a single-phase alloy, also shows failures under SCC,[241] cathodic hydrogen charging,[254,255] and gaseous hydrogen[256] test conditions. The extent of embrittlement, incidentally, depended upon thermal history and microstructure.[255] It is interesting that recent Auger spectroscopy of the grain boundary surfaces in K-500 has identified considerable sulfur segregation.[257] A parallel case exists for Co additions to Ni. These have been found to accelerate crack growth rate in hydrogen,[258] and some heat treatments of Ni–Co alloys can cause S segregation at grain boundaries, thereby accentuating hydrogen embrittlement.[246,258]

There are three major classes of commercial single-phase alloys in addition to the Ni–Cu system, all of which are primarily used in elevated-temperature service. These are the high-Ni alloys containing Cr (Inconel alloys are typical, with the 600-series alloy numbers indicating a (normally) non-age-hardening composition); the intermediate-Ni alloys, generally containing appreciable Fe, such as the Incoloy series; and the Mo-containing alloys. These are discussed separately below.

Alloying Cr with Ni produces considerable resistance to oxidation and general corrosion, particularly for about 20% Cr. The Ni–20% Cr binary alloy, however, exhibits a transition to intergranular fracture from ductile rupture in the presence of hydrogen,[259] and loses considerable ductility[109,259,260]; the SFE of Ni–20% Cr is considerably lower than that of Ni,[259,261] which suggests that slip would be somewhat more planar. The commercial alloys with compositions similar to Ni–20% Cr are Inconel 625 and Inconel 600, the latter having the higher SFE due to lower Cr content and appreciable Fe. These alloys are highly resistant to chloride environment SCC at temperatures below 375 K,[262] but as temperature is raised, SCC is observed.[241,262–264] Inconel 600 is also rather susceptible in fluoride environments[241] and in polythionic acid ($H_2S_xO_6$, where $x = 3, 4$, or 5) and other sulfide-bearing environments.[241,262] It should be noted, however, that the polythionic acid results have been described as "stress-assisted intergranular corrosion" rather than SCC in one review.[241]

Inconel 600 and 625 also show intergranular cracking in some other environments, including high-purity water and caustic solutions at ~ 600 K.[241,262,264,265] These failures occur in both aerated and degassed water. Crevices are probably necessary in the aerated case[264]; the degassed case seems less well understood. As in other cases, intergranular corrosion is poorly correlated with SCC,[264] but sulfur segregation to grain boundaries does appear to exacerbate SCC. It was suggested[264] that sulfur tends to be bound into carbides when sensitization occurs, thus explaining the observation[266] that the *least* sensitized grain boundaries crack most easily in deaerated water. This suggestion, however, is essentially opposite to the postulated behavior of carbides in steels,[12] and more work to examine this behavior would be welcome. Finally, it should be mentioned that addition of "massive amounts"[264] of Pb produces a unique, transgranular SCC in Inconel 600, but there is no explanation for this beyond the possibility that precipitated Pb exerts a galvanic effect on the Inconel.[264]

Binary alloys of Ni and Fe show decreasing susceptibility to hydrogen-induced ductility loss as Fe content is increased,[108,109] particularly for Fe contents between 20 and 50%. This effect is of interest because of the behavior of alloys with 20–30% Fe in addition to 20% Cr. These Ni–Cr–Fe ternary alloys, such as Ni-o-nel, Incoloy 800, and Incoloy 804, exhibit susceptibility to SCC in certain environments,[241,262,265–268] and in some circumstances perform less well than the alloys based on the Ni–20Cr composition.[241] Furthermore, improved SCC performance is obtained by successive replacement of Fe with Ni, in going from Incoloy 800 (33% Ni) to Incoloy

825 (42 % Ni) and to Inconel 625 (61 % Ni).[66,67,241,267,269] SCC failures can be produced in all these alloys, however, and Inconel 625 and Hastelloy X have been shown to be adversely affected by high-pressure hydrogen.[39,84,122,270] One conclusion, in the absence of systematic work, would be that Fe is damaging in ternary and more complex alloys, perhaps because of yet-to-be identified synergistic effects which overwhelm the behavior shown in binary alloys. But the 800, 804, and 825 (and even 625) alloys could be aged to produce γ' strengthening (see below) of varying amounts, as the compositions in Table 7 indicate; several of the studies cited provide no thermal history, so the microstructure tested is not known. Thus, comparisons of these alloys to alloys which cannot develop γ', such as Inconel 600 and Hastelloy X, may not be warranted. Further work in this area, with appropriate control of single-phase structure, seems needed.

A more convincing case can be made for Mo additions. The alloys based on Ni–Mo and Ni–Cr–Mo compositions, such as Hastelloy B and C, are fairly good performers in SCC environments.[67,241,267] On the other hand, small additions of Mo to Inconel 600-type alloys appear detrimental under SCC conditions,[241] and it is possible that an intermediate composition of several percent Mo gives rise to minimum SCC resistance. Alternatively, as mentioned for Fe, it is possible that synergistic effects are operating. The situation for both Fe and Mo might be clarified with relatively little additional work, although the good resistance of the single-phase alloys to corrosion and SCC[241,262,263] has resulted in their extensive use in challenging environments, for example in pressurized water nuclear reactors.[263]

Precipitation-Strengthened Alloys

Alloys of this class make up the majority of the high-temperature materials suitable for aircraft gas turbines and other demanding applications. The literature discussed below, however, relates primary to ambient temperature behavior; these alloys are interesting at low temperatures because they can exhibit yield strengths in excess of $1100 \, \text{MN/m}^2$ (160 ksi). The microstructure which makes this possible is relatively simple. It is a γ (fcc) solid solution containing coherent particles of γ', usually $Ni_3(Al, Ti)$, and a small volume fraction of fine carbides.[271,275] Neglecting the carbides, the ordered $(L1_2) \gamma'$ dominates, and different alloys contain different γ' compositions, since not only Al and Ti but also Nb (and to a lesser extent, V, Mo, Ta, and W also) can participate in the γ' phase.[274,276] The precipitation sequence normally is[123,126,272,274]

$$\text{solid solution} \rightarrow \gamma' \rightarrow \eta(Ni_3Ti)$$

where the η has the DO_{24} structure and is incoherent, but in a few alloys, notably Inconel 718, there is a simultaneous (and more important) sequence[277]:

$$\text{solid solution} \rightarrow \gamma'' \rightarrow \delta(Ni_3Nb)$$

Here the ordered γ'' has a body-centered tetragonal (DO_{22}) structure, while the δ is orthorhombic.[277] These phases are illustrated in Fig. 41.

Essentially all the alloys of this class which have been studied exhibit severe embrittlement by both internally charged and high-pressure hydrogen.[38,84,118,122,168,270,278,279] The most widely studied alloy is probably Inconel 718, which exhibits long life under some SCC conditions,[241,269] but typically shows intergranular fracture and a 97 % loss in RA in 69-MPa hydrogen[118]; in addition, $K_{Ienv} = 22$ MN/m$^{3/2}$ (20 ksi in.$^{1/2}$) in hydrogen, while $K_{Ic} \simeq 165$ MN/m$^{3/2}$ (150 ksi in.$^{1/2}$).[168] This is a larger loss in crack resistance than that shown for high-strength steels in Fig. 7. Other badly embrittled alloys include Ni–Cr–Al alloys,[241,280] René 41,[118,278,279] IN-100,[270] Waspaloy,[84,270] MAR-M 200,[270] and Udimet 700[279,281]; fractures are usually intergranular in hydrogen.

It was thought at one time that alloys of this type, such as René 41 and Inconel 718, were not susceptible to hydrogen embrittlement because even severe cathodic charging produced little ductility loss.[278,282] Cracking does occur, however, despite the modest losses in ductility.[283] These results may provide a link between the good SCC resistance[241,269] and very poor hydrogen resistance of Inconel 718, if it is assumed that combined dissolution and hydrogen embrittlement processes occur. The cathodic charging potential may lie in a permeation minimum, such as shown in Fig. 28, or the surface conditions may not permit significant hydrogen entry. The latter case would correspond to a low effective hydrogen fugacity; Inconel 718 is not embrittled by hydrogen below about 0.7 Pa (100 psi) pressure.[284] In addition, if the repassivation rates at the crack tip[99] do not permit crack resharpening by dissolution processes, SCC could not occur.

It is evident from the complex alloy compositions listed in Table 7 that correlation of hydrogen behavior with alloying elements would be difficult. Not only has there been no systematic work in this area, but many of the

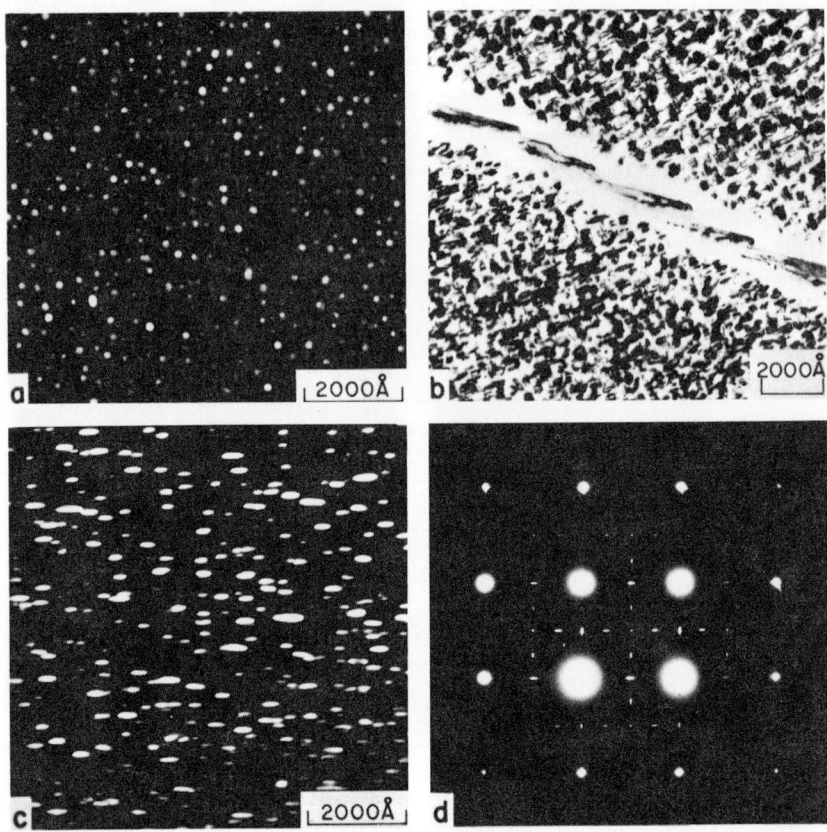

Fig. 41. Transmission electron microscopy of strengthening phases in nickel-base superalloys. (a) Dark field view of typical γ', here about 130 A in diameter. (b) Bright field of γ'' in Inconel 718, with δ particles at the grain boundary. (c) Dark field of one of the three variants of γ'' (disks seen edge-on). (d) Diffraction pattern of [100] matrix zone, showing γ'' reflections at $\{1\frac{1}{2}0\}$ positions. (Photographs courtesy of C. G. Rhodes.)

variations in alloy chemistry have their principal effect on the γ' phase.[274,276] Another factor is that these alloys vary widely in microstructure[274,285]; modern superalloys are usually divided into sheet, wrought, and cast alloys, which contain different amounts of γ'. Sheet alloys such as Waspaloy and René 41 contain less than 25% γ', while the strongest of the wrought alloys, Udimet 700, contains about 35% γ'. Cast alloys such as IN-100 and MAR-M 200 can contain as much as 55–65% γ'. All can be severely embrittled by hydrogen.[84,270]

Evidence is emerging, however, that microstructure does play a role in the hydrogen performance of these alloys. The essential element of microstructure to avoid is the presence of the equilibrium phase, usually η, at grain boundaries. This is true for η in A-286,[124] discussed earlier, and is also true for δ in Inconel 718. When prior history has brought about a relatively continuous layer of δ, Ni_3Nb, at grain boundaries, solution treatment at 1315 K (1900°F) is needed to dissolve it, since solution treatment above the γ'' solvus but below the δ solvus [e.g., at 1225 K (1750°F)] will not do so.[272] Use of the higher solution temperature for Inconel 718 has been found beneficial for hydrogen performance.[122,269,286] On the other hand, if prior processing has resulted in a dispersed or discontinuous δ structure, then the higher solution temperature is not needed and can be deleterious by causing grain growth.[122,286] Other changes in heat treatment (e.g., aging cycles) also affect performance of Inconel 718,[122] but in all cases the γ'' remains coherent,[277] contrary to some assertions.[286]

Thermomechanical history appears to be important in hydrogen performance of other superalloys[38,118,279,287]; for example, Fig. 42 shows the effects of heat treatment on René 41 sheet specimens,[279] thermally charged with hydrogen at 650°C for 1000 hr at 1-atm pressure. Note the deleterious effects of aging to form γ', and of furnace cooling from the solution temperature (perhaps by formation of η at grain boundaries; hydrogen fracture was intergranular[279]). Another study of thermomechanical effects,[288] in which Inconel 718 which had been HERF-forged prior to aging was compared to solution-treated and aged material, found little benefit: RA loss

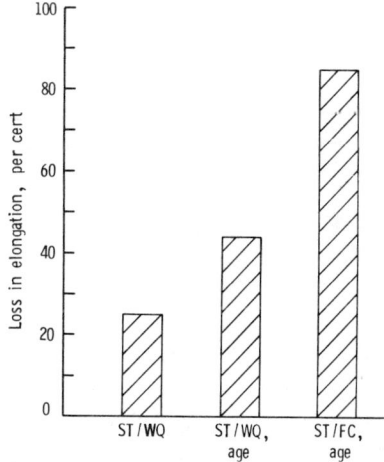

Fig. 42. Loss in elongation of hydrogen-charged sheet specimens of René 41 as a function of thermal history. Solution treated 30 min. at 1340 K (1950°F); aged 16 hr at 1035 K (1400°F) and air-cooled. ST, solution treat; WQ, water quench; FC, furnace cool. (Data are those of Gray.[279])

in 69-MPa hydrogen changed from 72% in the normally aged material, to 60% in the HERF material. Note the parallel between this result and the data shown in Fig. 21. Thus, formation of γ' or γ'' following TMT appears to be deleterious in these alloys, essentially to the same extent as in the absence of TMT. The refinement of substructure gained through aging *prior* to TMT appears critical in strengthening and SCC resistance,[160,289] and more sophisticated TMT sequences might well prove beneficial in nickel alloys.

It is evident that thermal and TMT history affects hydrogen performance in Ni. In combination with the several reviews on the physical metallurgy of these alloys,[123,126,271 277] it should be possible to determine in detail the microstructural origins of such behavior. But systematic characterization of microstructures studied in either SCC or hydrogen, particularly with thin-foil electron microscopy, is conspicuously lacking in nearly all the studies cited. Thus, the interaction of environmental behavior and metallurgical variables is to a large extent an unexplored topic. It is hoped that future work will begin to remedy this situation, particularly since it appears that both improved properties[286] and improved understanding can be expected.

It can be pointed out that these alloys are among the classic examples of dislocation cutting of precipitates, and such behavior is reasonably well understood.[123,126,285] The resulting planar slip may well be correlated with the poor hydrogen performance, as is the case in Fe-base alloys containing γ' precipitates.[124,125] The matrix-γ' misfit varies among these alloys,[274,276,285] and the matrix-γ'' misfit can be large.[277,290] Thus, the kind of relation shown in Fig. 22 may also apply to nickel-base alloys which have sufficiently large misfit,[126] provided that the deleterious grain boundary layers of η or δ are avoided. More work based on such concepts would appear valuable.

Dispersion-Strengthened Alloys

These alloys are strengthened by an inert oxide in the form of fine particles, which may be introduced by a variety of nonprecipitation processes.[291] The matrix of the alloy is typically unaffected in composition by the presence of the dispersoid, so the discovery by Thompson and Wilcox[238] that Ni–2ThO$_2$ (trade name TD Nickel) exhibited only moderate ductility losses when hydrogen-charged, rather than brittle intergranular fracture as in the case of Ni, came as something of a surprise. Extension of that work to Ni–20Cr–2ThO$_2$, at a much higher strength level, showed similar behavior,[259] although the thermal history of the material can be quite important.[259,279,292]

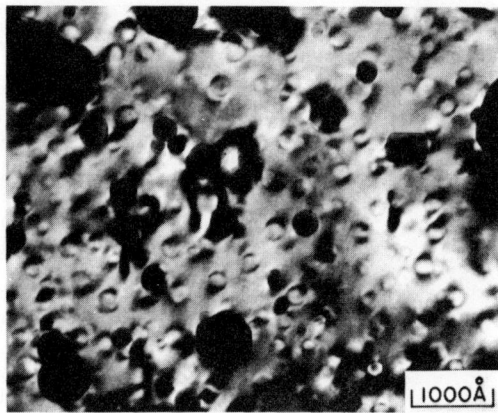

Fig. 43. Transmission electron micrograph of alloy MA 753, showing both yttria dispersoid (particles with sharp image) and ′ (particles showing strain contrast) in bright field; near matrix $\langle 110 \rangle$.

The reasons for behavior of these alloys appear to be that hydrogen is trapped at the particle–matrix interface and therefore does not reach the grain boundary. This behavior may saturate with an unlimited hydrogen supply,[293] but there are questions remaining on that point.[259]

An intriguing development in these alloys is the attempt to improve ambient temperature strength by superimposing strengthening by γ' precipitation upon that due to the dispersion.[291,294] One such alloy has been studied in the presence of hydrogen; it is basically Ni–20Cr with 2 vol % Y_2O_3 dispersoid and the addition of 2.2 % Ti and 1.1 % Al to produce γ' phase,[294] as shown in Fig. 43. This alloy, called Inconel MA 753, has a room temperature yield strength of about 900 MN/m^2 (130 ksi), and shows essentially no ductility losses in high-pressure hydrogen or when hydrogen charged.[259] This behavior is shown in Fig. 44.

Although these alloys are relatively costly, they offer good strength at both low and high temperatures[291] and seem to be compatible with hydrogen. Certain specialized applications in the future may well require use of the dispersion-strengthened nickel alloys.

Fig. 44. Ductility of MA 753 tested at two temperatures, 194 K and 295 K, for specimens tested in air when uncharged and when charged with hydrogen, and for tests in 69-MPa hydrogen. From Thompson.[259])

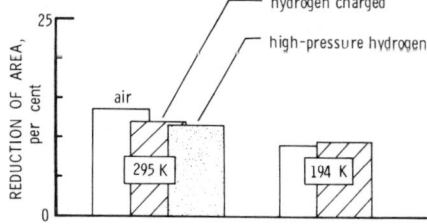

PARALLEL BEHAVIOR IN DIFFERENT ALLOY SYSTEMS

We have now discussed several different alloy systems in some detail. These included alloys of bcc, fcc, and hcp matrix crystal structures and a very wide variety of second phases. It may seem presumptuous to seek parallels in the SCC and hydrogen embrittlement behavior of such diverse materials, and certainly at a detail level, this would be the case. There are a number of significant, general parallels, however, and it seems worthwhile to point these out.

We have reviewed composition effects in each alloy system, but even though composition is a very useful variable in the hands of the alloy designer, it is in many cases one which yields relatively little information about mechanisms. The major exception is the effect which composition can exert on slip mode (see below). It is also difficult to generalize about composition in alloy systems of the diversity we have considered, but the poor environmental resistance conveyed by Mn additions in steels, austenitic stainless steels, A-286, and Ti alloys is noteworthy. Mn does seem, however, to play a beneficial role, albeit an indirect one, as a grain shape stabilizer in Al alloys. By contrast, Si seems beneficial in steels, austenitic stainless, A-286, and Al alloys, while possibly being poor in Ti alloys. The interstitial elements are potent strengtheners in many systems, but at certain concentrations can be deleterious to environmental behavior in steels, austenitic stainlesses, and Ti alloys. Most other elements act differently in each alloy system.

One particular solute element class deserves special mention: the hydrogen recombination poisons, such as S, As, Sb, and others. Segregation of these to grain boundaries appears to assist hydrogen-induced intergranular fracture, and since intergranular fracture is a common mode of environmental failure, careful attention to the presence and partitioning of these elements in future work is clearly warranted. The pioneering work of McMahon and others[11,12] in steels may presage similar work in austenitic stainless steels, and Al and Ti alloys, as is now beginning in Ni alloys.[246,257,264] These elements may also segregate at the interfaces of grain boundary intermetallic particles and worsen behavior at those interfaces, or the particles may absorb the poisons.[264]

There are also several parallels in the area of microstructure effects on hydrogen-induced environmental fracture. The most general of these parallels is the benefit obtained by refining microstructural size, whether it be grain size, martensite plate size, or the size of a precipitated phase such as Widmanstätten

α in Ti alloys. The same benefits are generally associated with considerations of strength, fracture toughness, and fatigue resistance, so that refined microstructures appear to be an instance of property improvement without obvious other penalties[64] (the most prominent exception, high-temperature creep resistance, is outside the scope of this chapter). Therefore, those environmental investigations which permit microstructural size to be an uncontrolled factor are ignoring a primary variable, even in cases where size effects do not exert the principal control over behavior.

Another interesting point is that equilibrium intermetallic compounds are deleterious to hydrogen-related performance, for steels, titanium alloys, nickel alloys, and in some cases for stainless steels. They play an indirect beneficial role in Al alloys, but they can play a damaging direct role. Since precipitation of these compounds can also negatively affect toughness and other properties, their avoidance seems generally helpful, except when necessary for strength.

Thermomechanical treatments or TMT have not been carefully investigated in any of the systems considered except in a preliminary way, but the ability to manipulate microstructure in this way seems promising in every case for which it has been tried. More work on such processing techniques seems potentially valuable, both for improvements in engineering materials and as a means to extend understanding of the microstructure variable. With respect to the latter point, it should be emphasized that use of thin-foil electron microscopy to characterize the microstructures obtained is an essential component of increased understanding; this point has begun to be appreciated by investigators working in this area.[101,160]

We believe that the most striking of the parallel types of behavior is that of slip mode effects. Planar slip can be brought about by a number of factors, including reduced SFE, low temperature, short-range and long-range order, clustering, and cutting of precipitates. All of these have been identified at various points in this chapter and in earlier reviews, and although the planar slip correlation with SCC and hydrogen behavior has been examined most completely and in detail for the austenitic stainless steels, it also applies to other austenites, to Al alloys, and to Ti alloys of both α and β structures; it may also apply to Ni alloys. The evident omission is the ferritic and martensitic steel family, but in that case relatively few investigations of slip mode have been made. We discuss below the possibility that slip mode may be unimportant in these alloys, but would also suggest that future work on them include slip mode characterizations, for example by the relatively simple slip line technique,[201] to see whether this correlation does extend to the bcc steels. The often-heard view that slip is always highly nonplanar in iron (and by extension in steels) is misleading. For example, temperature reduction leads to a marked increase in both planarity and coarseness of slip in pure iron,[295] and alloying elements tend to have similar effects[295]; the slip mode in highly alloyed bcc materials such as ferritic stainless steels is coarse and relatively planar.

We discuss possible reasons for the planar slip correlation in the next section, but whatever the explanation, it is an expression of some fundamental aspect of environmental behavior. Any model of hydrogen-induced SCC or embrittlement which cannot rationalize the importance of planar slip as a central element of behavior is significantly incomplete. At the same time, one must recognize that planar slip is not a sufficient condition, particularly under SCC conditions,[66,80,94,99] for environmental fracture to occur. Electrochemical factors, and metallurgical variables other than slip mode, must also be recognized and taken into account.

DISCUSSION

In the preceding section, several kinds of parallel behavior among alloy systems were described. We now turn to a discussion of some of these, and also to an overview discussion of mechanistic questions. Rather than attempting to analyze published mechanisms in detail, we seek to present the subject in a rather general way. The discussion first summarizes information on electrochemical factors, slip mode, and hydrogen effects; this information is then used to present a broad thesis about hydrogen behavior in materials which includes a number of new ideas in the hydrogen field. It is a thesis which we believe provides a unifying viewpoint for hydrogen phenomena, and which is consistent with the observations we have reviewed above. We also identify important areas of research in which work is needed.

Electrochemical Factors in Hydrogen-Induced Cracking

There are essentially three component processes of SCC in which electrochemical factors can play a predominant role when cracking behavior exhibits hydrogen-induced characteristics. These are (1) crack initiation, (2) crack resharpening, and (3) hydrogen production. (Anodic dissolution as a component of crack advancement is discussed below.) It is outside the scope of this review to discuss initiation behavior, process (1), in detail. It nevertheless seems clear that anodic dissolution of slip steps, chemically controlled pitting, passive film rupture, and other phenomena could singly or in combination provide both the initiation site and the stress concentration needed for cracking to begin (see Fig. 1). These particular phenomena may of course be quite different in different alloy systems.[2,66,99,186,296] Control of applied potential, anion concentration, or solution pH in a manner which retards the relevant phenomena would markedly delay initiation and therefore delay the onset of SCC, whatever the cracking process.

In ductile materials which resist crack extension by massive plastic flow and crack blunting, crack propagation is only possible if repeated crack initiation (i.e., resharpening) can occur. This can take place by the occurrence of phenomena similar to those of process (1), except that environmental chemistry within a crack tends to differ markedly from that at the external surface (e.g., pH may be reduced to very low values within the crack[60,61,175–178,297,298]). This effect occurs by hydrolysis of dissolved metal ions, M^{+y}, as follows[176]:

$$M^{+y} + xH_2O \rightarrow M(OH)_x^{y-x} + xH^+$$

Although this effect tends to reduce dependence on the external environment, as can the potential shift within a crack,[178] it is still true that the environment can play a profound role in the rate (or even the occurrence) of resharpening, just as is true for initiation.

Singly or in combination, then, the various electrochemical factors which influence the processes of initiation and resharpening can profoundly affect the rate of SCC, even when, as we are considering here, hydrogen plays at least a contributory role in the fracture. Furthermore, when second phases or other microstructural elements are favorable sites for attack, the crack path may be guided by the preferred location of initiation or resharpening. In many alloy systems, the presence of chloride ion is of particular importance[2,66,186,241]; a well-known example is the Williams and Eckel result for austenitic stainless steels (Fig. 45), which depicts a complex interaction of oxygen and chloride.

Finally, hydrogen production at or near the crack tip, process (3), also depends on environmental factors, although the occurrence of hydrolysis and of potential shifts as large as 1 V within the crack[178] means that the external environment is not always the controlling factor. The latter is of particular importance when SCC occurs under an applied anodic potential (relative to open-circuit conditions).

One way of integrating the various electrochemical factors is to examine the repassivation rate of the particular alloy system in the environment of

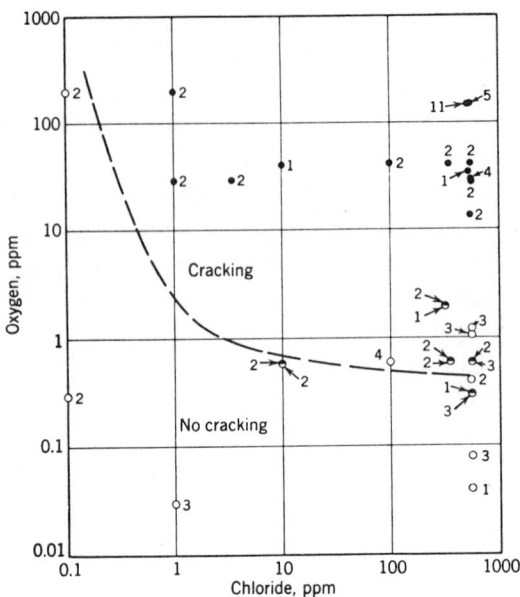

Fig. 45. Susceptibility of 18-8 type austenitic stainless steel to SCC failure as a function of oxygen and chloride content of environment (steam and intermittent wetting, 515–535 K, pH 10.6, 1–30 days exposure). ●, Failure; ○, no failure. Values inside figure denote number of specimens. (Data of Williams and Eckel[299]; figure from Uhlig.[300])

interest. The emergence of slip steps at the crack tip can rupture a passive film and permit localized dissolution, or pitting, and also accelerate the corrosion reactions which generate hydrogen; the repassivation rate is then an indication of the extent of these processes. Note that planar slip causes relatively larger and more intense slip steps, giving an effect of planar slip on SCC. As Staehle has shown,[99] the repassivation rate in many cases correlates well with SCC behavior. Such a correlation is thus a useful indicator of the interplay and net effect of the various electrochemical factors, although it does not, of itself, indicate the mechanism of cracking.

We mention the importance of electrochemical factors in order to demonstrate that comparison of hydrogen embrittlement and SCC results cannot be made without due consideration of the inherent differences between the two cases. This is not to say, however, that those differences preclude determination of a hydrogen role in some kinds of SCC. As was shown for most of the alloy systems considered in this review, hydrogen test results often are parallel to SCC results for cases in which hydrogen could be generated in the environment. Other reviewers have also pointed this out.[15,23,27,55,64,66,68,101,172,175,209]

One other example can be added to those already presented. Plots of crack velocity response to stress intensity, as in Fig. 2, are generally rather similar in certain SCC environments and in hydrogen gas, as illustrated in Figs. 46 and 47. An interpretation of the K-independent or region II behavior, under SCC conditions, has been that it is controlled by the diffusion rate of corrodents to the crack tip; this appears to be consistent with the temperature dependence of the behavior.[153,296] The existence of generally similar data in hydrogen gas, however, suggests that such an interpretation should be in terms of critical corrodents for the hydrogen production process. Theoretical efforts to model region II behavior in hydrogen terms are now appearing.[15,301]

We should also note that "activation energies" for region II (assuming that in this region a single thermally activated process controls the cracking

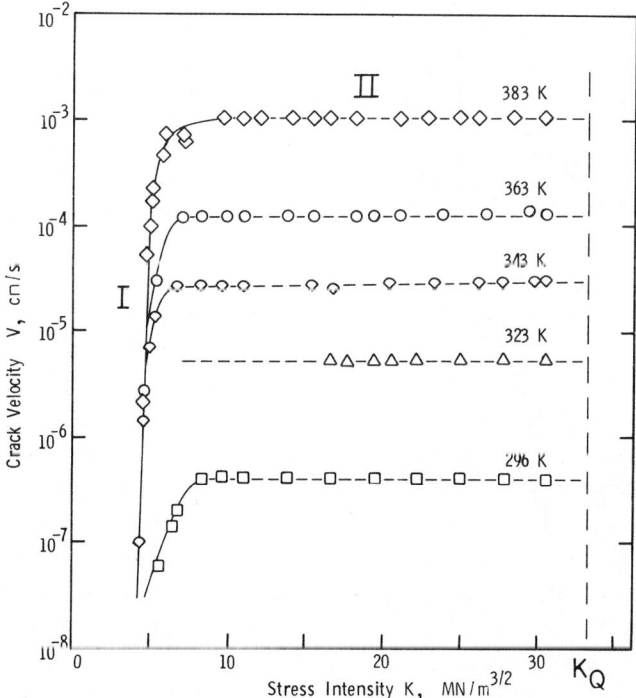

Fig. 46. Crack growth ($V–K$) curves as a function of temperature for 7039-T61 aluminum alloy, with crack plane perpendicular to short transverse direction, tested under open circuit conditions in $5\,M$ aqueous potassium iodide solution. (From Speidel and Hyatt.[2])

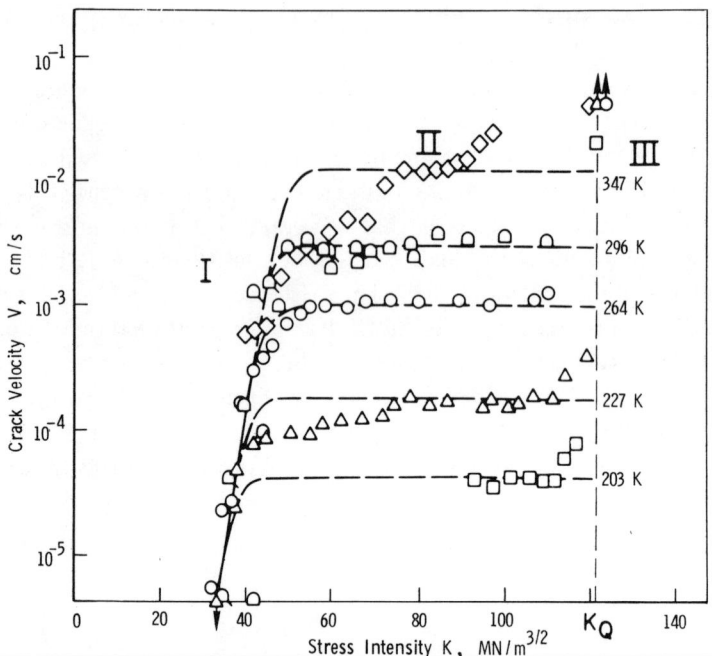

Fig. 47. Crack growth $(V-K)$ curves as a function of temperature for Ti–5Al–2.5Sn, tested in hydrogen gas at a pressure of 0.9 atm. (From Williams and Nelson.[207])

rate) vary widely, from the 85 kJ/mole (20.3 kcal/mole) in Fig. 46,[2] to less than 16 kJ/mole (3.8 kcal/mole)[2]. The value for the data of Fig. 47 was 23 kJ/mole (5.5 kcal/mole),[207] in agreement with SCC studies on Ti alloys[296]; values of about 38 kJ/mole have been reported for several material–environment combinations.[172] It is unclear exactly what, if anything, these various values mean, despite efforts at detailed interpretation.[2,172,207,296] In some cases the data seem to agree with published activation energies for hydrogen diffusion in the metals studied (we discuss diffusion below). Examples are the Ti data[207] and data in distilled water for alloys of Ni and Al and for high-strength steel[172]; the latter three have activation energies for hydrogen diffusion near 40 kJ/mole. We do not feel, however, that "energy" comparisons of this kind are particularly illuminating. The complexity of the various processes likely to be operating, several of which are thermally activated, means that "compound" activation energies would be expected in many cases

Before proceeding, we should refer again to the importance of testing methods and the data which they provide. The three-stage curve of

Wiederhorn,[4] shown in Fig. 2, gives both the region II cracking velocity V and the threshold K (K_{Iscc}), if one exists. Both these parameters are important, because they measure, respectively, the ease of moving a crack, and the ease of "initiating" measurable crack movement, particularly when region I has a high slope and K_{Iscc} is well defined. As we discussed in connection with Table 2, it is difficult to interpret observations unambiguously when only one of these parameters is available. Furthermore, the conclusions reached with one parameter may not be borne out by the other, as in the case of the Ti alloy Beta III, which shows no environmental susceptibility in terms of K_{Iscc}, but does show a marked susceptibility through increased crack velocity.[211-213] In many cases, it is also possible to correlate V values with fractographic observations.[186,212]

At the same time, we should emphasize that conventional smooth-specimen testing will continue to be important. For engineering tests on product forms such as sheet and wire, there is little alternative. There is also a very large body of information already available on the correlation of smooth specimen behavior and various service performance conditions. These data will only continue to be useful if both smooth specimens and fracture mechanics-type specimens are tested in the future.[6] For materials which are relatively tough, it can be difficult to perform TTF or V–K tests on precracked specimens, particularly when strengths are low and net section yielding intervenes early in the test. Care must also be used in interpreting the behavior of fracture mechanics specimens, which produce high strain rates at crack tips and which appear very sensitive to contamination effects.[302] Attention must be paid to these and similar points if the use of fracture mechanics techniques is not to become more fashionable than utilitarian.

Slip Mode as a Metallurgical Variable

In all the alloy systems we have surveyed, a distinct effect of metallurgical variables such as composition, presence and amount of second phases, texture, grain size, thermomechanical treatment, and slip mode was identified. Some of these variables, such as second phases which have distinct electrochemical differences from the matrix, can readily be incorporated into chemical and electrochemical models of SCC. Most, however, cannot, as several reviewers have concluded.[2,20,66,186] In addition, many of these variables affect hydrogen gas behavior in a similar way to their effect on hydrogen-induced SCC. In pursuing a general picture of hydrogen effects, then, we shall be attempting to rationalize the role of metallurgical variables in a quite general way. As we have stated, these have particular importance because they are accessible to manipulation by the alloy designer.

One of the most suggestive and pervasive of these variables is slip mode, which has a long history of interest by SCC investigators and more recently has attracted attention in hydrogen studies as well. The slip mode concept is a generalization first proposed[303] to describe the appearance of surface slip lines, as either being planar (i.e., rather straight and parallel) or wavy (i.e., generally curved and branching). These external slip line features generally correlate well with the structure of dislocation slip bands as seen in thin-foil electron microscopy, and thus this somewhat descriptive concept has become widely accepted. The terminology has not been without confusion, however, because slip which is either planar or wavy can also be either fine (referring to many slip steps of low height) or coarse (having large, infrequent steps).[201]

An example of this point can be found in the review of Blackburn et al.[186] They cite the alloy Ti–8Mo–8V–3Al–2Fe as an exception to the rule that alloys exhibiting planar slip are susceptible to SCC, since in the quenched condition this alloy is resistant. But the slip behavior they refer to is of the type shown in Fig. 48, which in reality is coarse, wavy slip.[201] This example is not intended to single out Blackburn et al.[186] for criticism, for theirs is an outstanding review; instead, it is to indicate that even these authors seem to have followed the conventional opinion that coarse slip is usually planar. Such an identification, although often true, is incorrect in general and in this specific case.[201]

The planar slip mode can be enforced by any variable which either makes dislocation cross-slip more difficult, or which perpetuates slip on those planes where slip initiates. Thus, not only the composition-dependent variable of

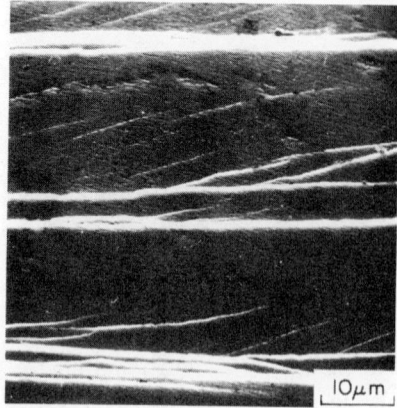

Fig. 48. Coarse, wavy slip in a Ti–40V (bcc) single crystal, tested at 300 K. (From Williams et al.[201])

SFE, but also temperature, and the microstructural variables of order, clustering, and coherent particle precipitation, can all affect slip mode, as we have shown. Many of the metallurgical variables considered in this chapter can therefore be subsumed under the slip mode heading (although there are certainly exceptions). We should also note that some of the cases which may appear to be exceptions in fact conform to the pattern. For example, grain-size refinement may be effective at least in part because a greater proportion of the specimen volume is required to deform by multiple slip at small strains,[304] and as we shall show, small strains are critical in a number of circumstances.

The Behavior and Effect of Hydrogen

In this section, we present a general view of a set of hydrogen processes which may be significant in both hydrogen-induced SCC and hydrogen embrittlement. It is not yet possible to be quantitative about many parts of this discussion, but it appears useful to present it in summary form and indicate the ways in which it can apply to the behavior reviewed earlier.

The Source of Hydrogen

Hydrogen may be present in a material before exposure to a service environment as a result of processing, such as the hydrogen retained upon solidification (in welds as well as in ingots) or as a result of heat treatment in a hydrogen atmosphere. But since such hydrogen sources are not fundamentally different from those occurring in service, they will be implicitly included in the following discussion.

The simplest case is hydrogen which is available as part or all of a gas phase, so that H_2 molecules adsorb onto the surface, dissociate there, and are absorbed into the metal as dissolved hydrogen or [H]. An important step here is dissociation, as has been shown by the accentuated hydrogen behavior when some of the H_2 molecules are dissociated by a hot filament before reaching the metal surface.[305] Figure 49a schematically shows such processes.

A second case is that of hydrogen which is available in solution as (hydrated) H^+ ions (e.g., in an acidic environment). Even high-pH solutions can produce rather acidic environments inside cracks and pits, typically with a pH in the range $1-3.5$[2.175–178.298.306]; the exact value depends on the solubility product of the hydrolysis reaction and on the particular Pourbaix diagram which applies.[176.307] In this case, as Fig. 49b shows, acquisition of an

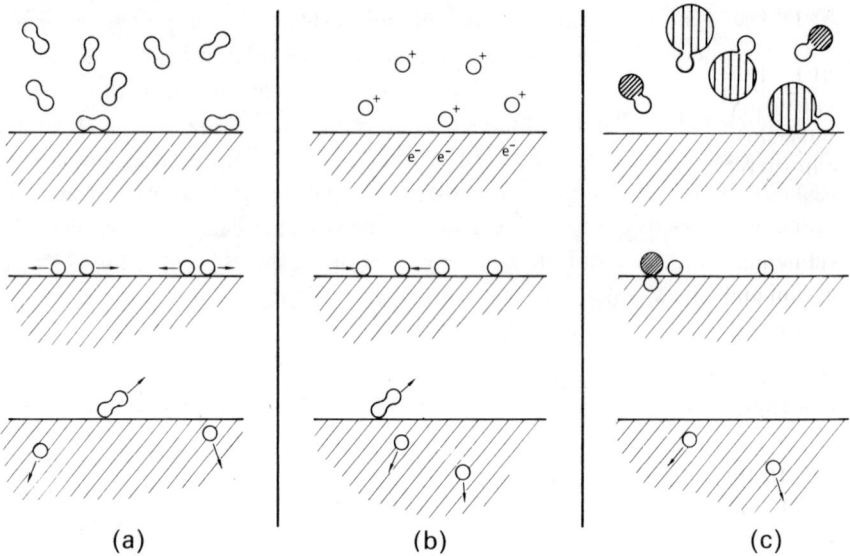

Fig. 49. Schematic depiction of hydrogen sources; sequence from top to bottom. (a) Hydrogen gas molecules adsorbing on metal surface, dissociating, and entering the metal as [H]. (b) Hydrogen ions combining with electrons at the metal surface to form [H]. (c) Hydrogen-bearing molecules reacting at metal surface to provide [H]. In all three cases, single hydrogen atoms at the surface can recombine to H₂ and desorb.

electron results in hydrogen atoms, which can either combine to form H_2 and bubble off, or can be absorbed into the metal as [H]. On a macroscopic scale, such a process can, of course, be prevented or reduced by an anodic potential, but the crack tip processes can be such as to alter local potentials and thereby give rise to significant hydrogen absorption.[178,297]

The third instance is a chemical reaction, in which a hydrogen-containing molecule reacts with the metal to release hydrogen, which can then enter the metal as [H] (see Fig. 49c). This can be true of many molecules, such as alcohols[186] and hydrides,[100] and can include complex ions in solution as well.

The presence of hydrogen recombination poisons such as S, As, Sb, and others, whether in the environment as ions or compounds such as H_2S, on the metal surface as part of a film, or in solid solution in the metal* can strongly

---

* Poisoning may not occur in solid solution if the valence state changes.

retard the $2H \rightarrow H_2$ reaction and thereby greatly increase the rate of [H] production in all three cases. An especially interesting case is the possibility that if these poisons are already segregated to grain boundaries (or other specific microstructural locations) prior to hydrogen entry, they can accelerate hydrogen entry at the intersections of those boundaries with the surface.[239] Critical experiments on this concept would be valuable.

Hydrogen Transport

The presence of [H] very near the metal surface may be sufficient for some kinds of hydrogen behavior, as we consider below, but other cases require hydrogen transport.

The most common of these transport processes would be diffusion through the lattice. Hydrogen diffuses fairly rapidly in most metals, particularly in bcc structures (steels and β-Ti alloys), and it is appropriate to compare cracking rates (e.g., in region II) to diffusion rates. It is common to make such a comparison on the basis of the activation parameters (e.g., activation energy), and agreement has been reported in a number of cases, specifically Ti alloys,[207] steels,[172,308,309] and others.[172] It should be pointed out, however, that it is usually not known if a single thermally activated process is occurring, and observation of an "activation energy" for a cracking process which approximates that for lattice diffusion demonstrates neither that diffusional transport alone is operative, nor that the critical step in the transport process is diffusional.[39,310] We do not doubt that some hydrogen cracking phenomena are diffusion-controlled, but proof of such control has not always been satisfactory.

In some cases, the diffusivity has been directly compared to the crack velocity. In several of these comparisons, it was concluded that diffusion was too slow to account for the observed cracking.[231] These observations, and other considerations described below, have led to interest in other mechanisms for hydrogen transport. Prominent among these is movement of hydrogen with dislocations, in the form of Cottrell atmospheres, a concept apparently first proposed by Bastien and Azou.[311–313] This movement can occur considerably faster than bulk diffusion,[314] because the association of hydrogen with the dislocation causes directed rather than random movement of the [H], often toward dislocation sinks in the material. For this to occur most effectively, it is required that "intermediate" values of E^β, the binding energy of hydrogen to dislocations, exist; high values could correspond to pinning of the dislocation, while low values could permit the dislocation to readily move away from the atmosphere.

Many investigators have concluded that dislocation transport of hydrogen occurred in their experiments, from Bastien's work on mild steel,[313] to a number of studies in austenitic stainless steel,[39,72,84,100,124] to work on Ni and Ni alloys,[108,238,253,259,293,315] and to a variety of other alloys, including those of aluminum.[68] The subject has been addressed in the case of Ti alloys also,[220] which is significant in view of a report[296] that cracking in these alloys occurs faster than hydrogen diffusion. Formation of Ti hydrides with stress assistance[224] would be consistent with dislocation transport, since hydrides form in slip bands rather than randomly in the matrix.[224,226,316] Recently, quantitative evaluations by Tien et al.[314,317] have shown that transport can be accelerated by factors of 10^2–10^4, and that because pickup and release times are of the order of microseconds, grain boundaries are not of major importance as barriers to dislocation transport. The latter point is in agreement with experimental evidence.[39,72,237,315]

Hydrogen transport can therefore occur by either lattice diffusion, dislocation motion, or short-circuit diffusion along grain boundaries or dislocations. Short-circuit diffusion of hydrogen, however, generally is scarcely more rapid than lattice diffusion. There may also be cases in which little long-range transport occurs, and the processes of Fig. 49 are sufficient to provide directly a supply of hydrogen. In most cases, however, some transport will occur in the ways we have described. We now consider the destinations of these transport processes.

Location of Critical Hydrogen Interactions

There are a wide variety of places in a material's microstructure at which the presence of hydrogen may be critical to fracture behavior. These include the lattice itself (hydrogen in solution), as well as grain boundaries, incoherent and coherent precipitates, voids, and dislocations. These are shown schematically in Fig. 50, which also emphasizes the role which may be played by certain solute elements, when solute–hydrogen pairs are formed in the lattice (Fig. 50b) or at grain boundaries.

On a larger scale, hydrogen diffuses to locations of maximum triaxial stress near crack tips.[318,319] Figure 51 shows that the location of such stresses is very close to the crack tip under conditions of plastic opening.[320] Hydrogen may accumulate, at any of the locations shown in Fig. 50, throughout the

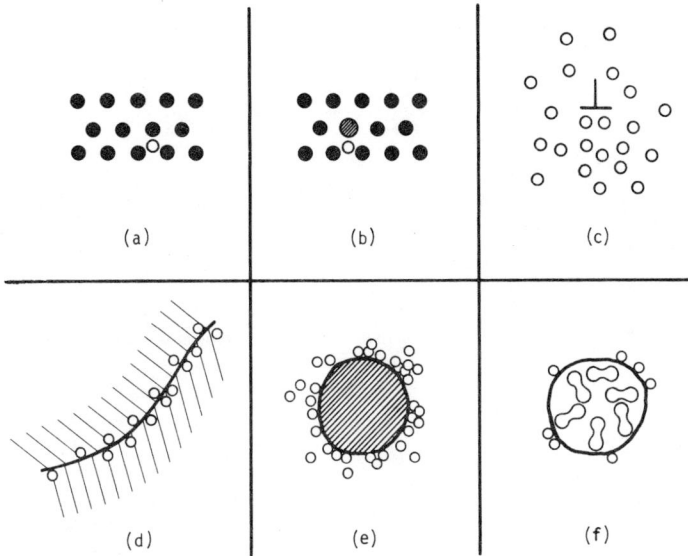

Fig. 50. Schematic view of destinations for hydrogen in a metal microstructure, discussed in the text. (a) Solid solution; (b) solute–hydrogen pair; (c) dislocation atmosphere; (d) grain boundary accumulation; (e) particle–matrix interface accumulation; (f) void containing recombined H_2.

plastic zone at the crack tip. In this zone, hydrogen transport may well occur predominantly by diffusion, particularly in steels,[318] because plastic zones are often small under the conditions of crack growth at high strength levels (cf. K_{Iscc} values in Fig. 7). Should lattice diffusion predominate in steels, slip mode correlations with the hydrogen component of environmental fracture would not be expected, although dislocation transport cannot be ruled out on the evidence now available. Dislocation transport of hydrogen is especially likely in other materials than steels, which have low hydrogen diffusivities.

An example of a hydrogen interaction, due to hydrogen in solution, is the case of cleavage cracking. Cleavage is usually thought of as a process of atomic separation intrinsic to the crystal lattice, although there is evidence[55] of intense, highly localized plastic flow on cleavage surfaces. Hydrogen "in solution" can also result in strain-induced hydride formation in Ti alloys,[224] and these hydrides may fracture by cleavage. These processes, and those described below, can be regarded as linked processes of transport followed by an increment of fracture, as is often observed in environmental cracking; many of the behavior types we are describing can result in such observations.

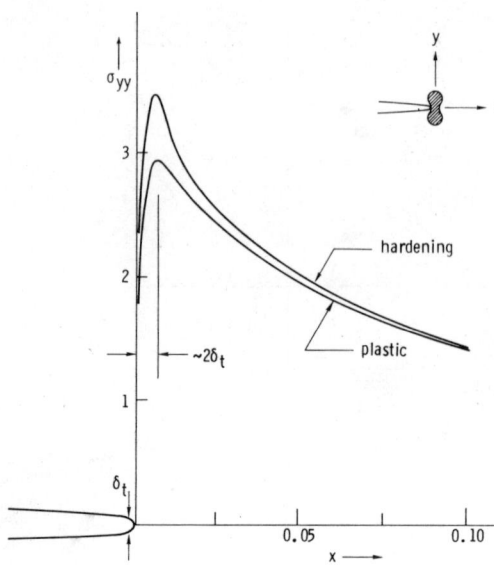

Fig. 51. Dependence of maximum stress, in units of yield stress σ_0, on distance ahead of a crack, in units of (K^2/σ_o^2). Two cases are shown: ideally plastic material, and power-law hardening material with exponent $n = 0.1$. Plane strain conditions and small scale yielding are assumed. Note the peak stress location in terms of plastic crack tip opening δ_t. (From Rice.[320])

This summary has indicated ways in which hydrogen from a variety of sources, transported by dislocations or lattice diffusion, can accumulate at any or all of a number of locations in the microstructure. The location most critically affected by hydrogen will form the fracture path; this may accentuate the fracture mode which occurs in the absence of hydrogen, or it may cause a different fracture mode. These possibilities are summarized in Fig. 52. Included in Fig. 52 is the occurrence of hydride formation and cleavage of the hydrides, although that is not strictly a behavior of the metal matrix. Figure 52 is a "roadmap" only for the *phenomenology* of hydrogen-induced fracture, and does not include any suggestion of the mechanism(s) of such fracture. We touch on the question of hydrogen mechanisms in a later section.

A General Viewpoint on Hydrogen-Induced Cracking

It is now appropriate to combine the electrochemical factors, the role of slip mode and other metallurgical variables, and the behavior of hydrogen to construct a general viewpoint on hydrogen-induced cracking. To be successful, such a rationale must be able to identify the common elements in such apparently diverse observations as these: ductile RA loss in an austenitic stainless steel tensile tested in high-pressure hydrogen gas, and cleavagelike fracture of a Ti alloy under sustained load in a methanolic chloride solution. It

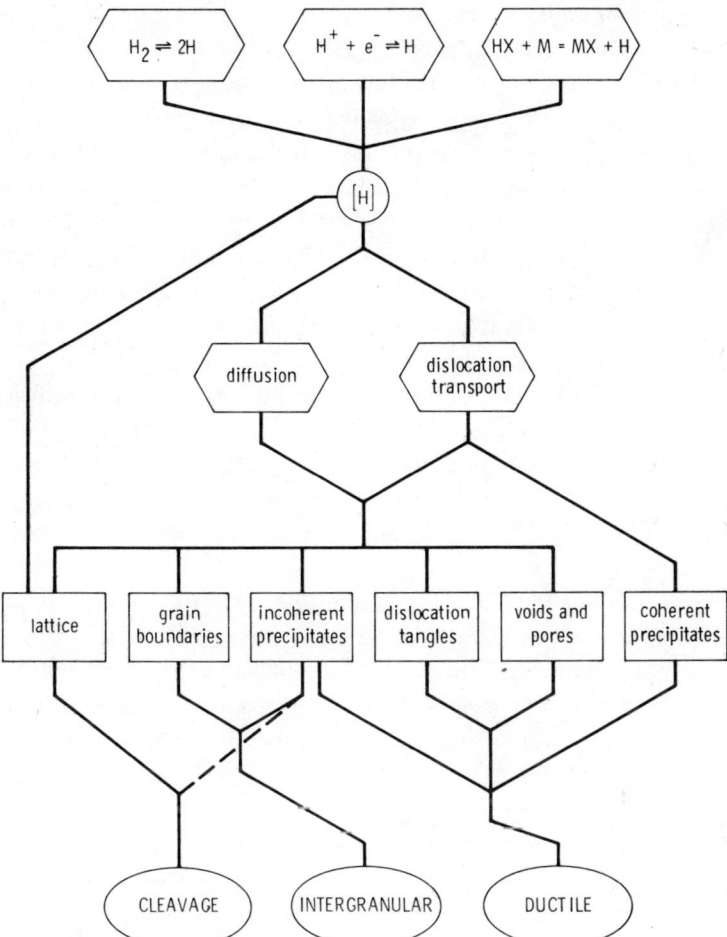

Fig. 52. Summary of hydrogen processes (hydrogen sources, hydrogen transport) and microstructural locations described in the text, with corresponding end results (hydrogen in solution, fracture). See the text for discussion and examples. Dashed line at the bottom refers to cleavage of hydrides.

must also be able to rationalize not only the occurrence of the "pure" processes of anodic dissolution and hydrogen embrittlement, but also the occurrence of mixed or multiple processes. We present below a qualitative description of at least a beginning for such a rationale, borrowing freely from earlier work.

A broad statement of the rationale should begin with a comment on the interplay between hydrogen embrittlement and anodic dissolution processes. Anodic dissolution, whether it occurs as a process mediated by competition between localized film rupture and repassivation,[99] as Logan originally suggested,[321] as a process assisted by yielding at the crack tip, in Hoar's formulation,[322] or by some other localized process, is a well-established phenomenon in SCC. In some systems (e.g., copper alloys), a process of the dissolution type appears to be the sole operative process.[323,324] On the other hand, cracking due to hydrogen absorption can be produced under suitable environmental conditions in all the alloy systems we have reviewed. One might conclude, then, that even though SCC usually requires a rather specific combination of alloy composition and microstructure, environmental composition, and other variables like potential range, one can produce cracking in an appropriately chosen system by either hydrogen or dissolution processes, *provided* that the environment (e.g., applied potential) is sufficiently modified.

In the great majority of cases, however, the situation is not clear-cut as to the operative process. These cases, which include many alloys under open-circuit conditions, might be called "intermediate conditions." There are two possibilities for intermediate conditions: (1) that anodic dissolution is the operative process, but at a rate greatly diminished from the bare-metal rate[325] because of altered repassivation, or (2) that mixed or multiple processes operate (e.g., dissolution and hydrogen processes together). It is difficult to believe that possibility (1) is *general*, for two reasons. First, it would require one to regard hydrogen embrittlement as a special case rather than as one end of a spectrum of behavior. It would be equally logical (or illogical) to regard dissolution as a special case. Second, it would require one to regard the many similarities in hydrogen and SCC dependencies on alloy composition, microstructure, thermal history, and other metallurgical variables as merely a large number of coincidences. We suggest, then, that possibility (2) is more general, keeping in mind that possibility (1) certainly occurs in some cases.

When multiple processes occur, they can be either sequential (i.e., required to operate in turn) or simultaneous (i.e., independent and perhaps additive). The distinction is important when the component processes differ markedly in rate, because if sequential, the slower one will control the phenomenon, while if simultaneous, the faster one will be dominant. It is implicit in possibility (2) that either sequential processes are involved, or that simultaneous processes are occurring under conditions of temperature, stress, strain rate, and so on, such that their respective contributions are comparable. There seems to be no evidence as to which type of multiple process occurs in hydrogen-induced SCC. One way to determine this would be to measure activation energies for cracking over a series of narrow temperature ranges, because the energy will tend to increase with temperature for independent processes and decrease for sequential ones,[326] provided that the temperature span includes the transition from dominance by one process to dominance by the other. It would also be required that the rate-controlling mechanism for each process remain the same (e.g., solution mass transport for dissolution or hydrogen absorption for hydrogen cracking) over the entire temperature range.

We would summarize this section as follows. There are cases in which either anodic dissolution or hydrogen cracking is entirely sufficient to explain an instance of SCC; an example of the former may be Ti alloys in N_2O_4 environments, while the latter seems to be represented by most high-strength steels in aqueous environments. But in most cases, both processes occur. This has been suggested for austenitic stainless steels in hot chloride

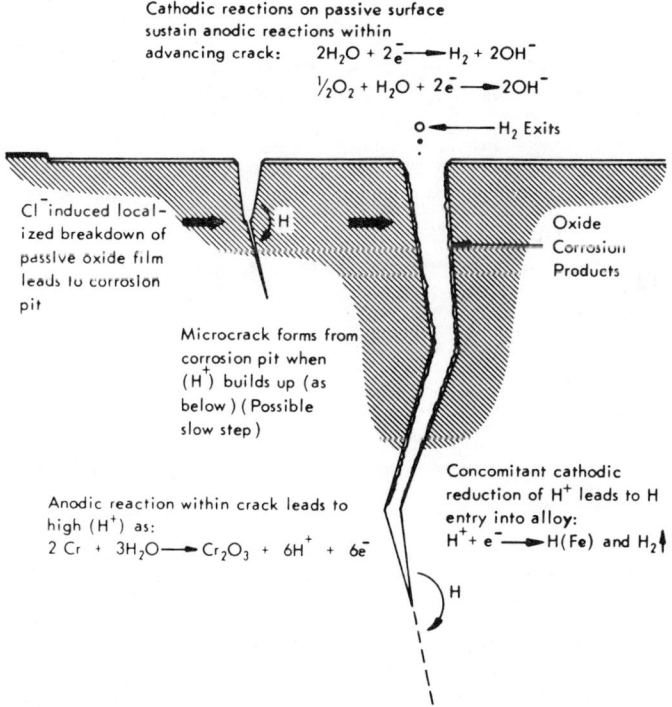

Fig. 53. Generalized model (chemical details for austenitic stainless steel in chloride environment) for crack initiation and propagation in SCC. (Modified from Rhodes.[327])

solutions,[64,327] for aluminum alloys as an outgrowth of Green's Mode I–Mode III experiments[179] and on other bases,[68,328,329] for Ti alloys,[235] and for nickel alloys. As summary illustration, we present Fig. 53; although the specific details are for austenitic stainless steels, the general picture corresponds both to Fig. 1 and to the general viewpoint we have been describing.

The Role of Metallurgical Variables in Fracture

We now wish to focus on the ways in which metallurgical variables affect hydrogen-induced fracture behavior. The discussion is predominantly phenomenological, although some comments on mechanisms are included. It is relevant here to state that we emphatically do not think there is one "hydrogen mechanism" operating in all cases; if there is, the evidence for it has not yet been identified. Instead, as Fig. 52 suggests, hydrogen seems capable of interacting with microstructure and fracture modes in a variety of ways.

The simplest kind of hydrogen fracture would occur from hydrogen in solution in the metal lattice, for example in the proposal which originated with Pfeil[330] in 1926 that hydrogen reduces the cohesive forces in the metal lattice,[318,331,332] particularly in the most highly stressed material near a crack tip. In such regions, thermodynamic arguments[319] suggest that hydrogen solubility will be increased. Since only a few atomic layers are stated to be affected under elastic conditions at a crack tip,[332] the situations in Fig. 49 could essentially provide the needed hydrogen without transport, for the case of a crack having direct access to the environment. This is indicated in Fig. 52 by a line bypassing the transport phenomena.

There are several difficulties with this specific model, including the conclusion[332] that plastic flow does not necessarily occur in high-strength materials under the stress singularity at the crack tip, and the dismissal of diffusion effects. Current fracture mechanics findings[320] indicate (Fig. 51) that the latter can be important because the maximum stress occurs very near the crack tip. A previous attempt to conduct "critical" experiments[333] in support of this model seems to have been unsuccessful.[310] The model also makes it difficult to rationalize intermittent cracking[318] and the role of metallurgical variables, except as they affect local solubilities. We therefore feel that this model, although possibly correct in principle and attractive in its clarity, is at present incomplete. The behavior it attempts to describe, of interatomic bonds being weakened by hydrogen, may well be at the root of many or most hydrogen embrittlement phenomena, but the model itself is not yet satisfactory. Additional work may place it on a firm footing and broaden its description of hydrogen behavior.

Grain boundaries are clearly an important microstructural location, since environmental fracture is often intergranular. Whether hydrogen accumulates at boundaries and weaken bonds, or reassociates as H_2, or has some other effect, is not known, although the H_2 damage possibility seems less likely. The presence of recombination poisons at boundaries should accentuate hydrogen accumulation and enhance fracture, as has been observed.[12,239,258] In the event that dislocation transport is important, microstructural features which reduce slip length within grains can thereby reduce hydrogen concentrations at boundaries. This is the explanation advanced for the effect of dispersoids in nickel alloys.[238,259] Another reason for intergranular fracture might be the presence of precipitates or inclusions at boundaries, as discussed next.

Hydrogen may accumulate at the interface between the matrix and a precipitate, particularly if it is incoherent. The presence of hydrogen may reduce the strength of the interface, thus aiding nucleation of cracking, or, if sufficient hydrogen is available, it may aid void growth at the interface by means of H_2 pressure. The latter case could occur in a dislocation transport situation if hydrogen is transported more rapidly to the particles than it can diffuse away. This viewpoint has been used to interpret instances of ductile fracture aided by hydrogen,[72,74,124] without specifying whether it was nucleation or growth of voids which was affected by hydrogen. However, the evidence that in many cases dimple sizes are reduced by testing in hydrogen[74,84,124] suggests that it is nucleation which is primarily altered. An example of this kind of observation is shown in Fig. 54. Particle effects have been suggested in other cases also,[63,334,335] and it appears important to evaluate further the role of inclusion type and orientation in the ferritic steels.[26,59] More work on this topic seems indicated, since ductile fracture is the mode of failure which would be desired for materials in service.

It is important to note that many inclusions and precipitates (particularly the overaged or equilibrium structure) are intermetallic compounds; these typically have very low solubility for elements other than those which participate in the compound. Formation of such precipitates would therefore be expected to result in rejection of other solutes at the interface (although an opposite proposal has been made[264] for the case of Inconel 600). The class of solutes which is of particular interest here is the recombination poisons.

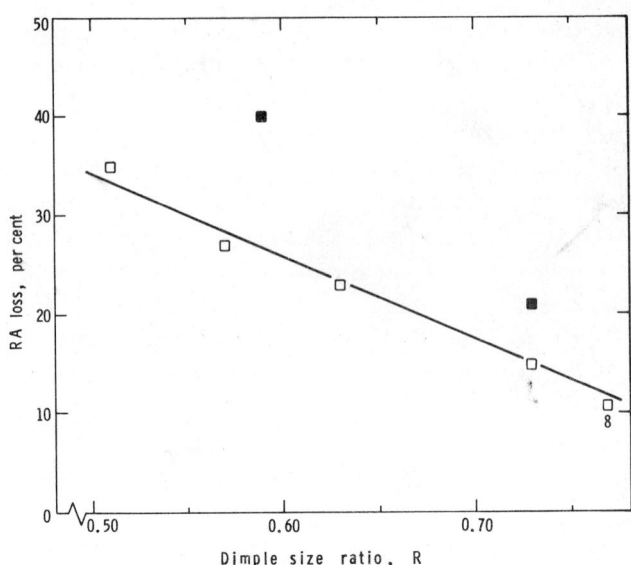

Fig. 54. Ductility (RA) loss of A-286-type alloys as a function of changes in fracture surface dimple size, expressed as the ratio R; R = (dimple size for hydrogen-charged material) ÷ (uncharged dimple size). Filled squares refer to partly intergranular fractures. For alloy 8, see Table 4 and Fig. 22. (From Thompson and Brooks.[124])

Presence of these at particle interfaces could aid particle–matrix interface fracture. These effects would be of greatest importance in precipitates of large surface area (e.g., cellular or pearlitic structures); it will be recalled that pearlite was found to be a poor microstructure in steels,[49] and cellular η was found deleterious[124] in A-286. The loss in SCC resistance with increasing grain size in pearlitic structures[49] is consistent with this interpretation, as the solute per unit area in the grain boundaries would be reduced, and thus available to pearlite interfaces, as grain size increases. A similar proposal for the behavior of carbides in steel was recently made by McMahon et al.[336] Finally, when these particles are at grain boundaries, a combined behavior of particle and grain boundary effects would be expected.

A final comment on slip mode is also in order, although it has been discussed at some length above. We have stated that slip planarity can affect electrochemical processes at crack tips or smooth surfaces by sharpening slip steps. It also increases the efficiency of dislocation transport of hydrogen,[314] of hydrogen storage at particles,[74,100,314] and of transport to grain boundaries. It can therefore affect both the dissolution and hydrogen components of SCC. At the same time, it cannot be regarded as a sufficient condition for environmental susceptibility, since some materials of planar slip character resist SCC fairly well[80,94,99] (this is not to say that a change in environmental composition could not produce SCC; one has only to provide the critical species or a sufficient local corrosion current, as the Inconel alloys illustrate[241,264]). In fact, planar slip is not always a necessary condition for susceptibility, since Nickel 200 exhibits SCC.[241] The intergranular fractures in Ni and Ni alloys may be aided by S segregation, an illustration that all relevant factors must be taken into account. These exceptions, however, do not obscure the widespread correlation between slip planarity and environmental susceptibility which we have identified in this chapter; they merely demonstrate that other factors can be sufficiently effective to intervene.

We conclude this section with illustrations of the above discussion for several alloy systems. These are not intended to be exhaustive, but rather to serve as examples.

Austenitic Stainless Steels

These alloys are in some ways the simplest case, since their hydrogen behavior is relatively easily correlated with their simple physical metallurgy. Here it has been pointed out that there is suggestive, although not yet conclusive, evidence that the ductility losses due to hydrogen occur at inclusions and precipitates.[72,74,87] The sequence of behavior is thought to be as follows. Dislocations carrying hydrogen tangle or pile up at the particles during deformation, so that a locally high hydrogen concentration can be created dynamically.[314] Some of this hydrogen can be released as dislocation stress fields overlap, and more will be released due to trapping at the inclusion.[314] When necking begins, hydrogen is available locally, either to lower the strength of the particle–matrix interface; to stabilize small voids, or cracks which form in particles; or to enter the growing void around the particle and aid its growth by the internal H_2 pressure developed. Note that the latter interaction begins only upon the initiation of necking. All these processes would aid and accelerate normal ductile rupture and cause it to occur at a smaller strain, which in turn corresponds to reduced ductility and an RA loss, or accelerated cracking in SCC. The overall sequence can be followed in Fig. 52.

The role of slip character in this process is that planar slip enhances dislocation transport of [H][314] and prevents the dislocations from cross-slipping around small particles and escaping. To the extent that SFE controls slip character, Fig. 12 indicates the stainless steel compositions for which SFE is high and for which hydrogen susceptibility should be low. For example, 310 (Table 3) has high SFE, and it usually exhibits little or no ductility loss in hydrogen.[278] But with higher hydrogen content[337] or when tested at lower temperature,[84,337] which increases slip planarity, 310 *does* show increased RA losses. This illustrates again our point that SFE is only one of the variables which affect planarity. On the basis of SFE alone, however, one can observe in Figs. 14 and 16 that ductility losses begin to be evident as SFE falls below about 40 mJ/m^2, as in 309S stainless.[74] Further, this correlation is consistent with the largely transgranular SCC cracking observed at low SFE levels.[78]

There are significant unanswered questions about austenitic stainless steel behavior. For example, are the effects of large Si or Ti additions due to structural changes (i.e., δ ferrite stabilization) or to SFE effects of the solutes in solution? As we have stated, we favor the former view, but SFE effects cannot be ruled out in this case at present.[68,94] The role of δ ferrite in SCC conditions is also unresolved, as to whether it resists cracking because it is tough and ductile, or because its electrochemical behavior inhibits crack tip resharpening. Finally, the role of Mn in environmental fracture should be examined in detail, because of increasing interest in the substitution of Mn for Ni and Cr as the latter elements become increasingly scarce and expensive. It is possible that additions of Mn + Si, or other combinations, could prove to be more effective substitutes.

Precipitation-Strengthened Alloys

An interesting modification of the rationale discussed above can be made for the precipitation-strengthened iron-base and nickel-base alloys. These are strengthened by coherent precipitates, whose interfaces may be less attractive locations for hydrogen accumulation than incoherent ones, but the loss of coherency at low strains[124,126] due to dislocations entering the interface appears to be an important step in hydrogen behavior, as noted above. When coherency is lost, hydrogen transported to the interface can be accumulated, so that ductile fracture is apparently affected by hydrogen in the same way as in austenitic stainless steels; that is, increased ductility loss is correlated with decreased fracture surface dimple size. It can be seen in Fig. 54 that this correlation is fairly good. It even includes two alloys which exhibited partly intergranular fracture, when the R measurements were made within the dimpled areas of the fracture surface; the low ductility of the intergranular portion presumably accounts for the increased RA loss compared to the trend for wholly ductile fractures. This kind of relationship may also exist for other materials such as the nickel-base alloys, and it is to be hoped that future work on these materials will include fractographic observations of the kind shown in Fig. 54. The special role of dislocation transport in this case is emphasized in Fig. 52.

We have noted the importance of slip planarity in both Fe-base and Ni-base alloys, caused by precipitate cutting, and more work on the dependence of such planarity on metallurgical variables and thermal history would appear most valuable. Coherency loss observations as a function of the same variables would also be of value as an aid in understanding behavior such as Fig. 54; it may also be of importance in understanding intergranular fracture, provided that such fracture results from dislocation transport of hydrogen to grain boundaries.[259] And the deleterious effects of the η and δ phases in these alloys deserve closer attention to see whether hydrogen recombination poisons are present at their interfaces, or whether the phases themselves are embrittled by hydrogen.

Alpha Titanium Alloys

The correlation of slip mode and environmental cracking in these alloys has been described, and it is at least possible that the correlation is related to dislocation transport effects of the kind discussed above. The role of hydrides in these alloys must also be considered, and several authors have discussed the interaction of hydrides with fracture, most notably Scully and his co-workers.[231,338] Hydride formation in slip bands[222,224,316,339] may imply a connection with dislocation transport, as may the observation that SCC cracking rates are faster than lattice diffusion.[231,296] Slip mode may also enter the mechanism through restriction of slip due to hydrides,[340] although attempts to model the "cleavage" behavior of Ti alloys in this and other ways have so far been unsuccessful.[229,340,341] It now appears that the observation[226] of (1017) hydrides may resolve this problem as a simple case of hydride cleavage.

The hydrogen-induced fracture of Ti alloys (which the results of Nelson[209] and Green[179] indicate may include many instances of SCC) might then be rationalized in terms of the relative amount of hydrogen available to the alloy. Relatively low concentrations would be expected under SCC conditions on the basis of the low hydrogen fugacity implied by Fig. 34. "Dissolved" hydrogen would also be at low levels in view of the usual alloy solubilities.[224] Thus, slow strain-rate embrittlement may occur at low [H] levels[339] through dislocation transport[342] (as affected by slip mode) and strain-induced hydride formation in slip bands. Fracture would then occur by cleavage of the hydrides. On the other hand, high [H] levels giving rise to profuse preexisting hydrides would cause a different fracture mode,[221] predominantly of the kind found at high strain rates. Further work to examine such a division of Ti alloy behavior[302] should encompass both the complex effects of hydride formation[224,226] and of hydrogen lattice sites in alloys.[343]

Ferritic and Martensitic Steels

There is relatively little microstructural evidence on the scale of transmission electron microscopy which bears on environmental fracture in these materials, particularly in mechanistic terms and in comparison to the other systems in this chapter. The high-strength martensitic steels exhibit behavior which is dominated by considerations rooted in fracture mechanics,[15,16,22,344] and may well be controlled by hydrogen diffusion ahead of cracks.[318] Alone of the alloy systems we have reviewed, therefore, slip mode effects and dislocation transport of hydrogen may be absent in steels, particularly those of high strength. As we have reviewed, however, there are compositional and microstructural effects in these steels which demand explanation, and dislocation transport may be involved. One of the pivotal questions is the behavior of hydrogen recombination poisons, since the partitioning of these in the microstructure depends on the tempering temperature.[12,19] Partitioning of poisons to grain boundaries or interphase interfaces may shift the critical fracture path (Fig. 52). Thus, efforts to vary yield strength by varying tempering conditions, for example, may not be simply interpretable without parallel investigation of the effect of tempering on the intergranular chemistry and the resultant crack path.[64,336] There can be little question, however, of the dominant role of hydrogen in most SCC failures in martensitic steels. This has been shown by a variety of means,[15,20,22,27,55,175,231,297,345,346] including Mode I–Mode III tests.[180,181]

Turning to ferritic microstructures, there are a number of marked similarities[39] between plain carbon and low-alloy steels, and austenitic stainless steels, in hydrogen; among these are observations of ductile fracture modes with losses in RA.[24,40] There are also many dissimilarities; but the similarities suggest that the kind of behavior described for austenitic steels may also occur in ferritic steels, and work to investigate this would be of value. Diffusion is far more rapid in the bcc lattice than in stainlesses, however, so careful experiments would be needed to establish that dislocation transport was indeed important.

We should reiterate that comprehensive work on the role played by Ni, Cr, and Mo in all the steels is urgently needed in view of the importance and extensive use of these solutes. Work to date has been no more than preliminary.

Aluminum Alloys

We have discussed the evidence, including Green's Mode III tests,[179] that hydrogen may be an important participant in aluminum alloy SCC. We shall now consider the processes by which this might occur. We have reviewed elsewhere[68] the status of the existing microstructural models for SCC in these alloys; those based on anodic dissolution paths or on preferential slip in precipitate-free zones (PFZ) in grain boundaries, of the sort illustrated in Fig. 55, were concluded to be inadequate. There are two other rationales, one of which correlates SCC resistance with the advent of nonplanar slip,[153,155] and the other of which correlates SCC resistance with increased coverage of grain boundaries by precipitates[347] (see Fig. 55). There is experimental support for each of these,[68] but careful experiments on commercial alloys showed that correlation coefficients were highest for the matrix slip mode concept,[348] in agreement with earlier work,[349,350] and in contradiction to the grain boundary precipitate concept. This appears true for 6000 alloys[137] as well as for 7000 alloys; we should also mention the success of TMT experiments premised on slip mode control.[160] Evidence in favor of the grain boundary precipitate proposal has in specific cases also seemed strong,[351,352] particularly when matrix slip character was maintained constant and only the boundary precipitates varied.

We propose[68] that both these concepts are in fact important. Slip mode would be a significant parameter because hydrogen transport by dislocations should be occurring; if this occurs during cutting of strengthening precipitates,

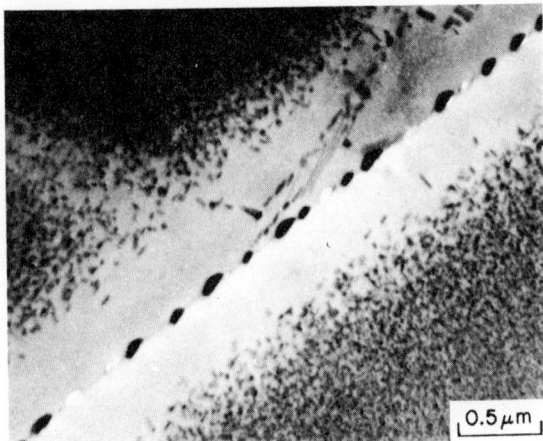

Fig. 55. Grain boundary precipitates, matrix precipitates, and PFZ in Al–Zn–Mg alloy, aged 500 min at 165 C. Matrix contains $\eta' + \eta$. (From Poulose *et al.*[352])

slip is highly planar and considerable hydrogen can reach grain boundaries. The effect of this hydrogen will then depend upon the precipitate character at the boundaries, since the precipitates will serve as hydrogen accumulators and thus as fracture nuclei.[173,328,353] Thus, we believe that the competition between these rationales, which depend, respectively, upon the character of grain interior and grain boundary precipitates, may simply reflect two aspects of the same phenomenon, provided that hydrogen is indeed involved. We therefore support those who have ascribed aluminum alloy behavior under most SCC conditions in part to hydrogen.[169–173,179,183,328,329,354–358]

We should mention an attempt to study the effect of hydrogen recombination poisons at grain boundary precipitates in aluminum alloys,[359] which found that these elements modified the precipitate behavior so strongly that any role as poisons was overshadowed; and the observation that solute depletion near grain boundaries affects initiation (and possibly resharpening) behavior in SCC.[360] There has also been an attempt[325,361,362] to compare quantitatively the dissolution and hydrogen processes in a model Al–7% Mg alloy. That study showed that under sufficiently anodic conditions and high stresses, the crack growth rate was faradaically equivalent to the dissolution rate.[362] At less anodic potentials or lower stresses, the comparison is flawed because the hydrogen model used was an equilibrium one, the behavior of Fig. 28 appears not to be taken into account, and it was assumed without justification that the rate of hydrogen absorption is rate-controlling. The work does show, however, as we have suggested, that in at least some systems, conditions can be adjusted to give a "pure" process, in this case anodic dissolution, even when mixed behavior appears relevant to open-circuit conditions.[179]

Implications for Hydrogen Embrittlement Mechanisms

We close the discussion with brief comments on the implications of this chapter for mechanisms of hydrogen embrittlement. We have already mentioned the problems with the Pfeil–Troiano–Oriani cohesive strength mechanism; these problems do *not* necessarily mean that the mechanism is fundamentally incorrect, only that it is not now in satisfactory form. Efforts to develop the underlying premise in different terms[363] are therefore of great interest. The second mechanism usually cited is the surface energy proposal of Petch and Stables.[364] Oriani has argued persuasively[331] that this is merely a different statement of the cohesive strength mechanism, provided that the cracking process can be treated as reversible.

A third mechanism is Zappfe's suggestion[365] that hydrogen accumulates in internal voids and cracks, and exerts gas pressure there to assist fracture. The suggestion was intended to explain hydrogen damage during charging, and is undoubtedly correct for some cases of that kind,[62] but is not general in its original form.[309,318] The suggestion has more recently been modified to suggest that hydrogen pressures are in many cases only developed during deformation, and have no significant effect prior to necking.[72,74,100,124] In combination with increased understanding of nonequilibrium hydrogen transport by dislocations[314] and of hydrogen effects on nucleation of ductile fracture,[366] it now appears that this modified pressure theory may have some applicability. It has been used to rationalize some of the material in the present review.

A fourth mechanism, suggested by Beachem,[55] is that hydrogen assists deformation in some unspecified way and thereby assists fracture. Although reduced flow stresses due to hydrogen have been observed,[55,367] the proposal is at present too unspecific to assess in detail; more work is called for. There have also been efforts to devise highly generalized theories of hydrogen embrittlement,[368,369] based in large part on the phenomenology of fracture in the presence of hydrides. As with Beachem's suggestion, these will have to be made more specific before they can be tested under a range of metallurgical conditions.

The present review suggests that none of the extant mechanisms is entirely general for the wide range of metallurgical variables examined; others have also concluded this.[370] Barring the possibility of an undiscovered mechanism, it seems unavoidable at this point to state that there is no underlying mechanism of complete generality for hydrogen embrittlement. The same appears true, incidentally, for the many proposals regarding anodic dissolution mechanisms, although we have not reviewed those here.

CONCLUDING REMARKS

There are several themes which have appeared repeatedly in this chapter. The most fundamental is the parallelism of hydrogen-induced SCC and hydrogen embrittlement. We have not claimed that these are the same phenomenon; anodic dissolution processes and crack tip resharpening, to name only two, are absent from hydrogen tests and inevitably alter the phenomenology of SCC processes by comparison. And there is certainly an ongoing dispute (e.g., in austenitic stainless steels[61,327,371–374]) as to the role of hydrogen in SCC. This review, emphasizing the existing and particularly the new data in both SCC and hydrogen embrittlement, has led us to suggest that the boundaries between these two phenomena cannot and should not be very sharply drawn. More work is needed, and will no doubt be performed,* to address this question.

The various suggestions we have made about hydrogen sources, transport processes, and locations of hydrogen damage, summarized in Fig. 52, comprise a broad and versatile viewpoint on hydrogen participation in environmental fracture. In combination with the view that mixed processes of anodic dissolution and hydrogen embrittlement frequently occur, a view which has been advocated by many others,[172,179,231,327,328,372] we believe the picture developed does help account for behavior in the most common service environments. Efforts to put it on a firmer foundation and make it quantitative are needed.

Fractography is an important aspect of any investigation of fracture behavior. It has been at the center of many disputes about mechanism, and at certain points in this review, we too have used it to discriminate among

---

* Even in copper alloys, there are SCC experiments in progress to evaluate a possible role of hydrogen.[375]

alternative explanations. Yet in many cases we have omitted detailed comparison and discussion of fracture surface appearance. The reason is that too often, no fractography (or worse, inconclusive or unconvincing fractography) has been performed in an investigation. We would urge greater use of fractography at lower magnifications (e.g., $200-1000\times$) when the fracture mode is mixed or complex, as it often is. And such cases should include some estimate of the proportions of different fracture modes which occur,[55,124] as well as (where possible) quantitative data on such features as secondary cracking, cleavage facet size, and dimple size. Finally, more use of Auger electron spectroscopy, when it can be applied, may well prove of great interest.

The ceaseless pattern of engineering materials use is of ever greater demands for increased resistance to environmental failure, frequently in environments which contain or can generate hydrogen. It is common to meet such demands with ad hoc simulation of the service environment, which leads to selection of the best material tested. But the alloy designer or materials engineer needs better guidance than service simulation to cope with the demands of new environments. In assessing the role of metallurgical variables in environmental fracture, we have attempted to emphasize those factors which can be manipulated to provide better materials. It is our hope that this review may serve in tandem with engineering tests as a guide to the improvement of old alloys and the development of new ones, as environments continue to become more demanding.

ACKNOWLEDGMENTS

We would like to thank A. G. Bell and the corporation which bears his name, for the development of the long-distance communication device which has made this collaboration possible. We appreciate discussions with J. C. Williams and critical comments by D. O. Sprowls, M. J. Blackburn, F. B. Mansfeld, and R. W. Staehle on parts of the manuscript, and are also grateful for the assistance of G. Pressouyre and R. Garber, who made preliminary searches of parts of the literature. J. C. Williams, N. E. Paton, and C. G. Rhodes kindly provided some of the photographs used. This work was carried out under Rockwell International's Independent Research and Development Program (AWT) and under the sponsorship of the Office of Naval Research (IMB).

EPILOGUE

This chapter was written in the spring of 1975 and somewhat revised in response to the reviewers in the spring of 1976. As we now read it for the last time prior to publication, it has been some 30 months since that last revision. It has been gratifying in this interim to observe that there are no major, and few minor aspects of the review which need to be modified. There have, however, been a number of subsequent publications which bear on the content of the review, and these are briefly cited below.

There have been a number of interesting observations on stainless steels, including additional work on sensitization[376] and on sustained load cracking in 1-atm hydrogen gas.[377] The latter experiments are of particular interest because they indicate that the exacerbative role of strain-induced martensite can be pivotal in providing a means of keeping crack tips sharp even though, as mentioned in the text, the martensite does not appear to play a primary or causative role. An additional finding of interest is that tensile tests in high-pressure hydrogen gas show a trend at pressures above 10,000 psi (69 MPa) toward intergranular fracture,[378] a finding which appears consistent with a role of dislocation transport in the embrittlement,[314] since grain boundaries can serve as hydrogen accumulators.[376]

There have also been several interesting reports on steels. In one, it was stated that decreasing grain size *lowered* K_{th}[379]; previous data have always shown the opposite. However, the evidence, Fig. 5 of the paper,[379] is not unequivocal. Additional work to examine the role played by grain size in steels of varying strength level would be useful, particularly since there is now evidence[380] that decreasing grain size *increases* K_{th} when impurity elements in the steel are reduced to a low level. Work on SCC of 4340-type steels[381] suggests again that a hydrogen role is central. Work on lower-strength (ca. 700 MPa or 100 ksi) plain carbon steels of varying Mn content[382] showed that Mn content does not affect hydrogen-induced ductility loss but does affect SCC behavior in pearlitic microstructures. In spheroidized carbide microstructures, however, Mn content did not adversely affect SCC performance.[382] Contrary to some assertions,[383] therefore, microstructure does affect the role of Mn at strength levels below 690 MPa. At the same time, the predominant role of nonmetallic inclusions[383,384] in hydrogen damage should be reiterated. Finally, it cannot be emphasized too strongly that attempts to evaluate thermomechanical processing and microstructure effects without controlling strength level or cooling rate from heat treatments[385] cannot give meaningful results, particularly when neither microstructural nor fractographic information is presented. Thermomechanical processing does appear promising for high-strength steels[386] in adequately documented studies, as the text discusses.

Valuable studies on sustained load cracking in Ti–6Al–4V have been published by Boyer and Spurr.[387,388] Their results on the temperature dependence of the phenomena strongly suggest a hydride-related mechanism of embrittlement,[387] in agreement with earlier suggestions.[224] They also reemphasize the importance, in Ti–6Al–4V, of the oxygen content, texture, and presence of α_2 in relation to SCC resistance.[388] Other work has also supported the hydride viewpoint,[389–392] including the mechanistic suggestions of Nelson[393] and Margolin[394]; regarding the latter work, it now appears[392] that hydrogen fractures occur entirely in α or interface regions, not along interfaces.

Aluminum alloy SCC data continue to be analyzed by some authors solely in terms of dissolution mechanisms,[395–397] but new evidence increasingly favors a major hydrogen contribution to cracking.[398–404] Microstructure studies in 7075 have shown[405] that hydrogen embrittlement has the same dependence on microstructure as does SCC, with underaged material the most sensitive and overaged (T73) less sensitive than T6. These observations are consistent with, although they do not prove, a major role of hydrogen in SCC of 7075 alloys. Attention continues to be devoted to solute element segregation effects in aluminum alloys, and the possible role of such segregation on SCC and hydrogen embrittlement[402,406,407]; the importance of such effects is not yet clear.

Extensions of the phenomenological descriptions in the present review have been published by the present authors and others[401,408–414] and in some cases offer additional detail. Significantly new discussions of fracture mode behavior,[403,409,415–417] mechanistic calculations of a new kind,[418,419] and of mechanisms in general[401,403,408,409,416,420-424] are also now available. Of particular interest are Fujita's calculations on a "hydridelike" mechanism in iron, involving disks of accumulated hydrogen on {100} planes[418]; Heady's reassessment of the Petch–Stables surface energy mechanism[425] and Hirth and Johnson's observation[408] that this mechanism is a special case of the Pfeil–Troiano–Oriani mechanism *only* when cracking is reversible; Heady's partial rationale[419] for the Beachem suggestion[55]; and continued interest in cooperative anodic dissolution-hydrogen embrittlement processes for SCC.[401,420–422,426–429] It still appears that no single mechanism can serve to rationalize all the data on metallurgical and other variables,[408,409,416,421-424] nor can any of the five current classes of mechanism[424] be ruled out as yet.[416,419,421-424]

In the area of dislocation transport, Johnson and Hirth[430] have demonstrated that mutual annihilation of hydrogen-carrying dislocations cannot of itself result in local hydrogen enrichment; this is a process not evaluated in earlier work.[311-315] Jensen and Tien have now shown,[431] however, that the pivotal point here is the existence of a binding energy for hydrogen to accumulation sites such as inclusions; insertion of such an energy into the Johnson–Hirth analysis gives rise to significant hydrogen accumulations, while removal of the binding energy from the Tien *et al.* model[314,432] results in no significant accumulation. Since it is generally found that metallurgical locations such as particles and grain boundaries do possess such a binding energy,[433,434] it would be concluded that both published transport analyses[314,430] can predict hydrogen accumulations at such sites. Experimental information has continued to suggest[429,435-438] that dislocation transport does occur in practice; on the other hand, a number of investigators[421,424,439-441] have increasingly felt that such transport perhaps plays a somewhat less central role in embrittlement phenomena than described in the text.

In summary, we feel that the great majority of the correlation, analysis, and discussion of the literature we prepared in 1975 continues to be both accurate and relevant. Our hope continues, however, that our comments on needed critical experiments will help stimulate new work which can lead to refinement and extension of the understanding of hydrogen-assisted fracture.

REFERENCES

1. B. F. Brown, in *Stress Corrosion Cracking in High Strength Steels and in Titanium and Aluminum Alloys* (B. F. Brown, ed.), pp. 2–16, Naval Research Laboratory, Washington, D.C. (1972).
2. M. O. Speidel and M. V. Hyatt, in *Advances in Corrosion Science and Technology* (M. G. Fontana and R. W. Staehle, eds.), Vol. 2, pp. 115–335, Plenum Press, New York (1972).
3. M. V. Hyatt and M. O. Speidel, in *Stress Corrosion Cracking in High Strength Steels and in Titanium and Aluminum Alloys* (B. F. Brown, ed.), pp. 147–244, Naval Res. Lab., Washington, D.C. (1972).
4. S. M. Wiederhorn, in *Materials Science Research* (W. W. Kriegel and H. Palmour III, eds.), Vol. 3, pp. 503–28, Plenum Press, New York (1966).
5. H. R. Smith and D. E. Piper, in *Stress Corrosion Cracking in High Strength Steels and in Titanium and Aluminum Alloys* (B. F. Brown, ed.), pp. 17–77, Naval Research Laboratory, Washington, D.C. (1972).
6. A. J. Jacobs and H. L. Marcus, *Corrosion* **30**, 305–319 (1974).

7. J. H. Hoke, *Corrosion* **26**, 396–397 (1970).
8. I. I. Vasilenko, G. V. Karpenko, A. B. Kuslitskii, G. A. Khasin, G. D. Stefanov, and S. N. Chubatina, *Sov. Mater. Sci.* **5**, 247–248 (1970).
9. C. B. Gilpin and N. A. Tiner, *Corrosion* **22**, 271–279 (1966).
10. E. Snape, *Corrosion* **24**, 261–282 (1968).
11. U. Q. Cabral, A. Hache, and A. Constant, *C. R. Acad. Sci. Paris* **260**, 6887–6890 (1965).
12. K. Yoshino and C. J. McMahon, *Met. Trans.* **5**, 363–370 (1974).
13. R. P. M. Proctor and H. W. Paxton, *Trans. ASM* **62**, 989–999 (1969).
14. J. B. Greer, Corrosion/73 preprint no. 55, NACE, Houston (1973).
15. W. W. Gerberich, in *Hydrogen in Metals* (I. M. Bernstein and A. W. Thompson, eds.), pp. 115–47, ASM, Metals Park, Ohio (1974).
16. A. S. Tetelman, in *Fundamental Aspects of Stress Corrosion Cracking* (R. W. Staehle, ed.), pp. 446–460, NACE, Houston (1969).
17. C. S. Carter, *Corrosion* **25**, 423–431 (1969).
18. B. B. Rath and I. M. Bernstein, *Met. Trans.* **2**, 2845–2851 (1971).
19. J. A. Marquez, I. Matsushima, and H. H. Uhlig, *Corrosion* **26**, 215–222 (1970).
20. I. M. Bernstein and A. W. Thompson, *Int. Met. Rev.* (Rev. 212) **21**, 269–287 (1976).
21. G. Sandoz, *Met. Trans.* **2**, 1055–1063 (1971).
22. G. Sandoz, *Met. Trans.* **3**, 1169–1176 (1972).
23. G. Sandoz, in *Stress Corrosion Cracking in High Strength Steels and in Aluminum and Titanium Alloys* (B. F. Brown, ed.), pp. 79–133, Naval Research Laboratory, Washington, D.C. (1972).
24. H. H. Uhlig, *Trans. AIME* **158**, 183–201 (1944).
25. R. R. Fessler, Applications of Stress-Corrosion Cracking Research to the Pipeline Problem, *5th Symposium on Line Pipe Research*, American Gas Association (November 1974).
26. P. C. Hughes, I. R. Lamborn, and B. B. Liebert, *J. Iron Steel Inst.* **203**, 728–731 (1965).
27. J. W. Kennedy and J. A. Whittaker, *Corros. Sci.* **8**, 359–375 (1968).
28. A. E. Scheutz and W. D. Robertson, *Corrosion* **13**, 437t–458t (1957).
29. L. W. Vollmer, *Corrosion* **8**, 326–332 (1952).
30. NACE Tech. Pract. Comm. 1-G, *Corrosion* **8**, 351–354 (1952).
31. G. M. Pressouyre and I. M. Bernstein, *Met. Trans.* **9A**, 1571–1578 (1978).
32. J. D. Hobson and C. Sykes, *J. Iron Steel Inst.* **169**, 209–220 (1951).
33. H. R. Copson, in *Physical Metallurgy of Stress Corrosion Fracture* (T. N. Rhodin, ed.), pp. 247–267, Interscience, New York (1959).
34. H. H. Uhlig, K. E. Perumal, and M. Talerman, *Corrosion* **30**, 229–236 (1974).
35. L. M. Long and H. H. Uhlig, *J. Electrochem. Soc.* **112**, 964–967 (1965).
36. I. M. Bernstein, *Met. Trans.* **1**, 3143–3150 (1970).
37. W. Hoffman and W. Rauls, *Welding J.* **44**, 225s–230s (1965).
38. R. P. Jewett, R. J. Walter, W. T. Chandler, and R. P. Frohmberg, Hydrogen Environment Embrittlement of Metals, NASA Report CR-2163, Rocketdyne Division, Rockwell International, Canoga Park, Calif. (1973).
39. M. R. Louthan, in *Hydrogen in Metals* (I. M. Bernstein and A. W. Thompson, eds.), pp. 53–78, ASM, Metals Park, Ohio (1974).
40. J. B. Seabrook, N. J. Grant, and D. Carney, *Trans. AIME* **188**, 1317–1321 (1950).
41. I. M. Bernstein, *Mater. Sci. Eng.* **6**, 1–19 (1970).
42. E. A. Steigerwald and W. D. Benjamin, *Met. Trans.* **2**, 606–608 (1971).
43. M. F. Baldy and R. C. Bowden, *Corrosion* **11**, 417t–422t (1955).
44. C. M. Hudgins, R. L. McGlasson, P. Mehdizadeh, and W. M. Rosborough, *Corrosion* **22**, 238–251 (1966).
45. R. A. McCoy and W. W. Gerberich, *Met. Trans.* **4**, 539–547 (1973).

46. R. R. Vandervoort, A. W. Ruotola, and E. L. Raymond, *Met. Trans.* **4**, 1175–1177 (1973).
47. R. T. Ault, K. O. McDowell, P. L. Hendricks, and T. M. F. Ronald, *Trans. ASM* **60**, 79–87 (1967).
48. C. N. Ahlquist, *Met. Eng. Quart.* **8**(4), 52–56 (1968).
49. M. Henthorne and R. N. Parkins, *Br. Corros. J.* **2**, 186–192 (1967).
50. R. H. Cavett and H. C. Van Ness, *Welding J.* **42**(7), 316s–319s (1963).
51. J. D. Hobson and J. Hewitt, *J. Iron Steel Inst.* **173**, 131–140 (1953).
52. R. A. Davis, G. A. Dreyer, and W. C. Gallaugher, *Corrosion* **20**, 93t–103t (1964).
53. J. Legrand, C. Couderc, and J.-P. Fidelle, in *L'Hydrogène dans les Métaux*, Vol. 2, pp. 316–318, Éditions Science et Industrie, Paris (1972).
54. J. E. Truman, R. Perry, and G. N. Chapman, *J. Iron Steel Inst.* **202**, 745–756 (1964).
55. C. D. Beachem, *Met. Trans.* **3**, 437–451 (1972).
56. P. Lillys and A. E. Nehrenberg, *Trans. ASM* **48**, 327–355 (1956).
57. W. W. Kirk, R. A. Covert, and T. P. May, *Met. Eng. Quart.* **8**(4), 31–37 (1968).
58. I. M. Bernstein and B. B. Rath, *Met. Trans.* **4**, 1545–1551 (1973).
59. R. A. Davis, *Corrosion* **19**, 45t–55t (1963).
60. I. M. Bernstein and H. W. Pickering, *Corrosion* **31**, 105–107 (1975).
61. H. H. Uhlig and R. T. Newberg, *Corrosion* **28**, 337–339 (1972).
62. A. S. Tetelman, in *Fracture of Solids*, pp. 671–708, Gordon and Breach, New York (1963).
63. K. Ohnishi, *Trans. Jpn. Inst. Met.* **12**, 329–336 (1971).
64. I. M. Bernstein and A. W. Thompson, in *Alloy and Microstructural Design* (J. K. Tien and G. S. Ansell, eds.), pp. 303–347, Academic Press, New York (1976).
65. *Metals Handbook*, p. 1261, ASM, Cleveland, Ohio (1948).
66. R. M. Latanision and R. W. Staehle, in *Fundamental Aspects of Stress Corrosion Cracking* (R. W. Staehle, ed.), pp. 214–296, NACE, Houston (1969).
67. A. J. Sedriks, *J. Inst. Met.* **101**, 225–232 (1973).
68. A. W. Thompson and I. M. Bernstein, *Rev. Coatings Corros.* **2**, 3–44 (1975).
69. A. W. Loginow and J. F. Bates, *Corrosion* **25**, 15–22 (1969).
70. J. E. Truman and R. Perry, *Br. Corros. J.* **1**, 60–66 (1966).
71. R. Lagneborg, *J. Iron Steel Inst.* **207**, 363–366 (1969).
72. A. W. Thompson, in *Hydrogen in Metals* (I. M. Bernstein and A. W. Thompson, eds.), pp. 91–101, ASM, Metals Park, Ohio (1974).
73. D. V. Neff, T. E. Mitchell, and A. R. Troiano, *Trans. ASM* **62**, 858–868 (1969).
74. A. W. Thompson, *Mater. Sci. Eng.* **14**, 253–264 (1974).
75. R. E. Schramm and R. P. Reed, *Met. Trans.* **6A**, 1345–1351 (1975).
76. C. G. Rhodes and A. W. Thompson, *Met. Trans.* **8A**, 1901–1906 (1977).
77. J. D. Harston and J. C. Scully, *Corrosion* **26**, 387–395 (1970).
78. P. R. Swann, *Corrosion* **19**, 102t–112t (1963).
79. L. M. Brown, *Phil. Mag.* **10**, 441–466 (1964).
80. J. Bade and R. A. Dodd, in *Fundamental Aspects of Stress Corrosion Cracking* (R. W. Staehle, ed.), pp. 342–350, N.A.C.E., Houston (1969).
81. S. Barnartt, R. Stickler, and D. van Rooyen, *Corros. Sci.* **3**, 9–16 (1963).
82. L. K. Zamiryakin, *Sov. Mater. Sci.* **7**, 9–13 (1971).
83. M. Kowaka and H. Fujikawa, in *Localized Corrosion* (R. W. Staehle, B. F. Brown, and J. Kruger, eds.), pp. 437–446, NACE, Houston (1974).
84. M. R. Louthan, G. R. Caskey, J. A. Donovan, and D. E. Rawl, *Mater. Sci. Eng.* **10**, 357–368 (1972).
85. H. H. Uhlig and R. A. White, *Trans. ASM* **52**, 830–847 (1960).
86. H. H. Uhlig, R. A. White, and J. Lincoln, *Acta Met.* **5**, 473–475 (1957).
87. F. S. Lang, *Corrosion* **18**, 378t–382t (1962).

88. J. F. Eckel and G. S. Clevinger, *Corrosion* **26**, 251–255 (1970).
89. M. Kowaka and H. Fujikawa, *Trans. Jpn. Inst. Met.* **12**, 243–249 (1971).
90. J. Hochman and J. Bourrat, *C. R. Acad. Sci. Paris* **255**, 3416–3417 (1962).
91. S. Matsushima and J. Ishiwara, *Trans. Jap. Inst. Met.* **14**, 20–25 (1972).
92. H. Okada, Y. Hosoi, and S. Abe, *Corrosion* **27**, 424–433 (1971).
93. J. M. Genin, G. Le Caer, P. Maitrepierre, and B. J. Thomas, *Scripta Met.* **8**, 15–22 (1974).
94. D. L. Douglass, G. Thomas, and W. R. Roser, *Corrosion* **20**, 15–27 (1964).
95. D. Dulieu and J. Nutting, in *Metallurgical Developments in High Alloy Steels* (Spec. Rep. 86), pp. 140–45, Iron and Steel Institute, London (1964).
96. J. W. Flowers, F. H. Beck, and M. G. Fontana, *Corrosion* **19**, 186t–198t (1963).
97. A. J. West and J. A. Brooks, in *Effect of Hydrogen on Behavior of Materials* (A. W. Thompson and I. M. Bernstein, eds.), pp. 686–697, AIME, New York (1976).
98. B. C. Odegard, The Effect of δ-Ferrite on the Hydrogen Sensitivity of 309S Stainless, Report SCL-TM-720367, Sandia Laboratories, Livermore, Calif. (October 1972).
99. R. W. Staehle, in *Theory of Stress Corrosion Cracking in Alloys* (J. C. Scully, ed.), pp. 223–286, NATO, Brussels (1971).
100. A. W. Thompson, *Met. Trans.* **4**, 2819–2825 (1973).
101. M. R. Louthan, J. A. Donovan, and D. E. Rawl, *Corrosion* **29**, 108–111 (1973).
102. H. Fujita and S. Ueda, *Acta Met.* **20**, 759–767 (1972).
103. P. L. Mangonon and G. Thomas, *Met. Trans.* **1**, 1577–1586 (1970).
104. A. W. Thompson and O. Buck, *Met. Trans.* **7A**, 329–331 (1976); see also T. Suzuki and J. C. Shyne, in *New Aspects of Martensitic Transformation*, pp. 305–308, Japan Institute of Metals, Sendai (1976).
105. E. G. Coleman, D. Weinstein, and W. Rostoker, *Acta Met.* **9**, 491–497 (1961).
106. V. L. Barnwell, J. R. Myers, and R. K. Saxer, *Corrosion* **22**, 261–264 (1966).
107. A. W. Thompson, Science Center, Rockwell International, Thousand Oaks, Calif., unpublished research (1974).
108. G. C. Smith, in *Hydrogen in Metals* (I. M. Bernstein and A. W. Thompson, eds.), pp. 485–511, ASM, Metals Park, Ohio (1974).
109. P. Blanchard and A. R. Troiano, *Mem. Sci. Rev. Met.* **57**(6), 409–414 (1960).
110. A. Holden, J. D. Bolton, and E. R. Petty, *J. Iron Steel Inst.* **209**, 721–728 (1971).
111. J. F. Breedis and L. Kaufman, *Met. Trans.* **2**, 2359–2371 (1971).
112. C. H. White and R. W. K. Honeycombe, *J. Iron Steel Inst.* **200**, 457–466 (1962).
113. R. A. McCoy, in *Hydrogen in Metals*, (I. M. Bernstein and A. W. Thompson, eds.), pp. 169–178, ASM, Metals Park, Ohio (1974).
114. A. W. Thompson, in *Hydrogen in Metals* (I. M. Bernstein and A. W. Thompson, eds.), pp. 179–182, ASM, Metals Park, Ohio (1974).
115 R. B. Benson, in *Hydrogen in Metals* (I. M. Bernstein and A. W. Thompson, eds.), pp. 183–195, ASM, Metals Park, Ohio (1974).
116. A. W. Thompson, *Metallography* **5**, 366–369 (1972).
117. A. W. Thompson and W. N. Posey, *J. Test. Eval.* **2**, 240–242 (1974).
118. R. J. Walter and W. T. Chandler, Effects of High-Pressure Hydrogen on Metals at Ambient Temperatures, Report R-7780-1 (Access No. N70-18637), Rocketdyne Div., Rockwell International, Canoga Park, Calif. (February 1969).
119. A. W. Thompson, in *Fracture 1977: Proceedings of the Fourth International Conference on Fracture* (D. M. R. Taplin, ed.), Vol. 2, pp. 237–242, Univ. Waterloo Press, Waterloo, Ont. (1977).
120. J. Papp, R. F. Hehemann, and A. R. Troiano, in *Hydrogen in Metals* (I. M. Bernstein and A. W. Thompson, eds.), pp. 657–668, ASM, Metals Park, Ohio (1974).

121. J. LeGrand, M. Caput, C. Couderc, R. Broudeur, and J.-P. Fidelle, *Mem. Sci. Rev. Met.* **68**, 861–869 (1971).

122. R. J. Walter and W. T. Chandler, Influence of Gaseous Hydrogen on Metals, Report NASA CR-124410, Rocketdyne Div., Rockwell International, Canoga Park, Calif. (October 1973).

123. A. Kelly and R. B. Nicholson, *Prog. Mater. Sci.* **10**, 149–391 (1963).

124. A. W. Thompson and J. A. Brooks, *Met. Trans.* **6A**, 1431–1442 (1975).

125. A. W. Thompson, *Met. Trans.* **7A**, 315–318 (1976).

126. L. M. Brown and R. K. Ham, in *Strengthening Methods in Crystals* (A. Kelly and R. B. Nicholson, eds.), pp. 12–135, Wiley, New York (1971).

127. H. Y. Hunsicker, in *Aluminum* (K. R. van Horn, ed.), Vol. 1, pp. 109–162, ASM, Metals Park, Ohio (1967).

128. D. O. Sprowls and R. H. Brown, in *Fundamental Aspects of Stress Corrosion Cracking* (R. W. Staehle, ed.), pp. 466–506, NACE, Houston (1969).

129. E. A. Starke, *J. Met.* **22**(1), 54–63 (1970).

130. L. F. Mondolfo, *Met. Mater.* **5**, 95–124 (1971).

131. J. T. Staley, *Met. Eng. Quart.* **13**(4), 52–57 (1973).

132. W. A. Anderson, U.S. Patent 3,232,796 (February 1, 1966).

133. R. B. Niederberger, J. L. Basil, and G. T. Bedford, *Corrosion* **22**, 68–73 (1966).

134. R. W. Rogers, W. D. Vernam, and M. B. Shumaker, Test and Exploratory Development of an Optimum Aluminum Alloy System for Ship Structures, Final Report on Contract N00024-72-C-5571, Alcoa Laboratories, Alcoa Center, Pa. (July 1974).

135. Y. Baba, M. Hagiwara, and J. Hamada, *J. Inst. Met.* **100**, 309–312 (1972).

136. Y. Baba, *J. Jpn. Inst. Met.* **36**, 341–346 (1972).

137. J. M. Dowling and J. W. Martin, in *Proceedings of the Third International Conference on the Strength of Metals and Alloys*, Vol. 1, pp. 170–174, Institute of Metals, Cambridge, England (1973).

138. J. Nock, in *Aluminum* (K. R. van Horn, ed.), Vol. 1, pp. 303–336, Metals Park, Ohio (1967).

139. J. M. Silcock, T. J. Heal, and H. K. Hardy, *J. Inst. Met.* **82**, 239–248 (1953–54).

140. J. M. Silcock, *J. Inst. Met.* **89**, 203–210 (1960–61).

141. M. O. Speidel, personal communication (1974).

142. W. A. Dean, in *Aluminum* (K. R. van Horn, ed.), Vol. 1, pp. 163–208, ASM, Metals Park, Ohio (1967).

143. H. Y. Hunsicker, J. T. Staley, and R. H. Brown, *Met. Trans.* **3**, 201–209 (1972).

144. E. H. Dix, *Trans. ASM* **42**, 1057–1127 (1950).

145. R. H. Brown, D. O. Sprowls, and M. B. Shumaker, in *Stress Corrosion Cracking of Metals—A State of the Art*, ASTM STP 518, pp. 87–118, ASTM, Philadelphia (1972).

146. C. A. Grove and G. Judd, *Met. Trans.* **4**, 1023–1027 (1973).

147. C. Chen and G. Judd, SCC Susceptibility of Al–Zn–Mg–Ti Alloys, Report No. 6 on Contract N00014-67-A-0117-0009, Rensselaer Polytechnic Inst. (September 1973); see also microstructural study, *Met. Trans.* **9A**, 553–559 (1978).

148. H.-J. Engell, W. Neth, and A. Šuchma, *Z. Metallkd.* **61**, 261–266 (1970).

149. J. T. Staley, Evaluating the New Aluminum Aerospace Forging Alloys, Alcoa Laboratories, presented at WESTEC, Los Angeles, Calif. (March 1974).

150. I. J. Polmear, *J. Met.* **20**(6), 44–51 (1968).

151. J. T. Staley, R. H. Brown, and R. Schmidt, *Met. Trans.* **3**, 191–199 (1972).

152. P. R. Swann, in *Electron Microscopy and Strength of Crystals* (G. Thomas and J. Washburn, eds.), pp. 131–179, Interscience, New York (1963).

153. M. O. Speidel, in *The Theory of Stress Corrosion Cracking in Alloys* (J. C. Scully, ed.), pp. 289–341, NATO, Brussels (1971).

154. G. M. Ugiansky, L. P. Skolnick, and S. W. Stiefel, *Corrosion* **25**, 77–86 (1969).

155. M. O. Speidel, in *Fundamental Aspects of Stress Corrosion Cracking* (R. W. Staehle, ed.), pp. 561–573, NACE, Houston (1969).

156. B. W. Lifka and B. M. Ponchel, Fundamental Concept of an SCC Control Procedure for 7075-T73 Alloy, Alcoa Laboratories, presented at WESTEC, Los Angeles, Calif.(March 1973).

157. M. S. Hunter, G. R. Frank, and D. L. Robinson, in *Proceedings of the Second International Congress on Metallic Corrosion*, NACE, Houston (1963).

158. A. J. McEvily, J. B. Clark, and A. P. Bond, *Trans. ASM* **60**, 661–671 (1967).

159. V. A. Pavlov, I. Y. U. Filippov, and S. A. Frizen, *Phys. Met. Metallogr. (USSR)* **20**, 770–774 (1965).

160. N. E. Paton and A. W. Sommer, in *Proceedings of the Third International Conference on the Strength of Metals and Alloys*, Vol. 1, pp. 101–108, Institute of Metals, Cambridge, England (1973).

161. M. Conserva, M. Buratti, E. DiRusso, and F. Gatto, *Mater. Sci. Eng.* **11**, 103–112 (1973).

162. E. DiRusso, M. Conserva, F. Gatto, and H. Markus, *Met. Trans.* **4**, 1133–1144 (1973).

163. U. R. Evans, in *Stress Corrosion Cracking and Embrittlement* (W. D. Robertson, ed.), pp. 158–162, Wiley, New York (1956).

164. T. Broom and A. J. Nicholson, *J. Inst. Met.* **89**, 183–190 (1960–61).

165. A. Hartman, *Int. J. Fract. Mech.* **1**, 167–187 (1965).

166. F. J. Bradshaw and C. Wheeler, *Appl. Mater. Res.* **5**, 112–120 (1966).

167. R. P. Wei, in *Fundamental Aspects of Stress Corrosion Cracking* (R. W. Staehle, ed.), pp. 104–111, NACE, Houston (1969).

168. P. M. Lorenz, Effect of Pressurized Hydrogen upon Inconel 718 and 2219 Aluminum, Report D2-114417 (Access No. N69-19152), Boeing Co., Seattle, Wash. (February 1969).

169. R. J. Gest and A. R. Troiano, in *L'Hydrogène dans les Métaux*, Vol. 2, pp. 427–432, Editions Science et Industrie, Paris (1972).

170. R. J. Gest and A. R. Troiano, *Corrosion* **30**, 274–279 (1974).

171. J. Albrecht, B. J. McTiernan, I. M. Bernstein, and A. W. Thompson, *Scripta Met.* **11**, 893–897 (1977).

172. M. O. Speidel, in *Hydrogen in Metals* (I. M. Bernstein and A. W. Thompson, eds.), pp. 249–273, ASM, Metals Park, Ohio (1974).

173. L. Montgrain and P. R. Swann, in *Hydrogen in Metals* (I. M. Bernstein and A. W. Thompson, eds.), pp. 575–584, ASM, Metals Park, Ohio (1974).

174. J. Berggreen, Ph.D. thesis, University of Erlangen-Nürnberg (1973); J. Berggreen and H. Kaesche, unpublished research (1973); quoted in Ref. 172.

175. B. F. Brown, in *Theory of Stress Corrosion Cracking in Alloys* (J. C. Scully, ed.), pp. 186–203, NATO, Brussels (1971).

176. B. F. Brown, C. T. Fujii, and E. P. Dahlberg, *J. Electrochem. Soc.* **116**, 218–219 (1969).

177. B. E. Wilde, *Corrosion* **27**, 326–333 (1971).

178. B. G. Ateya and H. W. Pickering, in *Hydrogen in Metals* (I. M. Bernstein and A. W. Thompson, eds.), pp. 207–222, ASM, Metals Park, Ohio (1974).

179. J. A. S. Green, H. W. Hayden, and W. G. Montague, in *Effect of Hydrogen on Behavior of Materials* (A. W. Thompson and I. M. Bernstein, eds.), pp. 200–215, TMS–AIME, New York (1976).

180. H. W. Hayden and S. Floreen, *Corrosion* **27**, 429–433 (1971).

181. C. St. John and W. W. Gerberich, *Met. Trans.* **4**, 589–594 (1973).

182. J. A. S. Green and H. W. Hayden, in *Hydrogen in Metals* (I. M. Bernstein and A. W. Thompson, eds.), pp. 235–245, ASM, Metals Park, Ohio (1974).

183. E. F. Smith, R. Jacko and D. J. Duquette, in *Effect of Hydrogen on Behavior of Materials* (A. W. Thompson and I. M. Bernstein, eds.), pp. 218–228, TMS–AIME, New York (1976).

184. R. I. Jaffee, *Prog. Met. Phys.* **7**, 65–163 (1958).
185. M. K. McQuillan, *Met. Rev.* **8**, 41–104 (1963).
186. M. J. Blackburn, J. A. Feeney, and T. R. Beck, in *Advances in Corrosion Science and Technology*, Vol. 3, pp. 67–292, Plenum Press, New York (1973).
187. J. C. Williams, in *Titanium Science and Technology* (R. I. Jaffee and H. M. Burte, eds.) Vol. 3, pp. 1433–1494, Plenum Press, New York (1973).
188. J. Brettle, *Met. Mater.* **6**, 442–451 (1972).
189. S. R. Seagle, R. R. Seeley, and G. S. Hall, in *Applications Related Phenomena in Titanium Alloys*, ASTM STP 432, pp. 170–188, ASTM, Philadelphia (1967).
190. R. E. Curtis, R. R. Boyer, and J. C. Williams, *Trans. ASM* **62**, 457–469 (1969).
191. M. J. Blackburn and J. C. Williams, in *Fundamental Aspects of Stress Corrosion Cracking* (R. W. Staehle, ed.), pp. 620–636, NACE, Houston (1969).
192. R. A. Wood, J. D. Boyd, D. N. Williams, and R. I. Jaffee, The Effect of Alloy Composition on SCC of Titanium Alloys in Aqueous Environments, NASA CR-131872 (Access No. N73-22475), Battelle Mem. Inst., Columbus, Ohio (June 1972).
193. R. E. Adams and E. von Tiesenhausen, in *Fundamental Aspects of Stress Corrosion Cracking* (R. W. Staehle, ed.), pp. 691–699, NACE, Houston (1969).
194. M. J. Blackburn and J. C. Williams, *Trans. ASM* **62**, 398–409 (1969).
195. J. C. Williams, A. W. Sommer and P. P. Tung, *Met. Trans.* **3**, 2979–2984 (1972).
196. E. A. Metzbower, *Met. Trans.* **2**, 3099 (1971).
197. R. Haynes, *J. Inst. Met.* **101**, 97–102 (1973).
198. R. G. Berryman, F. H. Froes, J. C. Chesnutt, C. G. Rhodes, J. C. Williams, and R. F. Malone, High Toughness Titanium Alloy Development, Final Engineering Report on Contract N00019-73-C-0335, Science Center, Rockwell International, Thousand Oaks, Calif. (July 1974).
199. I. R. Lane and J. L. Cavallaro, in *Applications of Related Phenomena in Titanium Alloys*, ASTM (STP 432), pp. 147–169, ASTM, Philadelphia (1968).
200. G. Lütjering and S. Weismann, *Acta Met.* **18**, 785–795 (1970).
201. J. C. Williams, A. W. Thompson, and R. G. Baggerly, *Scripta Met.* **8**, 625–630 (1974).
202. J. L. Cavallaro and R. C. Wilcox, *Corrosion* **27**, 157–163 (1971).
203. J. C. Williams, in *Fundamental Aspects of Stress Corrosion Cracking* (R. W. Staehle, ed.), p. 700, NACE, Houston (1969).
204. D. N. Fager and W. F. Spurr, *Trans. ASM* **61**, 283–291 (1968).
205. J. C. Chestnutt and J. C. Williams, in *Scanning Electron Microscopy/1974* (O. Johari and I. Corvin, eds.), pp. 895–902, IITRI, Chicago (1974).
206. H. G. Nelson, D. P. Williams, and J. E. Stein, *Met. Trans.* **3**, 469–475 (1972).
207. D. P. Williams and H. G. Nelson, *Met. Trans.* **3**, 2107–2113 (1972).
208. H. G. Nelson, *Met. Trans.* **4**, 364–367 (1973).
209. H. G. Nelson, in *Hydrogen in Metals* (I. M. Bernstein and A. W. Thompson, eds.), pp. 445–464, ASM, Metals Park, Ohio (1974).
210. C. G. Rhodes and J. C. Williams, *Met. Trans.* **6A**, 2103–2114 (1976).
211. J. A. Feeney and M. J. Blackburn, *Met. Trans.* **1**, 3309–3323 (1970).
212. B. S. Hickman, H. L. Marcus, and J. C. Williams, paper presented at Stress Corrosion Mechanisms in Titanium Alloys, Atlanta, Ga., January 1971; to be published.
213. J. B. Guernsey, V. C. Peterson, and F. H. Froes, *Met. Trans.* **3**, 339–340 (1972).
214. J. A. Feeney and M. J. Blackburn, *Met. Trans.* **3**, 341–342 (1972).
215. M. J. Blackburn, W. H. Smyrl, and J. A. Feeney, in *Stress-Corrosion Cracking in High Strength Steels and in Titanium and Aluminum Alloys* (B. F. Brown, ed.), pp. 245–363, Naval Res. Laboratory, Washington, D.C. (1972).

216. N. E. Paton and J. C. Williams, in *Proceedings of the Second International Conference on Strength of Metals and Alloys*, Vol. 1, pp. 108–112, ASM, Metals Park, Ohio (1970).
217. C. J. McMahon and D. J. Truax, *Corrosion* **29**, 47–55 (1973).
218. V. A. Livanov, A. P. Bukhanova, and B. A. Kolachev, *Influence of Grain Size on Hydrogen Embrittlement of Ti Alloys and Commercially Pure Ti*, Moscow Aviation Technology Institute, Moscow (1966); see Access No. N67–27021.
219. D. A. Meyn and G. Sandoz, *Trans. AIME* **245**, 1253–1258 (1969).
220. D. A. Meyn, *Met. Trans.* **5**, 2405–2414 (1974).
221. D. A. Meyn, *Met. Trans.* **3**, 2302–2305 (1972).
222. T. S. Liu and M. A. Steinberg, *Trans. ASM* **50**, 455–477 (1958).
223. M. J. Blackburn and W. H. Smyrl, in *Titanium Science and Technology* (R. I. Jaffee and H. M. Burte, eds.), Vol. 4, pp. 2577–2609, Plenum Press, New York (1973).
224. N. E. Paton and J. C. Williams, in *Hydrogen in Metals* (I. M. Bernstein and A. W. Thompson, eds.), pp. 409–431, ASM, Metals Park, Ohio (1974).
225. N. E. Paton, B. S. Hickman, and D. H. Leslie, *Met. Trans.* **2**, 2791–2796 (1971).
226. N. E. Paton and R. A. Spurling, *Met. Trans.* **7A**, 1769–1774 (1976).
227. J. S. Bradbrook, G. W. Lorimer, and N. Ridley, *J. Nucl. Mater.* **42**, 142–160 (1972).
228. D. N. Williams, *Met. Trans.* **5**, 2351–2358 (1974).
229. R. J. H. Wanhill, *Acta Met.* **21**, 1253–1258 (1973).
230. J. A. S. Green and A. J. Sedriks, *Corrosion* **28**, 226–230 (1972).
231. J. C. Scully, in *Effect of Hydrogen on Behavior of Materials* (A. W. Thompson and I. M. Bernstein, eds.), pp. 129–147, TMS-AIME, New York (1976).
232. H. R. Gray, *Corrosion* **25**, 337–341 (1969).
233. R. S. Ondrejcin, *Met. Trans.* **1**, 3031–3036 (1970).
234. W. H. Smyrl and M. J. Blackburn, *Corrosion* **31**, 371–375 (1975).
235. R. J. H. Wanhill, *Corrosion* **31**, 143–145 (1975).
236. T. Bonizewski and G. C. Smith, *Acta Met.* **11**, 165–178 (1963).
237. B. A. Wilcox and G. C. Smith, *Acta Met.* **13**, 331–343 (1965).
238. A. W. Thompson and B. A. Wilcox, *Scripta Met.* **6**, 689–696 (1972).
239. R. M. Latanision and H. Opperhauser, *Met. Trans.* **5**, 483–492 (1974).
240. A. H. Windle and G. C. Smith, *Met. Sci. J.* **2**, 187–191 (1968).
241. W. K. Boyd and W. E. Bery, in *Stress Corrosion Cracking—A State of the Art* (ASTM STP 518), pp. 58–78, ASTM, Philadelphia (1972).
242. P. W. Palmberg and H. L. Marcus, *Trans. ASM* **62**, 1016–1018 (1969).
243. H. Gleiter and B. Chalmers, *Progress in Materials Science*, Vol. 16, p. 49, Pergamon Press, Elmsford, N.Y. (1972).
244. J. R. Rellick, C. J. McMahon, H. L. Marcus, and P. W Palmberg, *Met. Trans.* **2**, 1492–1494 (1971).
245. A. Joshi and D. F. Stein, *J. Test. Eval.* **1**, 202–208 (1973).
246. A. W. Thompson, in *Grain Boundaries in Engineering Materials* (Proc. 4th Bolton Landing Conf.), pp. 607–618, Claitor's, Baton Rouge, La. (1975).
247. P. D. Merica and R. G. Waltenberg, *Trans. AIME* **71**, 709–716 (1925).
248. S. Floreen and J. H. Westbrook, *Acta Met.* **17**, 1175–1181 (1969).
249. J. H. Westbrook and S. Floreen, *Can. Met. Quart.* **13**, 181–186 (1974).
250. J. E. Doherty, B. H. Kear, A. F. Giamei, and C. W. Steinke, in *Grain Boundaries in Engineering Materials* (Proc. 4th Bolton Landing Conf.), pp. 619–627, Claitor's Baton Route, La. (1975).
251. R. M. Latanision and H. Opperhauser, *Met. Trans.* **6A**, 233–234 (1975).
252. A. W. Thompson, in *Hydrogen in Metals* (I. M. Bernstein and A. W. Thompson, eds.), pp. 543–544, ASM, Metals Park, Ohio (1974).

253. J. D. Frandsen, N. E. Paton, and H. L. Marcus, *Met. Trans.* **5**, 1655–1661 (1974).
254. J. L. Mihelich and A. R. Troiano, *Nature* **197**, 996–997 (1963).
255. J. A. Harris, R. C. Scarberry, and C. D. Stephens, *Corrosion* **28**, 57–62 (1972).
256. J. D. Frandsen, W. L. Morris, and H. L. Marcus, in *Hydrogen in Metals* (I. M. Bernstein and A. W. Thompson, eds.), pp. 633–643, ASM, Metals Park, Ohio (1974).
257. H. L. Marcus, and J. M. Harris, Science Center, Rockwell International, Thousand Oaks, Calif., unpublished research (1975).
258. J. D. Frandsen, H. L. Marcus, and A. S. Teleman, in *Effect of Hydrogen on Behavior of Materials* (A. W. Thompson and I. M. Bernstein, eds.), pp. 299–306, AIME, New York (1976).
259. A. W. Thompson, *Met. Trans.* **5**, 1855–1861 (1974).
260. P. A. Blanchard and A. R. Troiano, Hydrogen Embrittlement of Several Face-Centered Cubic Alloys, Report WADC TR-49-172, Section III (Access No. AD 229 466), U.S. Air Force (1959).
261. P. C. J. Gallagher, *Met. Trans.* **1**, 2429–2461 (1970).
262. R. L. Cowan and C. S. Tedmon, in *Advances in Corrosion Science and Technology* (M. G. Fontana and R. W. Staehle, eds.), Vol. 3, pp. 293–400, Plenum Press, New York (1973).
263. J. A. Board, *J. Inst. Met.* **101**, 241–247 (1973).
264. D. van Rooyen, *Corrosion* **31**, 327–337 (1975).
265. P. D. Neumann and J. C. Griess, *Corrosion* **19**, 345t–353t (1963).
266. J. Blanchet, H. Coriou, L. Grall, C. Mahieu, C. Otter, and G. Turluer, in *Stress Corrosion Cracking and Hydrogen Embrittlement of Iron-Base Alloys* (R. W. Staehle *et al.*, eds.), pp. 1149–1160, NACE, Houston (1977).
267. R. C. Scarberry, D. L. Graver, and C. D. Stephens, *Mater. Prot.* **6**(6), 54–57 (1967).
268. A. J. Sedriks, *Corrosion* **31**, 339 (1975).
269. R. W. Staehle, *Trans. Inst. Chem. Engr.* **47**, T227 (1969).
270. J. A. Harris and M. C. Van Wanderham, Properties of Materials in High Pressure Hydrogen at Cryogenic, Room and Elevated Temperatures, Report FR-5768, Pratt and Whitney Aircraft, W. Palm Beach, Fla. (1973).
271. R. F. Decker and S. Floreen, in *Precipitation from Iron-Base Alloys*, pp. 69–128, Gordon and Breach, New York (1965).
272. D. R. Muzyka, *Met. Eng. Quart.* **11**(4), 12–19 (1971).
273. C. P. Sullivan and M. J. Donachie, *Met. Eng. Quart.* **11**(4), 1–11 (1971).
274. R. F. Decker and C. T. Sims, in *The Superalloys* (C. T. Sims and W. C. Hagel, eds.), pp. 33–77, Wiley, New York (1972).
275. P. J. Bridges, in *The Nimonic Alloys* (W. Betteridge and J. Heslop, eds.), pp. 36–62, Arnold, London (1974).
276. O. H. Kriege and J. M. Baris, *Trans. ASM* **62**, 195–200 (1969).
277. D. F. Paulonis, J. M. Oblak, and D. S. Duvall, *Trans. ASM* **62**, 611–622 (1969).
278. R. J. Walter, R. P. Jewett, and W. T. Chandler, *Mater. Sci. Eng.* **5**, 98–110 (1969–70).
279. H. R. Gray, Embrittlement of Nickel-, Cobalt-, and Iron-Base Superalloys by Exposure to Hydrogen, Report TN D-7805, NASA, Cleveland, Ohio (1975).
280. J. Munford, Sandia Laboratories, Albuquerque, N.M., unpublished research (1972).
281. W. C. Harrigan, C. R. Barrett, and W. D. Nix, *Met. Trans.* **5**, 205–216 (1974).
282. T. P. Groeneveld, E. E. Fletcher, and A. R. Elsea, A Study of Hydrogen Embrittlement of Various Alloys, Summary Report on Contract NAS8-20029 (Access No. N69-28526), Battelle Mem. Inst., Columbus, Ohio (1969).
283. D. L. Dull and L. Raymond, *Met. Trans.* **4**, 1635–1637 (1973).
284. R. J. Walter and W. T. Chandler, *Mater. Sci. Eng.* **8**, 90–97 (1971).

285. S. M. Copley and J. C. Williams, in *Alloy and Microstructural Design* (J. K. Tien and G. S. Ansell, eds.), pp. 3–63, Academic Press, New York (1976).
286. R. J. Walter and W. T. Chandler, in *Hydrogen in Metals* (I. M. Bernstein and A. W. Thompson, eds.), pp. 515–525, ASM, Metals Park, Ohio (1974).
287. R. J. Walter and W. T. Chandler, Rocketdyne Div., Rockwell International, Canoga Park, Calif., unpublished research (1974).
288. B. C. Odegard, Sandia Laboratories, Livermore, Calif., unpublished research (1972).
289. J. M. Oblak and W. A. Owczarski, *Met. Trans.* **3**, 617–626 (1972).
290. J. M. Oblak, D. F. Paulonis, and D. S. Duvall, *Met. Trans.* **5**, 143–153 (1974).·
291. B. A. Wilcox and A. H. Clauer, in *The Superalloys* (C. T. Sims and W. C. Hagel, eds.), pp. 197–230, Wiley, New York (1972).
292. J. A. Brooks and A. W. Thompson, in *Hydrogen in Metals* (I. M. Bernstein and A. W. Thompson, eds.), pp. 527–539, ASM, Metals Park, Ohio (1974).
293. J. D. Frandsen, N. E. Paton, and H. L. Marcus, *Scripta Met.* **7**, 409–414 (1973).
294. J. S. Benjamin, *Met. Trans.* **1**, 2943–2951 (1970).
295. A. S. Keh and S. Weismann, in *Electron Microscopy and Strength of Crystals* (G. Thomas and J. Washburn, eds.), pp. 231–299, Interscience, New York (1963).
296. T. R. Beck, in *The Theory of Stress Corrosion Cracking in Alloys* (J. C. Scully, ed.), pp. 64–83, NATO, Brussels (1971).
297. C. F. Barth and A. R. Troiano, *Corrosion* **28**, 259–263 (1972).
298. M. Marek and R. F. Hochman, *Corrosion* **26**, 5–6 (1970).
299. W. Williams and J. Eckel, *J. Am. Soc. Nav. Engr.* **68**, 93–107 (February 1956).
300. H. H. Uhlig, *Corrosion and Corrosion Control*, 2nd ed., p. 314, Wiley, New York (1971).
301. H. P. van Leeuwen, *Corrosion* **31**, 42–49 (1975).
302. J.-P. Fidelle, in *Hydrogen in Metals* (I. M. Bernstein and A. W. Thompson, eds.), pp. 748–749, ASM, Metals Park, Ohio (1974).
303. T. L. Johnston, R. G. Davies, and N. S. Stoloff, *Phil. Mag.* **12**, 305–317 (1965).
304. A. W. Thompson, M. I. Baskes, and W. F. Flanagan, *Acta Met.* **21**, 1017–1028 (1973).
305. H. G. Nelson, D. P. Williams, and A. S. Tetelman, *Met. Trans.* **2**, 953–959 (1971).
306. A. J. Sedriks, J. A. S. Green, and D. L. Novak, *Corrosion* **27**, 198–202 (1971).
307. M. Pourbaix, in *Localized Corrosion* (R. W. Staehle, B. F. Brown, and J. Kruger, eds.), pp. 12–32, NACE, Houston (1974).
308. E. A. Steigerwald, F. W. Schaller, and A. R. Troiano, *Trans. AIME* **215**, 1048–1052 (1959).
309. H. H. Johnson, in *Fundamental Aspects of Stress Corrosion Cracking* (R. W. Staehle, *et al.*, eds.), pp. 439–445, NACE, Houston (1969).
310. D. P. Williams and H. G. Nelson, *Met. Trans.* **2**, 1987–1988 (1971).
311. P. Bastien and P. Azou, *C. R. Acad. Sci. Paris* **232**, 1845–1848 (1951).
312. P. Bastien and P. Azou, in *Proceedings First World Metallurgical Congress*, pp. 535–552, ASM, Cleveland, Ohio (1951).
313. P. Bastien, in *Physical Metallurgy of Stress Corrosion Fracture* (T. N. Rhodin, ed.), pp. 311–339, Interscience, New York (1959).
314. J. K. Tien, A. W. Thompson, I. M. Bernstein, and R. J. Richards, *Met. Trans.* **7A**, 821–829 (1976).
315. J. A. Donovan, *Met. Trans.* **7A**, 145–149, 1677–1683 (1976).
316. J. D. Boyd, in *The Science, Technology, and Application of Titanium* (R. I. Jaffee and N. E. Promisel, eds.), pp. 545–555, Pergamon Press, Elmsford, N.Y. (1970).
317. J. K. Tien, R. J. Richards, O. Buck, and H. L. Marcus, *Scripta Met.* **9**, 1097–1101 (1975).
318. A. R. Troiano, *Trans. ASM* **52**, 54–80 (1960).
319. J. C. M. Li, R. A. Oriani, and L. S. Darken, *Z. Phys. Chem.* **49**, 271–290 (1966).

320. J. R. Rice, in *Stress Corrosion Cracking and Hydrogen Embrittlement of Iron-Base Alloys* (R. W. Staehle *et al* eds.), pp. 11–15, NACE, Houston (1977).

321. H. L. Logan, *J. Res. Nat. Bur. Stand.* **48**, 99–105 (1952).

322. T. P. Hoar, *Corrosion* **19**, 331–338 (1963).

323. H. W. Pickering and C. Wagner, *J. Electrochem. Soc.* **114**, 698–706 (1967).

324. E. N. Pugh, in *Theory of Stress Corrosion Cracking in Alloys* (J. C. Scully, ed.), pp. 418–441, NATO, Brussels (1971).

325. F. P. Ford and T. P. Hoar, in *Proceedings of the Third International Conference on Strength of Metals and Alloys*, Vol. 1, pp. 467–471, Institute of Metals, Cambridge, England (1973).

326. A. G. Evans and T. G. Langdon, in *Progress in Materials Science*, Vol. 21, pp. 171–441, Pergamon Press, Elmsford, N.Y. (1977).

327. P. R. Rhodes, *Corrosion* **25**, 462–472 (1969).

328. H. P. van Leeuwen, *Corrosion* **29**, 197–204 (1973).

329. H. P. van Leeuwen, J. A. M. Boogers, and C. J. Stentler, *Corrosion* **31**, 23–29 (1975).

330. L. B. Pfeil, *Proc. Roy. Soc. London* **112A**, 182–195 (1926).

331. R. A. Oriani, *Ber. Bunsenges. Phys. Chem.* **76**, 848–857 (1972).

332. R. A. Oriani and P. H. Josephic, *Acta Met.* **22**, 1065–1074 (1974).

333. C. F. Barth and E. A. Steigerwald, *Met. Trans.* **1**, 3451–3455 (1970).

334. J. Hewitt and J. D. Murray, *Br. Welding J.* **15**, 151–158 (1968).

335. V. P. Krylov and N. I. Vorob'eva, *Met. Sci. Heat Treat.* **15**, 402–408 (1973).

336. C. J. McMahon, K. Yoshino, and H. C. Feng, in *Stress Corrosion Cracking and Hydrogen Embrittlement of Iron-Base Alloys* (R. W. Staehle *et al.*, eds.), pp. 649–657, NACE, Houston (1977).

337. M. B. Whiteman and A. R. Troiano, *Corrosion* **21**, 53–56 (1965).

338. J. C. Scully and D. T. Powell, *Corros. Sci.* **10**, 719–733 (1970).

339. D. A. Mauney, E. A. Starke, and R. F. Hochman, *Corrosion* **29**, 241–244 (1973).

340. D. A. Mauney and E. A. Starke, *Corrosion* **25**, 177–179 (1969).

341. R. J. H. Wanhill and J. Brettle, *Met. Mater.* **8**, 514–515 (1972).

342. B. A. Kolachev, *Hydrogen Embrittlement of Nonferrous Metals*, Israel Program for Scientific Translations, Jerusalem (1968).

343. C. Korn, D. Zamir, and Z. Hadari, *Acta Met.* **22**, 33–45 (1974).

344. W. W. Gerberich and Y. T. Chen, *Met. Trans.* **6A**, 271–278 (1975).

345. C. F. Barth, E. A. Steigerwald, and A. R. Troiano, *Corrosion* **25**, 353–358 (1969).

346. V. J. Colangelo and M. S. Ferguson, *Corrosion* **25**, 509–514 (1969).

347. P. N. T. Unwin and R. B. Nicholson, *Acta Met.* **17**, 1379–1393 (1969).

348. H. P. van Leeuwen, L. Schra, and W. J. van der Vet, *J. Inst. Met.* **100**, 86–96 (1972).

349. L. Graf and W. Neth, *Z. Metallkd.* **60**, 789–799 (1969).

350. A. J. DeArdo and R. D. Townsend, *Met. Trans.* **1**, 2573–2581 (1970).

351. K. G. Kent, *J. Aust. Inst. Met.* **15**, 171–178 (1970).

352. P. K. Poulose, J. E. Morral, and A. J. McEvily, *Met. Trans.* **5**, 1393–1400. (1974).

353. W. E. Wood and W. W. Gerberich, *Met. Trans.* **5**, 1285–1294 (1974).

354. D. A. Vaughan and D. I. Phalen, in *Stress Corrosion Testing*, ASTM STP 425, pp. 209–227, ASTM, Philadelphia (1967).

355. G. P. Cherepanov, *Eng. Fract. Mech.* **5**, 1041–1046 (1973).

356. G. P. Cherepanov, *Corrosion* **29**, 305–309 (1973).

357. J. S. Enochs and O. F. Devereux, *Met. Trans.* **6A**, 391–397 (1975).

358. L. V. Corsetti and D. J. Duquette, *Met. Trans.* **5**, 1087–1093 (1974).

359. J. A. S. Green and W. G. Montague, *Corrosion* **31**, 209–213 (1975).

360. P. Doig and J. W. Edington, *Corrosion* **31**, 347–352 (1975).

361. F. P. Ford, Ph.D. thesis, Cambridge University (1973).

362. F. P. Ford and T. P. Hoar, submitted to *J. Electrochem. Soc.*

363. J. R. Rice, in *Effect of Hydrogen on Behavior of Materials* (A. W. Thompson and I. M. Bernstein, eds.), pp. 455–466, TMS-AIME, New York (1976).

364. N. J. Petch and P. Stables, *Nature* **169**, 842–843 (1952).

365. C. Zappfe and C. Sims, *Trans. AIME* **145**, 225–259 (1941).

366. A. W. Thompson, in *Effect of Hydrogen on Behavior of Materials* (A. W. Thompson and I. M. Bernstein, eds.), pp. 467–477, TMS-AIME, New York (1976).

367. I. M. Bernstein, *Scripta Met.* **8**, 343–349 (1974).

368. D. G. Westlake, *Trans. ASM* **62**, 1000–1006 (1969).

369. J. J. Gilman, *Phil. Mag.* **26**, 801–812 (1972).

370. M. R. Louthan and R. P. McNitt, in *Effect of Hydrogen on Behavior of Materials* (A. W. Thompson and I. M. Bernstein, eds.), pp. 496–506, TMS-AIME, New York (1976).

371. N. A. Nielsen, *J. Mater.* **5**, 794–829 (1970).

372. B. E. Wilde and C. D. Kim, *Corrosion* **28**, 350–356 (1972).

373. A. A. Seys, M. J. Brabers, and A. A. van Haute, *Corrosion* **30**, 47–52 (1974).

374. M. L. Mehta and J. Burke, *Corrosion* **31**, 108–110 (1975).

375. E. N. Pugh, University of Illinois, personal communication (1976); see also J. M. Popplewell and J. A. Ford, *Met. Trans.* **5**, 2600–2602 (1974).

376. C. L. Briant, *Met. Trans.* **9A**, 731–733 (1978).

377. D. Eliezer, D. G. Chakrapani, C. Altstetter, and E. N. Pugh, *Met. Trans.* *10A* (1979), in press.

378. A. J. West, Sandia Laboratories, Livermore, personal communication (1978).

379. J. F. Lessar and W. W. Gerberich, *Met. Trans.* **7A**, 953–960 (1976).

380. S. K. Banerji, C. J. McMahon, and H. C. Feng, *Met. Trans.* **9A**, 237–247 (1978).

381. R. O. Ritchie, M. H. Castro Cedeño, V. F. Zackay, and E. R. Parker, *Met. Trans.* **9A**, 35–40 (1978).

382. R. Garber and I. M. Bernstein, in *Environmental Degradation of Engineering Materials* (M. R. Louthan and R. P. McNitt, eds.), pp. 463–473, V.P.I. Press, Blacksburg, Va. (1977).

383. A. P. Coldren and G. Tither, *J. Met.* **28**(5), 5–10 (1976).

384. M. Iino, *Met. Trans.* **9A**, 1581–1590 (1978)

385. J. G. Morris and B. J. Roopchand, in *Environmental Degradation of Engineering Materials* (M. R. Louthan and R. P. McNitt, eds.), pp. 475–485, V.P.I. Press, Blacksburg, Va. (1977).

386. J. W. Munford, H. J. Rack, and W. J. Kass, in *Environmental Degradation of Engineering Materials* (M. R. Louthan and R. P. McNitt, eds.), pp. 501–510, V.P.I. Press, Blacksburg, Va. (1977).

387. R. R. Boyer and W. F. Spurr, *Met. Trans.* **9A**, 23–29 (1978).

388. R. R. Boyer and W. F. Spurr, *Met. Trans.* **9A**, 1443–1447 (1978).

389. R. Dutton, K. Nuttall, M. P. Puls, and L. A. Simpson, *Met. Trans.* **8A**, 1553–1562 (1977).

390. G. H. Koch, A. J. Bursle, and E. N. Pugh, *Met. Trans.* **9A**, 129–131 (1978).

391. I. W. Hall, *Met. Trans.* **9A**, 815–820 (1978).

392. K. A. Peterson, J. C. Schwanebeck, and W. W. Gerberich, *Met. Trans.* **9A**, 1169–1172 (1978).

393. H. G. Nelson, *Met. Trans.* **7A**, 621–627 (1976).

394. H. Margolin, *Met. Trans.* **7A**, 1233–1235 (1976).

395. P. Doig, P. E. J. Flewitt, and J. W. Edington, *Corrosion* **33**, 217–221 (1977).

396. R. Haag, J. E. Morral, and A. J. McEvily, in *Environmental Degradation of Engineering Materials* (M. R. Louthan and R. P. McNitt, eds.), pp. 19–40, V.P.I. Press, Blacksburg, Va. (1977).

397. F. P. Ford, in *Mechanisms of Environment Sensitive Cracking of Materials*, pp. 125–142, Metals Society, London (1977).

398. G. M. Scamans, R. Alani, and P. R. Swann, *Corros. Sci.* **16**, 443–459 (1976).
399. G. M. Scamans, *J. Mater. Sci.* **13**, 27–36 (1978).
400. R. J. Jacko and D. J. Duquette, *Met. Trans.* **8A**, 1821–1827 (1977).
401. R. M. Latanision, O. H. Gastine, and C. R. Compeau, in *Environment-Sensitive Fracture of Engineering Materials*, pp. 48–70, TMS-AIME, New York (1979).
402. G. M. Scamans and C. D. S. Tuck, in *Mechanisms of Environment-Sensitive Cracking of Materials*, pp. 482–491, Metals Society, London (1977).
403. A. J. Bursle and E. N. Pugh, in *Mechanisms of Environment-Sensitive Cracking of Materials*, pp. 471–480, Metals Society, London (1977).
404. T. Klimowicz and R. M. Latanision, *Met. Trans.* **9A**, 597–599 (1978).
405. J. Albrecht, A. W. Thompson, and I. M. Bernstein, *Met. Trans.* (1979), in press.
406. J. M. Chen, T. S. Sun, R. K. Viswanadham, and J. A. S. Green, *Met. Trans.* **8A**, 1935–1940 (1977).
407. E. C. Pow, J. C. Schwanebeck, and W. W. Gerberich, *Met. Trans.* **9A**, 1009–1011 (1978).
408. J. P. Hirth and H. H. Johnson, *Corrosion* **32**, 3–26 (1976).
409. A. W. Thompson and I. M. Bernstein, in *Fracture 1977* (D. M. R. Taplin, ed.) Vol. 2, pp. 249–254, University of Waterloo Press, Waterloo, Ont. (1977).
410. A. W. Thompson and I. M. Bernstein, *Int. J. Hydrogen Energy* **2**, 163–173 (1977).
411. A. W. Thompson, *Int. J. Hydrogen Energy* **2**, 299–307 (1977).
412. A. W. Thompson, in *Transmission and Storage of Hydrogen*, Vol. II of *Hydrogen: Its Technology and Implications* (K. E. Cox and K. D. Williamson, eds.), 85–124, CRC Press, Cleveland, Ohio (1977).
413. A. W. Thompson, *Plat. Surf. Finish.* **65**(9), 36–44 (1978).
414. A. W. Thompson, *Am. Pet. Inst., Sec. 3* **57**, 317–326 (1978).
415. A. W. Thompson, in *Fracture 1977* (D. M. R. Taplin, ed.), Vol. 2, pp. 237–242, University of Waterloo Press, Waterloo, Ont. (1977).
416. A. W. Thompson and I. M. Bernstein, in *Hydrogen in Metals* (Proc. 2nd Int. Congr., Paris), Vol. 10, paper 3A-6, Pergamon Press, Elmsford, N.Y. (1977).
417. Y. Kikuta, T. Araki, and T. Kuroda, in *Hydrogen in Metals* (Proc. 2nd Int. Congr., Paris), Vol. 10, paper 3A-4, Pergamon Press, Elmsford, N.Y. (1977).
418. F. E. Fujita, in *Hydrogen in Metals* (Proc. 2nd Int. Congr., Paris), Vol. 5, paper 2B-10, Pergamon Press, Elmsford, N.Y. (1977).
419. R. B. Heady, *Corrosion* **34**, 303–306 (1978).
420. S. R. Bala and D. Tromans, *Met. Trans.* **9A**, 1125–1132 (1978).
421. I. M. Bernstein and A. W. Thompson, in *Mechanisms of Environment-Sensitive Cracking of Materials*, pp. 412–426, Metals Society, London (1978).
422. A. W. Thompson, in *Environmental Degradation of Engineering Materials* (M. R. Louthan and R. P. McNitt, eds.), pp. 3–17, V.P.I. Press, Blacksburg, Va. (1977).
423. R. A. Oriani and P. H. Josephic, *Acta Met.* **25**, 979–988 (1977).
424. A. W. Thompson, in *Environment-Sensitive Fracture of Engineering Materials*, pp. 379–410, TMS-AIME, New York (1979).
425. R. B. Heady, *Corrosion* **33**, 441–447 (1977).
426. J. C. Scully, in *Mechanisms of Environment-Sensitive Cracking of Materials*, pp. 1–18, Metals Society, London (1977).
427. R. M. Parkins and I. H. Craig, *Mechanisms of Environment-Sensitive Cracking of Materials*, pp. 32–50, Metals Society, London (1977).
428. F. Perdieus, M. Brabers, and A. Van Haute, *Mechanisms of Environment-Sensitive Cracking of Materials*, pp. 53–61, Metals Society, London (1977).
429. D. J. Duquette, *Mechanisms of Environment-Sensitive Cracking of Materials*, pp. 305–320, Metals Society, London (1977).

430. H. H. Johnson and J. P. Hirth, *Met. Trans.* **7A**, 1543–1548 (1976).

431. R. Jensen and J. K. Tien, Columbia University, unpublished research (1977).

432. J. K. Tien, in *Effect of Hydrogen on Behavior of Materials* (A. W. Thompson and I. M. Bernstein, eds.), pp. 167–175, TMS-AIME, New York (1976).

433. T. Asaoka, G. Lapasset, M. Aucouturier, and P. Lacombe, *Corrosion* **34**, 39–44 (1978).

434. B. D. Craig, *Acta Met.* **25**, 1027–1030 (1977).

435. J. E. Smugeresky, *Met. Trans.* **8A**, 1283–1289 (1977).

436. C. G. Rhodes and A. W. Thompson, *Met. Trans.* **8A**, 949–954 (1977).

437. A. Atrens, N. F. Fiore, and K. Miura, in *Environmental Degradation of Engineering Materials* (M. R. Louthan and R. P. McNitt, eds.), pp. 487–494, V.P.I. Press, Blacksburg, Va. (1977).

438. R. J. Richards, S. Purushothaman, J. K. Tien, J. D. Frandsen, and O. Buck, *Met. Trans.* **9A**, 1107–1111 (1978).

439. A. J. West, J. A. Brooks, and S. L. Robinson, Sandia Laboratories, Livermore, Calif. personal communication (1978).

440. J. A. Donovan and G. R. Caskey, Savannah River Laboratory, E. I. du Pont, personal communication (1978).

441. M. R. Louthan, Virginia Polytechnic Institute and State University, personal communication (1978).

EMF MEASUREMENTS AT ELEVATED TEMPERATURES AND PRESSURES

J. V. Dobson

Department of Physical Chemistry
University of Newcastle upon Tyne
Newcastle upon Tyne
England

INTRODUCTION

The types of electromotive force (emf) measurements covered in this chapter include galvanic, concentration with and without transport, and glass electrode cells, over wide ranges of temperature and pressure. Measurement of rest potentials and the instantaneous, but continual comparison of working electrode potentials with reference electrodes in potentiostatic work are basically emf measurements but are generally avoided and are obviously dealt with more fully in other papers devoted to electrode kinetics. Emf studies on nonisothermal, polarographic cells are not discussed for similar reasons. The present chapter attempts to demonstrate, by reference to recently published work, factors important in the measurement of cell emf. Moreover, by devoting various sections to specific quantities, such as pH, K_a, K_w, $\gamma_\pm$, dE/dP, dE/dT, E/T, and E/P, the versatility of this type of electrochemical measurement is highlighted. From the aspect of review, this chapter only discusses in detail information published since 1965, although comparisons with earlier work are made occasionally. Because the range of applications of emf measurements is very large and not concerned in general with the continuation of a particular problem, it is difficult to prepare a review without discontinuities. Apologies are also given by the writer for any studies unintentionally overlooked.

RECENT APPLICATIONS OF EMF MEASUREMENTS

An idea of the variation in applications of emf measurements to academic and industrial problems may be obtained by reiterating the reasons the authors of some recently published emf work have given for the need and justification of their particular studies. According to Schöön[1] (Sweden), for example, the most important factors in the technical sulfite process, in addition to temperature and cooking time, are the pH and the HSO_3^- concentration in the cooking liquids. Both of these, he demonstrates, may be determined satisfactorily by emf measurements. The measurement of pH is necessary in the control of reactions in hydrolysis and esterification in the chemical industry. Fiker and Santova[2] (Czechoslovakia) have used emf measurements to measure pH under plant conditions with through-flow sensing elements at high temperatures. Measurement of pH by emf measurements in superheated water contained in borax springs in Kamchatka, USSR, were needed by Kryukov et al.[3] (USSR) in recent studies connected with geothermal fissures and calcification. Emf–pH measurements at elevated pressures were also used in a microbiological study by Smith and Lauro[4] (USA).

Emf measurements also provide an experimental background to current theories of the origin and composition of seawater; mean ion activities are needed for the major components over wide temperature and pressure ranges. Whitfield[5] (Australia), in his emf studies on the ionic product of water (K_w), has pointed out that the moderate range in temperature and high pressures are also of direct interest to the marine scientist. Advances in emf techniques now make it possible to collect all the data necessary in the oceanographically significant temperature and pressure ranges. Whitfield also states that fresh K_w data are also important, because of persistence by authoritative sources in requoting old or incorrect values. For example, Refs. 6–8 quoted values of K_w calculated by Owen and Brinkley[9] on the basis of a standard partial molal volume change which has since been shown to be $2 \, cm^3/mole$ in error. Accurate values of K_w as functions of temperature and pressure are also required when setting up thermodynamic diagrams and pH scales for kinetic studies involving hydrogen and hydroxyl ions.

Thermodynamic information, possibly derived by emf measurements, are of particular interest in boiler corrosion and fuel cell studies. In order to produce information in the study of corrosion of materials for use in nuclear reactor systems, Indig and Vermilyea[10] (USA) were required to measure and control the potentials of electrodes in water at approximately 290°C. The choice, by Mesmer et al.[11] (USA), of boric acid/borate buffer mixtures for cell electrolytes at elevated temperatures was made because they serve as pH standards, occur in natural aqueous systems and in detergent solutions, and are used as burnable nuclear poisons in the coolants of pressurized water nuclear reactors. Every and Banks[12] (USA) have demonstrated in a patent on a practical application of anodic protection at elevated temperatures that precise emf measurements are required continuously. Finally, Staehle and Cowan[13] (USA), in the course of an electrode kinetic study, have shown concern, and the need for reliable emf data, with respect to determination of standard electrode potentials at elevated temperatures.

These few random examples do not attempt to be exhaustive, and others could have been cited (e.g., hydrothermal synthesis, water pollution, desalination). However, the author hopes that the examples chosen have indicated the interdisciplinary role that this type of electrochemical measurement possesses.

HISTORICAL BACKGROUND

A major part of early work was focused on practical problems revolving around fabrication of pressure vessels, cell container design, electrical and pressure seals, and materials of construction. All problems have not been solved. An indication of the interest that is still apparent is suggested by the large number of papers devoted to such matters given at the NACE High Temperature High Pressure Electrochemistry Conference (HTHPE), 1973, Guildford, U.K.

Detailed investigations with respect to stability, response, and measurements of standard electrode potentials of reference electrodes, as well as determinations of precise activity data and other thermodynamic quantities of electrolytes, have been, and still are, main preoccupations.

Several useful reviews on emf measurements, some containing bibliographies, have been published within the last decade, notably by Lietzke[14] (USA), Ives and Janz[15] (UK), Hills and Ovenden[16] (UK), de G. Jones and Masterson[17] (UK), and Hills[18] (UK). Particularly useful, but large, tables contained in the review by de G. Jones and Masterson listed all the work carried out between 1953 and 1966. The early work was discussed thoroughly in these reviews, but to provide a condensed historical background for the present chapter, the references, including some earlier studies not listed in the de G. Jones tables, are now compiled using the following code. The first symbol is the general classification of use: P represents pH studies; D, data; E, standard electrode potential; Ac, activity data; K, dissociation constant data. The range of pressure and temperature is then given. The third symbol given is the type of electrodes and electrolytes: H, platinum hydrogen electrode; G, glass electrode; A, silver–silver halide; Hg, mercury halide; other types named: C, chloride; B, bromide; I, iodide; SO_4, sulfate; Bu, mixed buffers; OH, hydroxide-bearing electrolyte. The final part of the code given is the year and reference number (in parentheses) in the general listing at the end of this chapter; for example, P25OHAC56(00) corresponds to a study which derived pH values up to 250°C using platinum hydrogen and silver–silver chloride electrodes in chloride-bearing electrolyte, published in 1956 and given in reference 00. These are: $D250Ag_2SO_4Hg_2SO_4SO_4$53 (19); E95HAC54 (20); D263AHgC54 (21); D250HAC55,56 (22); $Ac140PbO_2/PbSO_4Ag_2SO_4$57 (23); E275HAC60 (24); D275HAC60 (25); AcE200HAB60 (26), Ac150HAB63 (27); D225 Deuterium AC64 (28); D175HAC65 (29); EAc200HAI65 (30); and Ac125HAC55 (31).

All prefix (P) and temperature studies mainly at standard volume and pressure (SVP): 150HACBu30 (32); 250HAC56 (33); 160GCBu54,56 (34); 140GTlClBu59 (35); 240H/NiHg00H59 (36); 250HHCBu60 (37); 150Pb/HgABuC60 (38); 275HAC60 (39); 150Pb/HgGABuC62 (40); and 150HGThClBu66 (41).

All prefix (P) and pressure studies, maximum pressure in atm, around room temperature: 143GHgC54 (42); K1607GACBu59 (43); 1071GABu59 (44); K1071GABu62 (45); Kw2142GACOH63 (46); K1071GABu65 (47); D1000HHgC24 (48); $D1200Cd/Hg_2SO_4$32 (49); $D1500TlHgHg_2SO_4SO_4$23 (50); $D1500CdCd/Hg_2SO_4$20 (51); and $D1500Zn/HgHg_2SO_4SO_4$20 (52).

COMMENTS ON CELL EMF, THERMODYNAMIC RELATIONSHIPS, CELL DESIGN, MATERIALS, AND SOURCES OF ERROR

Cell EMF and Thermodynamic Relationships

It is highly probable that no electrochemical cell achieves and maintains ideal thermodynamic equilibrium for substantial periods of time, particularly at elevated temperatures. However, it is generally assumed for practical purposes to have reached a thermodynamic equilibrium when thermal and pressure conditions affecting solubility, density, vapor pressure, and so on, are relatively constant. Moreover, if emf is being measured, the values observed should have small variation over periods of 1 hr or so, the emf being taken over this period of time to represent precisely the change in free energy of the particular cell reaction which is involved. This is because an emf measurement is only a measurement of potential difference and is not considered to disturb the direction of the reaction involved. It is assumed that the measurement of emf is carried out by instruments which do not draw or pass current through the cell. Large variations of emf with time are generally taken to indicate, under constant temperature and pressure, that reactions other than those expected from the cell reaction are occurring.

The detailed and algebraic formal treatments of cell emf and their ability to produce precise thermodynamic data, and other quantities, are treated by

Table 1. Some Common Thermodynamic Relationships[a]

Temperature dependence at constant pressure	Pressure dependence at constant temperature
$\Delta G = -nFE = -RT \ln K$	$-nF\dfrac{dE}{dP} = \Delta V$
$\dfrac{d\Delta H}{dT} = \Delta C_p = f(T)$	$\dfrac{d\Delta S}{dP} = -\alpha V$
$\Delta H = \Delta H^\circ + \displaystyle\int \Delta C_p dT$	$T\left.\dfrac{\partial P}{\partial T}\right|_V = \dfrac{\partial U}{\partial V} + P = \dfrac{\alpha T}{\beta}$
$\dfrac{d\Delta G/T}{dT} = -\dfrac{\Delta H}{T^2}$	$\dfrac{dV}{dP} = -\beta V(KV)$
$\Delta S = -nF\dfrac{dE}{dT}$	$\dfrac{d\overline{V}_2}{dP} = \overline{K}_2$
$\Delta H = \dfrac{nF}{J}\left(E + T\dfrac{dE}{dT}\right)$	$nF\dfrac{dE}{dP} = \dfrac{RT\partial \ln K}{\partial P} = -\Delta \overline{V} = \sum_i \overline{V}_i$
$E_T^\circ = E_{25}^\circ + \dfrac{1}{nF}\left(\Delta S_{25}^\circ \Delta T\right.$	$nF\left.\dfrac{\partial(E - E^\circ)}{\partial P}\right|_m = 2RT\dfrac{\partial \ln \gamma_\pm}{\partial P} = \overline{V}_2 = \overline{V}_2$
$\left. -\displaystyle\int_{298}^{T} \Delta C_p^\circ dT + T\int_{298}^{T} \Delta C_p^\circ \dfrac{dT}{T}\right)$	
$pK = -\ln K = \dfrac{\Delta H}{RT} + \dfrac{\Delta C_p - \Delta s}{R} - \dfrac{\Delta C_p}{R}\ln T$	
$\dfrac{d\ln \gamma_\pm}{dT} = \dfrac{\overline{L}_2}{RT^2}$	

Liquid Junction Potential

$$E_j = \frac{RT}{F}\int_1^{11}\sum_i \frac{t_i}{z_i}\alpha \ln (m_i\gamma_i)$$

[a] All symbols have their usual significance.

any standard textbook. Little may be gained, therefore, from a detailed repetition of these well-known derivations, and only a few relevant comments will be made here. The original premise in the treatment of cell emf data relies on knowledge of a chemical reaction which may be represented by a combination of two electrodes, each being sensitive or reversible to its own particular ion, which is contained in the electrolyte of the cell. A number of the basic relationships are given in Table 1. It is possible therefore, from cell emf measurements (E), when made over ranges of temperature, pressure, and concentration, to derive the standard electrode potential $E°$, ΔG, ΔH, ΔV, ΔS, and ΔC_p for a cell reaction; the ionization constant of water K_w; and the dissociation constants, K, of weak and complex acids and bases. Other quantities may be found, such as transport numbers, t, pH, Soret coefficients, activity coefficients $\gamma_\pm$, and the concentration of ionic species other than hydrogen.

Cell Container Design

No comment will be made here on pressure vessel design, because this aspect is dealt with completely in other papers given at the HTHPE. Some comments will, however, be given on the cell containers which are housed inside the pressure vessels.

A number of different cell container designs have been used since the last half of the 1960s, each with its own particular advantages and disadvantages. In Fig. 1^{1-9} (pages 184 and 185) a number of interesting designs selected from the literature are reproduced, which are subdivided into two main classes. Two-phase cell systems are those which possess an aqueous vapor phase in equilibrium with the liquid electrolyte in the cell. These systems are generally operated at SVP, the pressure maintained by water contained in the pressure vessel surrounding the cell container (see Fig. 1, Type 1 or 2) or within the cell itself (Type 3). In the latter case calculated corrections are made to account for electrolyte concentration changes because of evaporation. Single-phase cell systems have the advantage that the pressure in the cell may be varied by oil or other liquids surrounding the sealed cell container in the pressure vessel, via a collapsible membrane (Type 8), bag (Type 6), piston (Type 9), or via dissimilar liquid interphase (Type 4 or 5). It is out of place here to discuss the various designs in detail, as all may be found in the papers quoted in the legend to Fig. 1; however, a few comments on salient features may be made in passing. Type 1 employs a long quartz capillary to reduce contamination of the bulk of the electrolyte with dissolved electrode materials. Types 2, 6 and 7, made almost completely of polytetrafluoroethylene (PTFE), use sintered PTFE at the entrances of electrode compartments to aid in the contamination problem. Loose-fitting palladium sheets were also used in Type 2 to assist in preventing hydrogen in solution from reaching the silver–silver chloride electrodes, palladium being used because of its property to absorb considerable quantities of hydrogen. Type 5 used a piece of nylon wrapped in the threads of a PTFE plug to give electrical connection between the two main compartments of the cell. Type 7 allows one, by a system of pistons and an electrically operated valve, to change the electrolyte in the cell. Type 8 has been used for nonisothermal studies[58]; the silver chloride electrodes at either end of the cell are maintained at known small differences in temperature.

Choice of Materials for Cell Containers

PTFE has become the generally used material for the construction of the cell containers for temperatures up to 300° C, and in preference to silica. Other materials, such as silicon hexaboride (Cerac, USA) and pure and doped alumina or zirconia (Sintox, UK), have been used for temperatures in excess of 300°C.

PTFE is easy to machine but has a tendency to "drill large," and suffers distortion if maintained at temperatures above 200°C and not supported. Early reports[17] of chemical attack on PTFE were due to the use of impure and cheap grades; top-quality PTFE is highly recommended. Refractory materials such as silicon hexaboride are not readily machinable and do suffer some slight chemical attack.[59]

Sources of Errors and Corrections

Two main sources of errors are those loosely described as uncontrolled and controlled variables. In the controlled category are those associated with cell design: calculated corrections, hydrogen pressure, for example; unwanted thermal gradients; and spurious emf's which are due to bad grounding or a cell resistance that is too high for the class of measuring instrument. All of these may be rectified by a subtle or a major variation in the system specifically or as a whole.

Uncontrolled variables are those associated with, and inherent in, the essential components of the system, particularly the electrodes: for example,

Two-Phase Systems

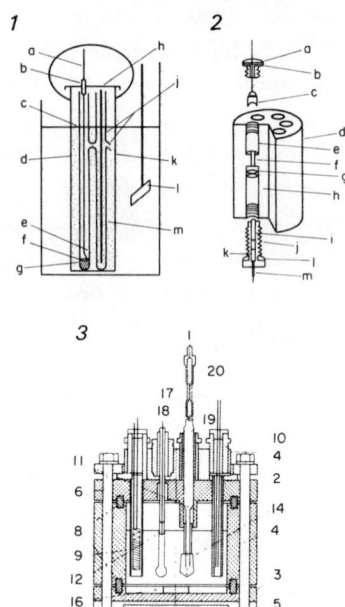

Single-Phase Systems

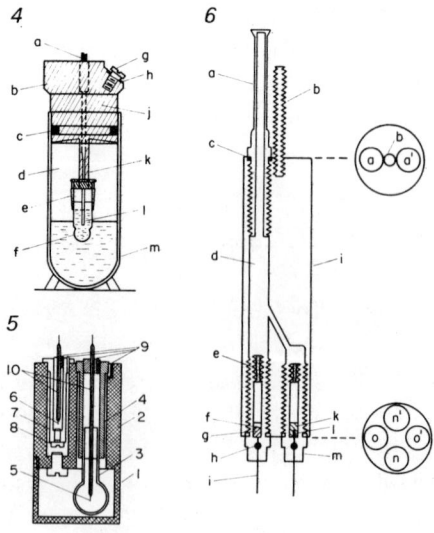

Fig. 1. See opposite page for legend.

Single-Phase Systems (cont'd.)

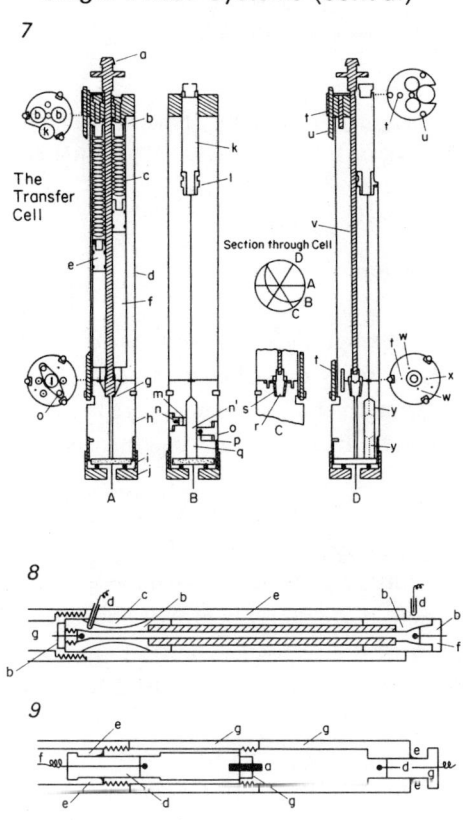

Fig. 1. Cell designs selected from the literature: Types 1 and 2 from Indig and Vermilyea[10] and Dobson and Thirsk[53]; Type 3 from Kryukov *et al.*[41]; Type 4 from Whitfield[5]; Type 5 from Kryukov *et al.*[41]; Type 6 from Dobson *et al.*[54]; Type 7 from Dobson *et al.*[57]; Type 8 from Dobson *et al.*[54]; Type 9 from Dobson.[55]

abrupt changes in $E°$ due to changes with temperature and hydration, as with the antimony oxide electrode[53]; solubility of electrode materials, causing changes in activity and creation of liquid junction potentials[54]; modifications of structure due to temperature, for example with silver–silver iodide electrodes[30]; spontaneous composition changes as by loss of hydrogen in the palladium hydride electrode[60]; effects due to gases, for example oxygen and hydrogen, on the silver–silver chloride electrode[33]; and chemical effects and changes due to attack of electrolyte on the electrode, for example the silver oxide electrode.[61] Generally, little can be done with these types of problems other than choosing a completely different electrode system, which may not always be possible.

CELL EMF'S, DEPENDENCE ON TIME, TEMPERATURE, ELECTROLYTE CONCENTRATION, AND DERIVATION OF STANDARD ELECTRODE POTENTIALS

Cell EMF—Time Dependence

This important characteristic of emf measurements is not often commented on in the various published papers dealing with experimental results. Usually, if referred to at all, statements are made to the effect that this or that cell was constant either over the time measurements reported or at some other temperature.

The manner in which a cell reaches its various equilibriums as a function of time, whether they be thermal, chemical, or solubility equilibriums, are decisive in understanding any anomalous behavior. Alternatively, a record, even by graphical means, particularly before or after the period of interest, of the emf as a function of time, is valid support for the various degrees of significance which a particular type of measurement is subsequently considered to have.

The importance of detailed observation of cell emf's with time has been shown from a study of some earlier work of Lietzke et al.[24] in connection with dissolution of cell container materials. A consideration of time and cell emf, because of other effects, allowed Dobson and Thirsk[53] to point out the advantages of using cell containers of small electrolyte volume.

Workers in the Lietzke school[23-28] (USA) and Isaki and Arai[62] (Japan) did not detect any drift in emf with time (Fig. 2) when employing cells using silver–silver chloride, and platinum–hydrogen electrodes, in cells with relatively large volumes of electrolyte. In the work of Dobson and Thirsk[53] (UK) considerable drifts were noted (Fig. 3) from cells using the same electrodes but electrolyte volumes probably at least 10 times smaller. Cells of small electrolyte volume magnify any change (e.g. concentration) and should be avoided unless there is a particular interest in these changes. For example, the behavior of the emf–time curves shown in Fig. 3 may be correlated to effects of hydrogen on the silver–silver chloride.[22]

The emf–time observations may be put to a different use. For example, the constancy of emf with time graphs, used in the work of Dobson et al.[54,63]

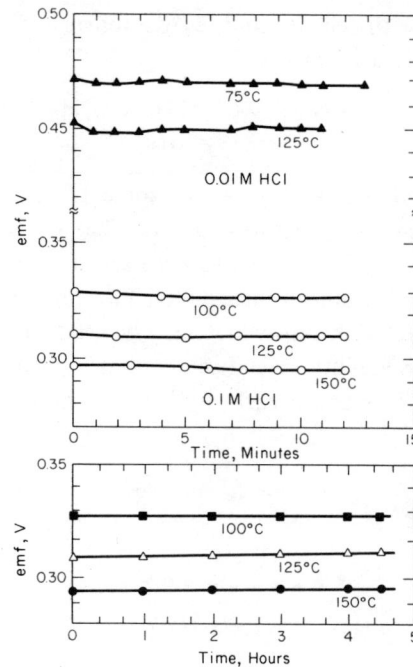

Fig. 2. EMF time curves of the couple of hydrogen electrode and Ag–AgCl electrode in 0.1 M HCl solution under 1 atm H_2. (From Isaki and Arai.[62])

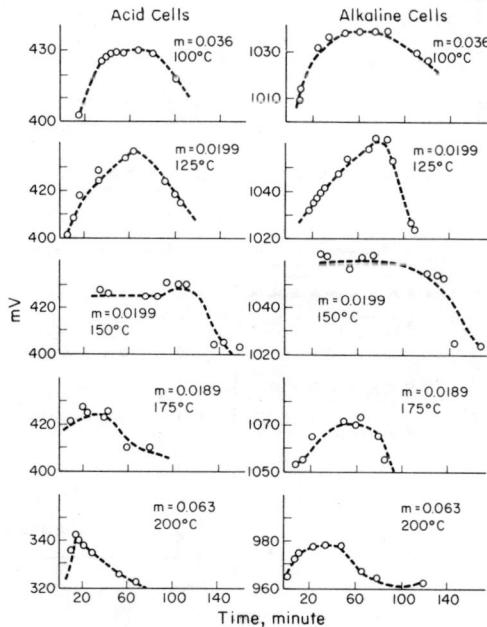

Fig. 3. Cell emf time and temperature relationships. (From Dobson and Thirsk.[53])

with cells (1) and (2), demonstrated stability and precision if their cells and measurements.

$$Ag, AgCl \,|\, XCl_{(m)} \,|\, Hg_2Cl_2, Hg \qquad Hg, Hg_2SO_4 \,|\, X_2SO_{4(m)} \,|\, Ag_2SO_4, Ag$$

$$(1) \qquad\qquad\qquad\qquad (2)$$

Shown in Fig. 4 are some plots of cell emf against time for both cells at temperatures up to 200° C. The demonstration of high stability, ± 0.05–0.1 mV, was necessary for the subsequent treatments of these precise data later in their paper. Two final examples now follow of cell emf and time relationships which may be used, but this time, employed as indicators of the reliability or endurance that a particular system may have. Fournie et al.[40] (France), as early as 1960, attempted to improve the performance and extend

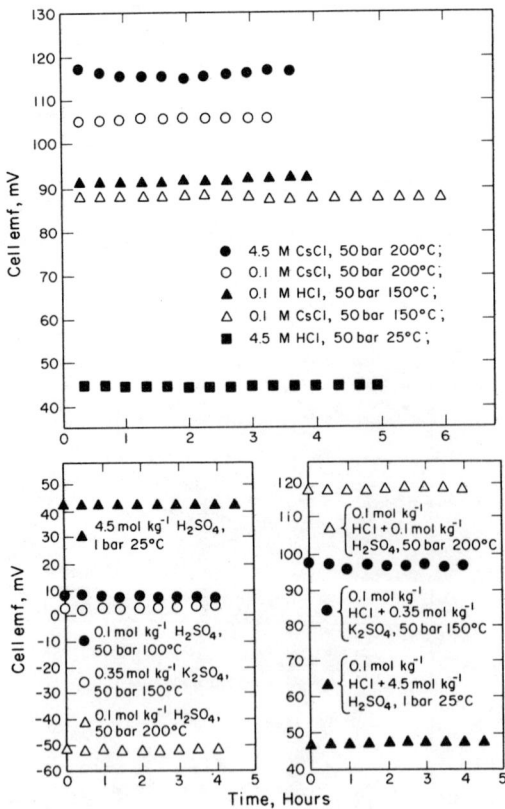

Fig. 4. Stability of cell emf with time. (From Dobson et al.[54,63])

the temperature range of glass electrodes. They chose a "Meci point noir" glass and filled the inside of the bulb of the electrode with lead amalgam instead of aqueous electrolyte. Figure 5 shows cell emf with time data using these electrodes in cells up to 150°C. In Fig. 6 similar types of plots are shown and were produced using a new type of pH/solid state electrode in an aqueous electrolyte cell system up to 200°C created and measured recently by Dobson and Firman.[64,65] In both of these examples the gradual appearance of an almost invariant cell emf with time demonstrates the aging processes, among others, involved with the electrodes at these elevated temperatures.

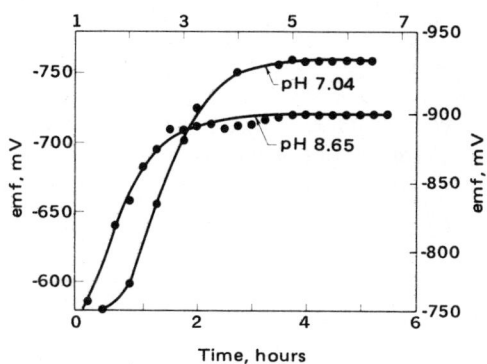

Fig. 5. Stabilization of glass electrodes with inner filling of lead amalgam at 150°C with time. (From Fournie *et al.*[40])

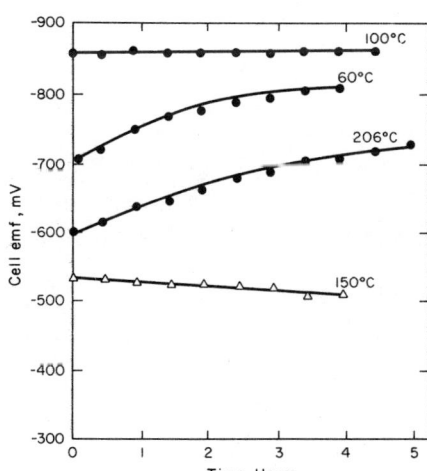

Fig. 6. Stability of cell emf with time at 50 bars between 60 and 206°C pH (solid state) (0.01$_m$ HCl/AgCl, Ag). ●, electrode 2; △, electrode 3. (From Dobson and Firman.[64,65])

Cell EMF—Temperature

In the second half of the 1960s, Körtum and Hausserman[30] (Germany) measured the emf of cell (3):

$$Pt, H_2 \mid HI_{(m)} \mid AgI, Ag$$

(3)

up to 200°C and derived standard electrode potential data for the silver–silver iodide electrode and activity data on hydroiodic acid (HI). Isaki and Arai,[62] using the same cell as Lietzke,[24,25] cell (4),

$$Pt, H_2 \mid HCl_{(m)} \mid AgCl, Ag$$

(4)

have also made measurements up to 200°C, but not as thoroughly as the earlier work of Lietzke. There is also some confusion in the Isaki paper with sign conventions, but he derives sensible $E°$'s for the silver–silver chloride electrode.

Lepeintre *et al.*[66] (France) have described in a patent what happens when cell (5),

$$Ag, AgCl \mid HCl_{(0.1m)} \mid K_2SO_4 \mid H_2SO_{4(0.005m)} \mid PbSO_4, Pb$$

(5)

is taken up to 250°C.

A number of cells involving oxides of platinum and other transitional metals, calomel, and mercurous sulfate electrodes up to 140°C are reported in a patent by Every and Banks[12] (USA). There is, of course, considerable cell emf and temperature data contained in the papers produced by Lietzke *et al.* during 1965–1971 from cells such as cell (6),

$$Pt, H_2 \begin{vmatrix} XCl_{(aq)} \\ XCl_{(aq)} \end{vmatrix} AgCl, Ag$$

(6)

where X may be an alkali metal or earth or even metal chloride, involving the production of activity data up to 275°C. These papers are referred to in detail in a later section of this chapter. Finally, there is the cell-temperature data produced in the papers of Dobson et al.[53,54,63] and associated with cells (**7a**) and (**7b**),

$$\text{Pt, H}_2 \left| \begin{array}{c} \text{NaCl} \\ \text{HCl or NaOH} \end{array} \right| \text{AgCl, Ag}$$
$$\quad\quad\text{(7a)}\quad\quad\text{(7b)}$$

and cells (**1**) and (**2**) up to 200°C.

An interesting comparison may be seen in Fig. 7, where a selection of the data produced by the workers mentioned in this section are plotted. Cell emf's taken at comparable ionic strengths of $0.1–0.2I$ are represented for clarity by curves that best-fit the data. Applying Stockholm sign conventions to the data, it is seen, with one exception, that the emf decreases with increasing temperature, the exception being cell (**1**), which uses calomel and silver–silver chloride electrodes, the variation being due to a less compensating or complementary effect of standard electrode potential temperature coefficients.

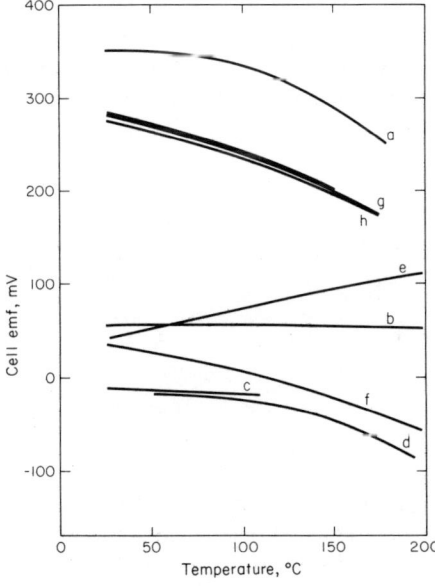

Fig. 7. Emf of various cells between 25 and 200°C: (a) Pt, H_2|HCl|AgCl, Ag (Isaki and Arai[62]); (b) Ag, AgCl|HCl|K_2SO_4|H_2SO_4 (0.05 m)|$PbSO_4$, Pb (Lepeintre et al.[66]); (c) Pt, PtO_2|NaOH|Hg_2Cl_2, Hg (Every and Banks[12]); (d) Pt, H_2|HI|AgI, Ag (Körtum and Hausserman[30]); (e) Ag, AgCl|XCl|Hg_2Cl_2 (Dobson et al.[54]); (f) Hg, Hg_2SO_4|X_2SO_4|Ag_2SO_4, Ag (Dobson and Firman[63]); (g, h, i) Pt, H_2|XCl|AgCl, Ag (Lietzke et al.[123,125,127]).

Cell EMF—Concentration and Nature of Electrolyte

A number of cells, for example cells (**1**) and (**2**), are, according to the formal cell reaction, independent of concentration of the electrolyte and nature of the cation. In recent papers Dobson *et al.*[54,63] have demonstrated that the observed emf of such cells depends considerably on concentration and type of cation. It will be shown later that pronounced differences are also observed for the effects of pressure on these cells, depending again on cation or concentration, at any particular temperature. Some examples of the cell emf dependence on concentration and cation are shown in Figs. 4, 8, and 9. Also plotted on Fig. 9 are some data from Lietzke and Vaughen's[21] earlier work on cell (**1**) and a calculated line derived from thermochemical data.[67]

Shown in Fig. 10 are data for cell (**2**) that are also compared with Lietzke's data and a calculated line. Lietzke and Stoughton in their papers attributed the deviations from the calculated line, for certain concentrations, to hydrolysis of calomel. This idea is based mainly on very old chemical data[68] and a little additional experimental evidence. Apart from some comments concerning mixed potentials, no clear reason was given to explain the deviations. Moreover, Dobson *et al.* were not able to confirm some of Lietzke's original findings concerning concentration, for example, that the cell fails at greater than 1 *M* KCl.

Undoubtedly, some hydrolysis of calomel may occur, particularly when the complex equilibria for calomel in terms of disproportionation are considered. A quantitative attempt is made to explain both concentration and cation effects in Dobson and Firman's paper,[63] in terms of enhanced solubility

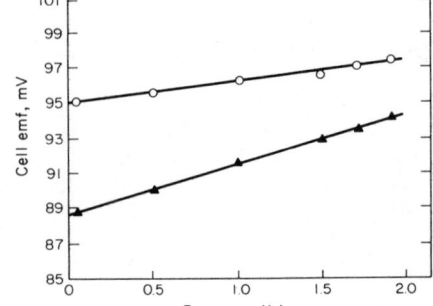

Fig. 8. Dependence of pressure and type of electrolyte of cell emf at 150°C. ○, Cell with 4.5 *M* CsCl: ▲, cell with 4.5 *M* KCl. (From Dobson *et al.*[54,63])

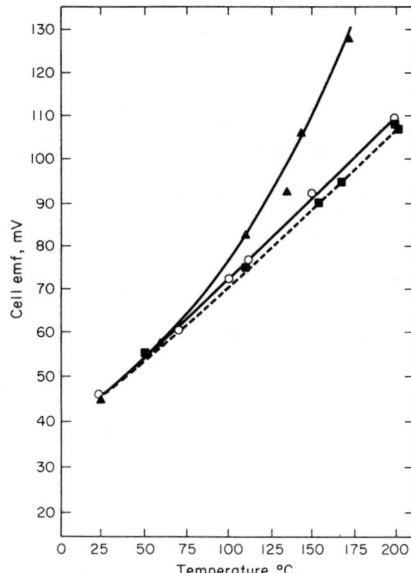

Fig. 9. Cell (1) emf as a function of temperature: ▲, Lietzke and Vaughen,[21] SVP 0.1 *M* HCl; ■, Lietzke and Vaughen,[21] SVP 1.0 *M* HCl; ○, Dobson *et al.*,[54,63] 50 bar 0.1 *M* HCl; –––, Kelly,[67] calculated (thermochemical data).

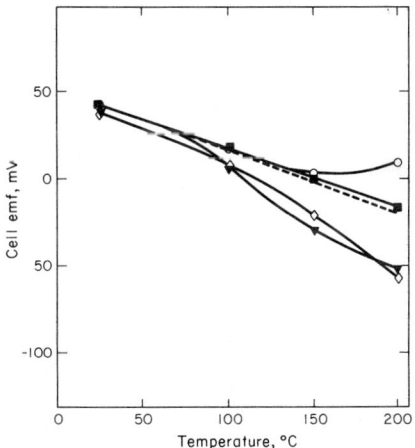

Fig. 10. Comparison of emf's of cell (2) as a function of temperature: ○, Dobson *et al.*,[54,63] 0.35 mol kg^{-1} K$_2$SO$_4$, 50 bar; ■, Dobson *et al.*,[54,63] 4.50 mol kg^{-1} H$_2$SO$_4$, 50 bar; ◇, Dobson *et al.*,[54,63] 0.10 mol kg^{-1} H$_2$SO$_4$, 50 bar; ▼, Lietzke and Stoughton,[19] 0.20 mol kg^{-1} H$_2$SO$_4$, S.V.P.; –––, Kelly,[67] calculated (thermodynamical data).

of calomel, for example, in concentrated solutions of the various alkaline metal chlorides. This enhanced solubility is related in their paper to concentration changes within the cell and a calculation of liquid junction potentials is made. As can be seen from plots of observed–calculated cell emf in Fig. 11, the apparent failure to account for the behavior of cells above 150°C is regarded by

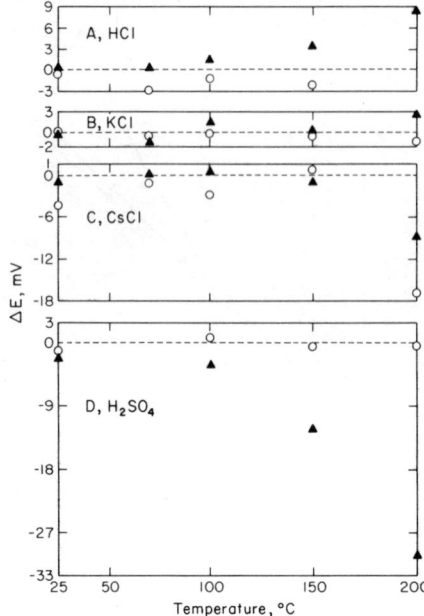

Fig. 11. Comparison of observed and calculated emf's of cells (1) and (2) ($\triangle E$), using HCl, KCl, CsCl (4), and H_2SO_4 between 25 and 200 C. ▲, $0.1\,mol\,kg^{-1}$; ○, $4.5\,mol\,kg^{-1}$. (From Dobson and Firman.[63])

the authors as being due to the lack of additional data and not to the original premise used in the formula of the equation, the equations involved being

$$E = E° = E_{x^+}\frac{RT}{F}\ln\frac{m''_{XCl}\,\gamma''_{\pm\,XCl}}{m'_{XCl}\,\gamma_{\pm\,XCl}}$$

$$-\sum\frac{E_{Ag(c)}}{2_{Ag(c)}}\frac{RT}{F}\ln\frac{m''_{Ag(c)}}{m'_{Ag(c)}}\frac{m''_{Cl^-}}{m'_{Cl^-}}\frac{\gamma''_{Ag(c)}\gamma''_{Cl}}{\gamma'_{Ag(c)}\gamma'_{Cl}}$$

$$-\sum\frac{E_{Hg(c)}}{2_{Hg(c)}}\frac{RT}{F}\frac{m''_{Hg(c)}}{m'_{Hg(c)}}\frac{m''_{Cl}}{m'_{Cl}}\frac{\gamma''_{Hg(c)}\gamma''_{Cl^-}}{\gamma'_{Hg(c)}\gamma'_{Cl^-}}$$

shortened to

$$E = E° = -\frac{RT}{F}E_{X^+}\ln\frac{m''_X + m''_{Cl}\,\gamma''^2_{\pm(XCl)}}{m'_X + m'_{Cl}\,\gamma'^2_{\pm(XCl)}}$$

The double and single primes refer to the right-hand and left-hand electrode compartments, respectively.

DETERMINATIONS OF STANDARD ELECTRODE POTENTIAL FROM EMF MEASUREMENTS AS A FUNCTION OF TEMPERATURE

There has been some discussion in the past[69,53] as to the significance of the values of $E°$. One reason for this doubt is the poor interagreement in $E°$ values at particular temperatures that have been found by various workers.

Recent evidence and controversy on hysteresis,[70,71] phase change,[30,33] and chemical attack[61,72] only support the earlier claims[69] that an experimental determination of the $E°$ of an electrode (or one of a batch of electrodes) that is to be used in a cell for a specific purpose is necessary. This single determination is to be preferred to accepting and using $E°$ values compiled by other workers, who may have used different preparations of the particular electrode. Single determinations of the $E°$ may not, of course, be practical, and the requirements of the precision must also be considered. While precision of $E° = \pm 0.1 \, \text{m V}$ is necessary for determination of a dissociation constant of a weak acid by emf methods, for an electrode kinetic study $E° \pm 10 \, \text{m V}$ would be sufficient in some cases. Moreover, the basis of the method suggested in Ref. 69 for a single determination of $E°$ relies on the availability of precise activity data on the electrolyte containing the ion to which the electrode is reversible.

The usual and detailed methods of the evaluation of $E°$'s from emf data are given in standard thermodynamic textbooks. Basically and generally, the method consists of measurement of cell emf over a range of concentrations of electrolyte whose activities are known, or may be represented by a proven empirical activity expression, and a short graphical extrapolation of equations and data in ionic strength to infinite dilution to give the $E°$ in the slope or intercept.

No detailed discussion is given here of the various studies and methods that have been made recently; however, a short compilation of $E°$ data and their sources is given in Table 2. Included are the experimental equilibrium or rest potentials for a number of common metals taken from the recent work of Wagramjam et al.[73] (USSR) and Indig and Vermilyea[10].

EMF Studies at Elevated Pressures

Only a small number of extensive cell emf and pressure studies were published in the last half of the 1960s. Disteche and Disteche[74] (Belgium) continued from 1965–1968 using cells involving glass electrodes to

Table 2. Recent Determinations of Standard Electrode Potentials and Equilibrium Potentials (Volts) of Some Common Electrodes and Metals at Selected Temperatures[a]

Type of electrode	Temperature (°C)					Source (reference number), remarks
	25	100	150	200	250	
$AgCl$, Ag	0.2223	0.1600	0.1032	0.0348	0.054	24, 62
AgI, Ag	−0.17195	−0.1926	−0.2375	−0.3094	—	30
Hg_2Cl_2, Hg	0.2680	0.2329	0.1959	0.1443	—	54
Hg_2SO_4, Hg	0.6125	0.5514	0.5106	0.4699	—	63
Ni	−0.68	−0.845	−0.875	−0.855	—	73
Co	−0.805	−0.805	−0.920	−0.920	—	Measured
Fe	−0.945	−1.110	−1.1105	—	—	with respect to Hg_2SO_4, Hg at same temperature in $MeSO_4$, pH 1.5

Type of electrode	pH 3, H_2SO_4	pH 3.1, 0.1 M $FeSO_4$	pH 4, H_2SO_4	pH 5.7, neutral	pH 10, NaOH	
$Pt + H_2$	−0.23	−0.28	−0.44	−0.37	−0.68	10
$Pt + O_2$			+0.59			Measured
$Inc + H_2$			−0.445			with respect to
$Inc + O_2$			+0.44			standard
Inc	−0.30					hydrogen
304 SS	−0.34	−0.31	+0.46		−0.70	electrode,
Fe		−0.36		−0.50		all at
Au				−0.41		289°C
Zircaloy 2					−1.0	

[a] Inc, Inconel 600; 304 SS, 304 stainless steel; Pt, platinum black; H_2 pressure, 1 atm at room temperature; oxygen pressure, unspecified.

complement their work on dissociation constants of weak acids at room temperature and several kilobars pressure. Ben Yaakov and Kaplan[75] (USA) reported the use of a glass electrode in combination with thermally coated silver–silver chloride electrodes for studies on seawater pH's at depths up to 280 m. Figure 12 shows a plot of emf data converted to pH as a function of depth and temperature. The differences between measurements made descending and those ascending are not due to the misbehavior of the electrode systems but are associated with true differences in water bodies. In

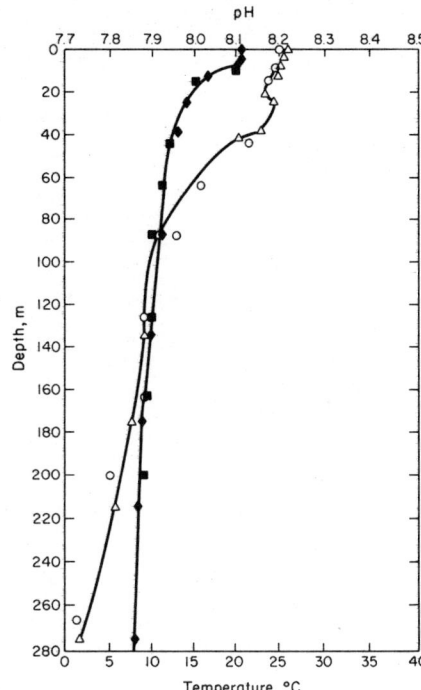

Fig. 12. Temperature and pH versus depth as recorded at Station II above the San Pedro Basin some 8.6 km from shore off southern California. $\bigcirc$, pH up; $\triangle$, pH down; $\blacksquare$, temperature up; $\blacklozenge$, temperature down. (From Ben Yaakov and Kaplan.[75])

1969, Kryukov et al.[76] (USSR) produced a design for a high-pressure cell containing glass electrodes and showed how the emf varied at 25°C with pressure of cell (8), in cell type 5 of Fig. 1 and in

$$\text{glass electrodes} \left|\begin{array}{c} \text{buffers} \\ \text{(a) \quad (b) \quad (c)} \end{array}\right| \text{AgCl, Ag}$$

$$(8)$$

where (a)–(c) are buffers containing (a) 0.01 M sodium tetraborate, (b) 0.025 M potassium dihydrogen and sodium phosphate, and (c) potassium hydrogen phthalate.

Figure 13 shows a plot of the difference $E_1 - E_p$ against pressure, where E_1 is the emf value (in mV) at 1 atm pressure and E_p is the emf value (in mV) at the pressure p (in kg/cm^2). Whitfield in 1969 and 1970[5,77] also used glass electrodes (see Type 4, Fig. 1) in cells, but involving measurements of the effects of membrane geometry on the performance at high pressures. Whitfield found that when electrodes are aged under "symmetrical conditions," flat membranes give more stable and reproducible results than those based on

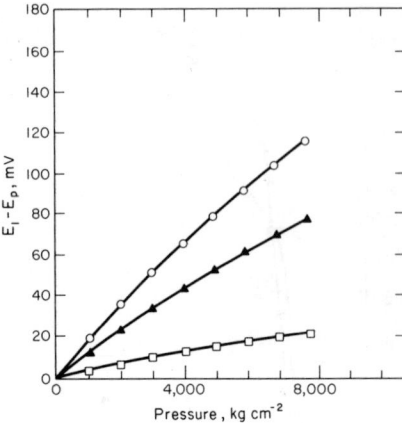

Fig. 13. Effect of pressure on emf at 25°C. O,
Sodium tetraborate (Borax), 0.01 mol kg^{-1};
▲, KH$_2$PO$_4$ + Na$_2$HPO$_4$, equimolal,
0.025 mol kg^{-1}; □, KHC$_8$H$_4$O$_4$ (phthalate),
0.05 mol kg^{-1}. (From Kryukov et al.[76])

conventional spherical geometry. Described in his paper is an electrode which
gives stable asymmetry potentials as low as 30 μV and is only slightly affected
by changes in environmental conditions such as pressure.

A plot of asymmetry potential as a function of pressure, taken from the
paper, is shown in Fig. 14; the letters refer to the various shapes of glass
electrodes; group (1) refers to general shapes, based on bulb design; group (2)
is based on a flat membrane.

Von Heusler and Gaiser[78] (Germany) studied cell (**7a**) as a function of
pressure for the first time at room temperature. From their emf measurements
at pressures up to 2.5 kbars (see Fig. 15 for $E_1 - E_p$ plots), the partial molal

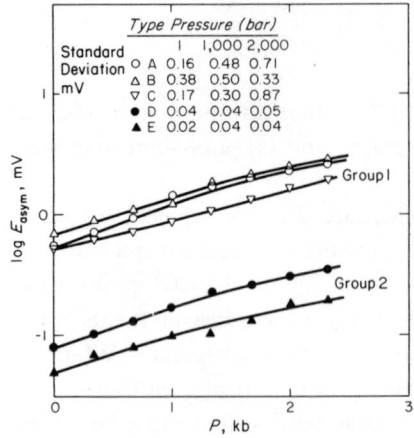

Fig. 14. Asymmetry potentials of glass
electrodes as a function of pressure at 20°C
(mean values plotted). (From Whitfield.[5,77])

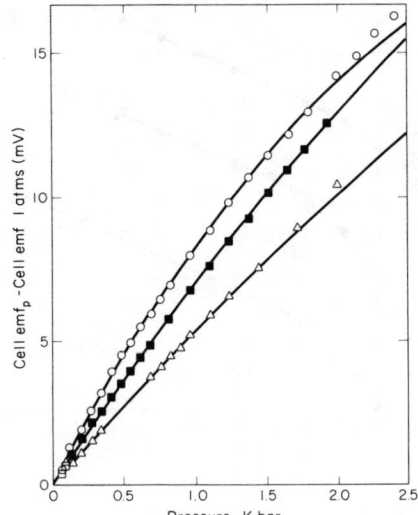

Fig. 15. Emf of cell (**7a**) at 25°C as a function of pressure.   ○, $0.2\,mol\,kg^{-1}$ HCl;   ■, $0.415\,mol\,kg^{-1}$ HCl; △, $0.80\,mol\,kg^{-1}$ HCl. (From Von Heusler and Gaiser.[78])

volumes and compressibilities of hydrogen have been calculated. Using the known partial molal volumes of hydrogen ions in the solution and of the electrons in the metal, the reaction volumes of the hydrogen electrode were estimated.[79] Incorporating overpotential measurements, von Heusler and Gaiser[78] were able to show that the rates of hydrogen evolution on copper, silver, and gold at constant overvoltage $\eta > 0.2$ V increase with pressure.

Shown in Fig. 16 are plots of current density against pressure, for copper, silver, and gold in various electrolytes, with transfer coefficients of $\alpha = 0.51$, 0.63, and 0.45 for copper, silver, and gold, respectively. It was further found that in acid solutions, activation volumes at high pressure were independent of pressure, pH, ionic strength, and electrode material. The increase in rate of hydrogen evolution with pressure at constant potential difference between the electrode and the solution is due to the more negative partial molal volume of the adsorbed, solvated proton in the transition state compared with the solvated proton in the interior of the solution.

Recent work by Dobson et al.[54,63] has shown the effects of pressure and time on cells (**1**) and (**2**); see Type 6, Fig. 1.

Shown in Fig. 17 are typical time transients in cell emf after changes in pressure. This type of behavior has been recently observed on other cells[80] and is due to the localized heating effects due to sudden pressure changes. The little temperature hysteresis that cells (**1**) and (**2**) possess is demonstrated in further

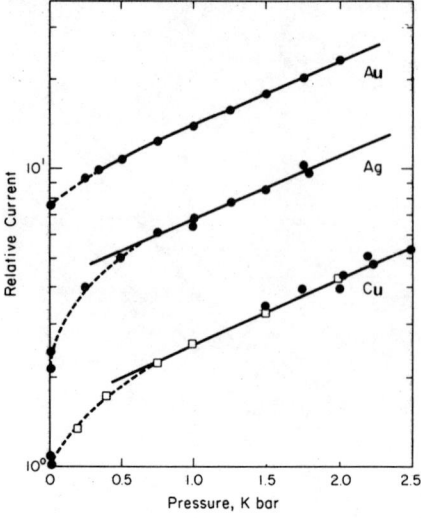

Fig. 16. Pressure dependence of the stationary rate of evolution of hydrogen at a constant overvoltage on copper, silver, and gold. ●, 1 mol kg^{-1} perchloric acid and 1 mol kg^{-1} sodium perchlorate; □, 0.01 mol kg^{-1} perchloric acid and 1.99 mol kg^{-1} sodium perchlorate. (From Von Heusler and Gaiser.[78])

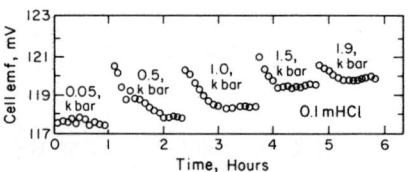

Fig. 17. Effect on cell (1) emf with change in pressure as a function of time. (From Dobson et al.[54,63])

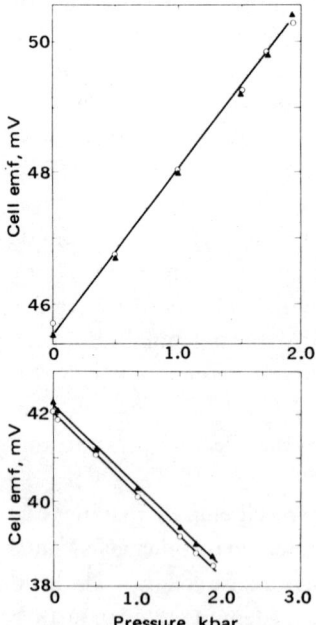

Fig. 18. (Top) Cell (1): Effect of temperature hysteresis on 4.5 M KCl cell. ○, Before temperature excursion to 90°C, measurements 25°C; ▲, after temperature excursion to 90°C, measurements, 25°C. (Bottom) Cell (2): Effect of temperature hysteresis on emf. ○, 4.5 mol kg^{-1} H$_2$SO$_4$ before temperature excursion; ▲, 4.5 mol kg^{-1} H$_2$SO$_4$ after temperature excursion to 200°C (3 days later). (From Dobson et al.[80])

plots of pressure given in Fig. 18 before and after temperature excursions. The difference in sign for dE/dP and very little temperature dependence is shown by

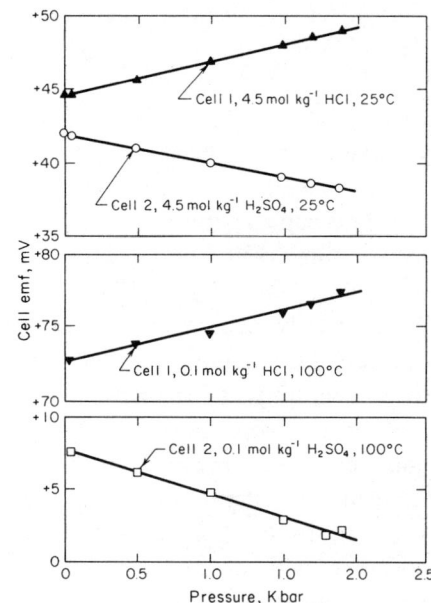

Fig. 19. Comparison of emf's of cells (**1**) and (**2**) as a function of pressure at 25° and 100°C. (From Dobson *et al.*[80])

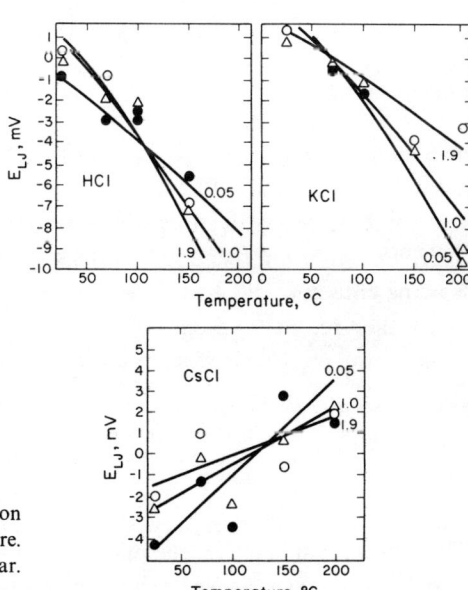

Fig. 20. Variation of apparent liquid junction potential with temperature and pressure. ●, 0.05 kbar; △, 1.00 kbar; ○, 1.90 kbar. (From Dobson *et al.*[80])

the plots for cells (1) and (2) in Fig. 19. Finally shown are the odd, and perhaps the not too convincingly demonstrated, plots of the variation of apparent liquid junction potentials that these cells possess when different cell electrolytes at 4.5 m are compared. In Fig. 20, note the change in slope of lines for CsCl solutions.

Temperature and Pressure Coefficients of Cell EMF

Two papers published in the last decade which were concerned with the temperature coefficients of cell emf's and standard electrode potentials are those of de Bethune and Salvi[81] (USA) and Lietzke et al.,[21,24,31] the former from a purely theoretical basis, the latter from experimental measurements. Considerable amounts of dE/dT, ΔH, and ΔC_p can be made readily available from a scrutiny of the many later papers that have been reviewed in this chapter. The more scarce dE/dP data and their availability, however, are now considered in some detail. In the mid-1960s Hills and Kinnibrugh[79] (UK) devoted their attention to the pressure coefficient of the hydrogen electrode reaction. In the course of their measurements, which were principally concerned with measurements of hydrogen overpotential on mercury surface, the equilibrium emf's of cell (9).

$$H_2, Pt \mid 0.1 \; M \; HCl \parallel M \; KCl \mid Hg_2Cl_2, Hg$$

(9)

as a function of pressure at 25°C were determined and found to be equal to 8.0×10^{-6} V/atm up to 1.5 k bars. The negative volume ΔV of $-3.4 \, cm^3/mole$ was not in accord with either of the two mechanisms usually proposed for the hydrogen evolution reaction.[82] They suggested that the rate-determining step was the emission and hydration of metallic electrons $e_{(metal)} \rightarrow e_{(aq)}$.

Other recent workers concerned with pressure coefficients have been Churagulov et al[83] (USSR), who made measurements on cell (10)

$$Cd \mid CdCl_{2(0.5 \, M)} \mid Hg_2Cl_2, Hg$$

(10)

This cell was studied over the pressure range of 1.5–10 kbars at 20°C, and they found that the increase (ΔE) in emf was a linear function of the square root of the pressure to within ± 0.1 mV. At a pressure of 10 kbars, $\Delta E = 51.65$ mV, which was 6.75% of the emf at atmospheric pressure, or a dE/dP of $+5.16 \times 10^{-6}$ V/atm. In further papers,[84] the same authors measured a similar cell at 25°C,

$$\text{Cd} \left| \begin{array}{c} \text{CdSO}_{4(m)} \\ \text{H}_2\text{O} \end{array} \right| \text{Hg}_2\text{SO}_4, \text{Hg}$$

(11)

again up to 10 kbars and were able to show that at about 6.4 kbars the ΔV was -5.7 cm^3/mole if the cell reaction was considered to involve $\text{CdSO}_4 \cdot (\frac{8}{3})\text{H}_2\text{O}$. Pogorelova[85] (USSR) has recently studied the saturated Clark cell,

$$\text{Zn} \left| \begin{array}{c} \text{ZnSO}_{4(\text{sat})} \\ \text{solid ZnSO}_4 \end{array} \right| \text{Hg}_2\text{SO}_4, \text{Hg}$$

(12)

at 25 and 50°C up to 10 kbars pressure. Again the effect of pressure is to increase the emf. Breaks in pressure isotherms were correlated to hydration transformation $\text{ZnSO}_4 \cdot 7\text{H}_2\text{O} \rightarrow \text{ZnSO}_4 \cdot 6\text{H}_2\text{O}$. The experimental findings were ultimately used to determine the solubility of ZnSO_4 in water at high pressures. A listing of a number of these and other cells giving dE/dP and ΔV's when available for the various reactions involved and their sources is given in Table 3. It is interesting to note the almost complete absence of dE/dP data as a function of temperature. Generally, it is assumed, because of this deficiency, that this coefficient is temperature-independent. This premise, because of Table 3, must only be a poor approximation, and in general the other coefficients, dE/dT, are not temperature-independent[60] when large ranges in temperature, $>100°$C, are considered. The approximation is generally acceptable when used as a correction factor to, say, experimental cell emf data, because of the small value of dE/dP when compared with the magnitude of the cell emf. It is not acceptable, however, to apply the assumed nonvariability of dE/dP with temperature to calculations of ΔV as is sometimes the case, because of the profound effect small variations of dE/dP have on the calculated values of ΔV.

Dissociation Constants and Effects of Pressure and Temperature

The Disteches,[74] continuing their earlier pressure work, reported in 1967 a study on carbonic acid using cells involving glass (Corning 015) electrodes. Some of their results, incidentally, have been subsequently confirmed

Table 3. Selected Values of Cell EMF Pressure Coefficients

Type of cell (number referred to in text)	Cell reaction	dE/dP (V/bar)	$\Delta_3 V$ (cm³/mole)	Reference number
25°C only				
(9)	$H_{2(g)} + Hg_2Cl_{2(s)} \rightarrow 2Hg_{(l)} + 2HCl_{(aq)}$	8×10^{-6} (1.5 kb)[a]	-3.4	79
(10)	$Cd_{(s)} + Hg_2Cl_{2(s)} \rightarrow CdCl_{2(s)} + Hg_{(l)}$	5×10^{-6} (1.0 kb)	—	83
(11)	$Cd_{(s)} + Hg_2SO_{4(s)} + \frac{8}{3}H_2O_{(l)} \rightarrow$	4.1×10^{-6} (1.0 kb)	—	49
	$CdSO_4 + \frac{8}{3}H_2O_{(s)} + 2Hg_{(l)}$	$\Delta G/\Delta P = -3.35$ (1.0 kb)	-5.7 (6.4 kb)	84
(12)	$Zn_{(s)} + Hg_2SO_{4(s)} \xrightarrow{N_2O} ZnSO_4(?)H_2O + 2Hg_{(l)}$	12×10^{-6} (1 kb)		52
		14×10^{-6} (1 kb)	-19.8–14.5	
		5.7×10^{-6} (10 kb)		85
(13)	$H_2CO_{3(aq)} \rightarrow HCO_{3(aq)}^- + H_{(aq)}^+$	$\sim 25 \times 10^{-6}$ (1 kb)	-26.6	74
	$HCO_{3(aq)}^- \rightarrow CO_{3(aq)}^{2-} + H_{(aq)}^+$	13×10^{-6} (1.5 kb)	-25.6	
Tl, Hg/Tl₂SO₄(sat)/Hg₂SO₄, Hg	$2Tl_{(s)} + Hg_2SO_{4(s)} \rightarrow Tl_2SO_{4(s)} + 2Hg_{(l)}$		—	50
Cd/CdSO₄/Cd, Hg₈%	$Cd_{(Hg)}^{2+} + SO_{4(aq)}^{2-} \rightarrow CdSO_{4(s)}$	1.85×10^{-6} (1.5 kb)	—	50
(7a)	$\frac{1}{2}H_{2(g)} + AgCl_{(s)} \rightarrow HCl_{(aq)} + \frac{1}{2}Ag_{(s)}$	7×10^{-6} (1.0 kb)	—	78
25–200°C				
(1)	$Ag_{(s)} + \frac{1}{2}Hg_2Cl_{2(s)} \rightarrow AgCl_{(s)} + Hg_{(l)}$	$2.1_9 \times 10^{-6}$ (1.0 kg) at 25°C	-2.0_9 (1 kb)	54
		2.35×10^{-6} at 100°C	-2.2_3	54
		1.3×10^{-6} at 200°C	-1.0	54
(2)	$Ag_2SO_{4(s)} + Hg_{(l)} \rightarrow Hg_2SO_{4(s)} + Ag_{(s)}$	-1.89×10^{-6} (1 kb) at 25°C	$+3.6$ (1 kb)	63
		$-2.0_3 \times 10^{-6}$ at 100°C	$+3.87$	63
		-1.5×10^{-6} at 200°C	$+2.86$	63

[a]kb, kilobar(s).

by Whitfield[5,77] as a function of pressure at 25°C. In view of the interest in precise knowledge of the effect of pressure on the dissociation of carbonic acid, for oceanographic work, Disteche's paper dealt with the determination, between 1 and 1000 atm, of the dissociation constants K and K_2 and the ionization functions $k_{(1)}$, $k_{(2)}$, and $k_{(3)} = [H^+][HCO_3^-]/[CO_2]$; $k_{(2)} = [H^+][CO_3^{2-}]/[HCO_3^-]$. The electrolyte of the cell also contained NaCl or KCl, over a wide ionic strength range of 0 to 0.8 at buffer ratios in the pH range 5.1–9.6. The effect of Ca^{2+}, Mg^{2+}, SO_4^{2-} ions, and boric acid was also studied in order to interpret the data in terms of natural seawater.

The emf of the Disteche cell [cell (13)]

$$\text{Ag,AgCl}\begin{vmatrix}\text{reference compartment} \\ \text{HCl}_{(0.01)}\text{MCl}_{(m_3-0.01)}\end{vmatrix}\begin{matrix}\text{glass} \\ \text{membrane}\end{matrix}\begin{vmatrix}x\text{ compartment} \\ \text{HR}_{(m_1)}\text{MR}_{(m_2)}\text{MCl}_{(m_2)}\end{vmatrix}\text{AgCl,Ag}$$

$$(13)$$

(where M = Na or K and HR = carbonic acid) at pressure P, with reference to the value at 1 atm, is given by

$$\frac{(E_1 - E_p)F}{2.3RT} = \log\frac{K_P^m}{K^m} + 2\log\frac{\gamma_{A_1}}{\gamma_{A_P}} + 2\log\frac{\gamma_{HCl}^{ref.\,0.1}}{\gamma_{HCl}^{ref.\,P}} - 2\log\frac{\gamma_{HCl}x_1}{\gamma_{HCl}x_P} \qquad (1)$$

when

$$K^m = \log\frac{m_H + m_1}{m_2} + 2\log\gamma_H$$

$$2\log\gamma_A = \log\frac{\gamma_{H^+}\gamma_{R^-}}{\gamma_{HR}}$$

Correction methods for asymmetry potential shifts with pressure were also considered in the paper and relied on earlier studies.[43,45,47] Shown in Fig. 21 (top) are the observed emf shifts produced by pressure as a function of ionic strength in NaCl and KCl, in the presence of Mg^{2+}, Ca^{2+}, and SO_4^{2-}, and in seawater. In Fig. 21 (bottom) the ionization function of the carbonic acid as a function of ionic strength at atmospheric pressure and at 1 kbar in similar solutions is shown. While this type of study is directly applicable to oceanographic study, its overall significance as a contribution to electrochemically derived thermodynamic data is limited because of the complex nature of the solutions. It would have been very instructive if the results were compared with data derived from solutions of a little less complex nature. The increase in emf of Whitfield's similar cell [cell (3)],

$$\text{Ag,AgCl}\begin{vmatrix}\text{HCl}_{(m_1)}\text{KCl}_{(m_2)}\end{vmatrix}\begin{matrix}\text{glass} \\ \text{membrane}\end{matrix}\begin{vmatrix}\text{KCl}_{(m_3)}\text{K}_2\text{CO}_{2(m_4)}\text{KHCO}_{3(m_4)}\end{vmatrix}\text{AgCl,Ag}$$

$$(14)$$

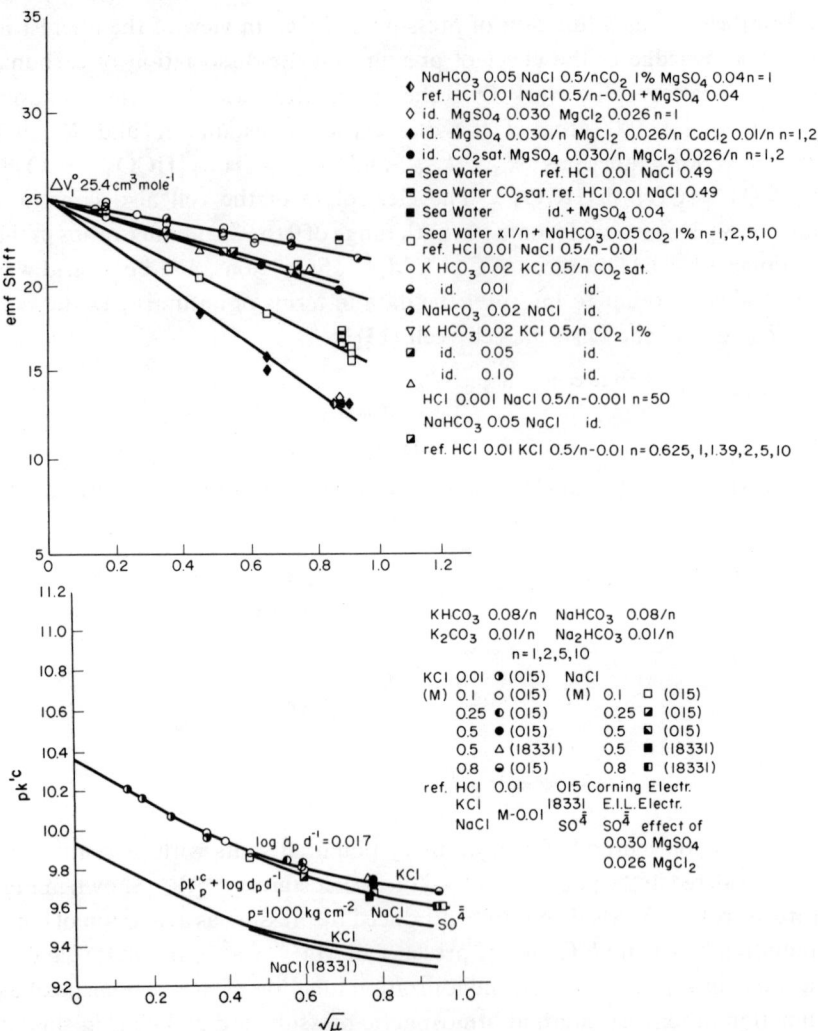

Fig. 21. (Top) Glass electrode emf shifts at 22°C produced by pressure (1000 kg cm^{-2}), in bicarbonate buffers, at different buffer ratios, as a function of $\mu^{1/2}$ in NaCl and KCl in the presence of Mg^{2+}, Ca^{2+}, and SO$_4^{2-}$ ions, and in seawater. (Bottom) Second ionization function (pk'^c) of carbonic acid as a function of $\mu^{1/2}$ at atm pressure and 100 kg cm^{-2} in NaCl and KCl at 22°C. Sulfate effect of 0.030 MgSO$_4$ + 0.026 MgCl$_2$. (From Disteche and Disteche.[74])

as a function of pressure, is shown in Fig. 22 and is in keeping with the findings of Disteche. A seeming hysteresis exhibited by cell (3) was traced according to Whitfield, although no detail was given to the "poor" performance of the

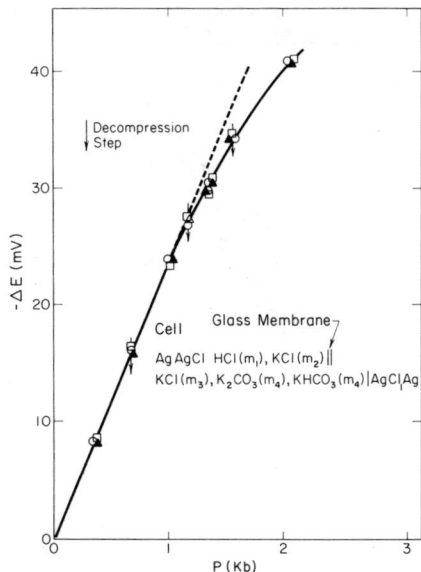

Fig. 22. Reproducibility of measurements on a carbonate/bicarbonate buffer. $\bigcirc$, Cell I; $\blacktriangle$, cell II; $\square$, cell III. $m_1 = 0.01\ M$; $m_2 = 0.76\ M$; $m_3 = 0.77\ M$; $m_4 = 0.01\ M$; $T = 20°C$. (From Whitfield.[5.77])

silver–silver chloride electrode. Cells I and II used calomel instead of silver–silver chloride electrodes.

Determinations of the acidity constants for sulfur dioxide and hydrogen sulfate ion in water solutions at different ionic strengths and at elevated temperatures have recently been carried out by Schöön and Wannholt[1] (Sweden). In this investigation the protolysis equilibria of sulfur dioxide and hydrogen sulfate ion were studied by emf measurements up to 150°C at ionic strengths between 0.25 and 3 M. Their cell comprised glass electrodes, type E1L GBQ28 or GHQ28, and a calomel electrode situated on the end of the salt bridge and mounted outside the pressure vessel. The calomel electrode, kept at 23°C, was maintained by a pressure balance system. They unwisely assumed[54,63,86] that the liquid junction potential and thermal effects between the salt bridge and the equilibrium solution contained in the autoclave were negligible. Schöön and Wannholt also somewhat haphazardly assumed that the activity coefficient for hydrogen ion is independent of low hydrogen concentration even at very high ionic strengths of 3 M. They justified this assumption by quoting some not wholly relevant work of Biedermann and Sillen.[87] The estimation, let alone determination, of dissociation constants of weak acids requires extremely precise and careful handling. It has been shown by a number of authors[53,88] how sensitive and how prone to error dissociation

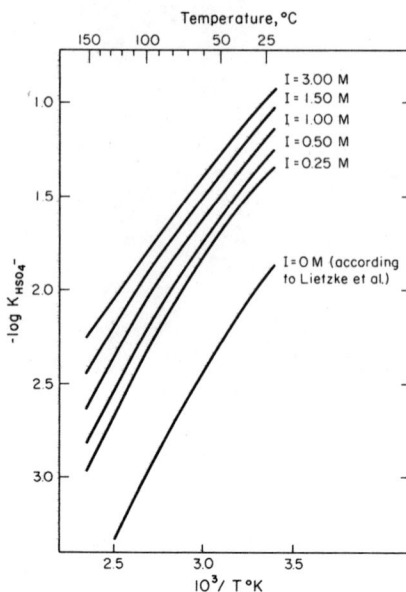

Fig. 23. The stoichiometric acidity constant for hydrogen sulfate ion at different ionic strengths. Ionic medium: $1\,M$ NaCl. (From Schöön and Wannholt[1].)

constants can be. Schöön and Wannholt also did not do justice to their apparently considerable amount of data, by not recording any experimental points or even smoothed data. Only smoothed lines of the final pK's are given, and then without line-fitting equations. If for one reason or another experimental data tables are not given, at least graphical representation of the raw data plus precise methods on how the data were smoothed should be given. Fundamental effects have been overlooked in the past[89] because of their negligence. The plot of $-\log K_{HSO_4}$ against the reciprocal of temperature shown in the Schöön and Wannholt paper and reproduced in Fig. 23 is in good agreement with a line produced by Lietzke and Vaughen[23] from solubility measurements.

A little less crude emf work was published in 1970 by a Polish worker, Letowski,[90] who was concerned with an investigation of nickel ammine equilibria up to 180°C. Following some earlier work[91] he was able to determine the successive stability constants

$$K_n = \frac{[Ni(NH_3)_2^{2+}]}{[Ni(NH_2)_{n-1}^{2+}][NH_3]}$$

using a $Ag/Ag(NH_3)_2^+$, NH_3 electrode in the cell.

$$Ag/Ag(NH_3)_2^+, NH_3 \overset{(1)}{\|} Ni^{2+}, NH_3, Ag(NH_3)_2^+ \overset{(2)}{\bigg|} Ag \text{ in } 2\,m \text{ } NH_4NO_3$$

(15)

The emf of this cell is

$$E = (2RT/F)\ln([NH_3]_1/[NH_3]_2) \tag{2}$$

when nickel ions are absent but changes to

$$E = (2RT/F)\ln C_{NH_3} - \ln[NH_3]$$

when nickel nitrate solutions are added to the cell. C_{NH_3} is the total ammonia concentration in the two half-cells and $[NH_3]$ the concentration of the free ammonia in the amminonickel equilibrium and is derived as a result of the emf measurement. The mean number of ligands ($\bar{n}$) of ammonia, NH_3, bound to a nickel ion is also given as $\bar{n} = C_{NH_3} - [NH_3]/C_{Ni}$. Shown in Table 4 and Fig. 24 are the values of successive stability constants and curves of formation and dependence of $\bar{n}$ on $-\log[NH_3]$ up to 200°C of the nickel ammoniates. These results demonstrate that the value of $-\log K$ decreases with increasing temperature and that the most stable ion complexes between 100 and 200°C are $Ni(NH_3)_2^{2+}$ ions.

Finally, in this section are reviewed the most recent of a number of acidity measurements carried out by Mesmer et al.[11] at Oak Ridge (USA) and are concerned with boric acid and borate equilibria. Their interest in this system stems from a desire to extend the temperature range of pH buffered mixtures and in estimating the pH and other properties of nuclear reactor coolants. They indicate in their paper that results from concentrated borate solutions would also establish more firmly the identity of the polyborate species which are formed in aqueous solutions.

Table 4. Stability Constants of Nickel Ammoniates between 25 and 200°C

t (°C)	$\log K_1$	$\log K_2$	$\log K_3$	$\log K_4$
25	2.61	2.15	2.03	1.56
50	2.31	1.98	1.57	0.56
100	1.67	1.64	(1.08)	—
150	0.97	1.29	—	—
200	(0.24)	(0.93)	—	—

[a] Values in parentheses are extrapolated.

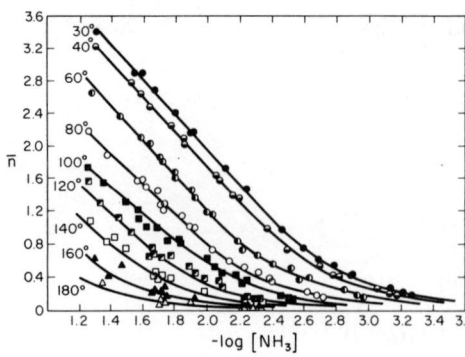

Fig. 24. Curves of formation of the nickel ammoniates in 2 moles/kg between 30 and 180°C: ●, 30°C; ◖, 40°C; ◑, 60°C; ○, 80°C; ■, 100°C; ◪, 120°C; □, 140°C; ▲, 160°C; △, 180°C. Plotted as the mean number of ligands bound to nickel ion as function of concentration of NH_3. (From Letowski.[90])

The emf measurements which Mesmer *et al.* carried out were on the concentration cell [cell (**16**)]

$$\text{H}_2,\text{Pt} \left| \begin{array}{c} a_m\,\text{KCl} \\ \sim 0.02a_m\,\text{B(OH)}_3 \\ \sim 0.01a_m\,\text{KOH} \end{array} \right| \left| \begin{array}{c} \text{Reference} \\ a_m\,\text{KCl} \\ 0.02a_m\,\text{KOH} \end{array} \right| \text{Pt,H}_2$$

(16)

over the temperature range 50–290°C. The equilibria in dilute boric acid solution

$$\text{B(OH)}_3 + \text{OH}^- = \text{B(OH)}_4 \tag{3}$$

was studied as a function of KCl concentration, by allowing a to vary from 0.13 to 1 m. The emf of this cell is given by

$$E = (RT/F)\ln\left([\text{OH}^-]/[\text{OH}^-]_{\text{ref}}\right) - D_{\text{OH}}([\text{OH}^-]_{\text{ref}} - [\text{OH}^-])$$

$$- \sum D_i([i]_{\text{ref}} - [i]) \tag{4}$$

where $[\text{OH}^-]$ and $[i]$ denote, respectively, the concentration of hydroxide ion and of each other's ionic species in solution. The liquid junction potential is given by the terms D_{OH} and D_i, which were calculated from the Henderson equation.[92] The applicability of the Henderson equation, which is a somewhat restricted representation of effects at liquid junctions, especially at high temperatures, is debatable. No indication of the values calculated and attributed to liquid junction potentials are given in Mesmer's paper, unfortunately. Criticism, similar to that given to the Schöön and Wannholt paper concerning data documentation, may also be applied to the work now under discussion, but perhaps not so vigorously.

Mesmer's paper is very long and has an in-depth discussion which is much too involved for detailed consideration in this chapter. However, the arguments and treatments are generally made with clarity. Figure 25 shows $\log Q_{11}$ against ionic strength at various temperatures, and is defined as

$$\log Q_{11} = \log \frac{\bar{n}}{(1 - \bar{n})\,[OH^-]} \tag{5}$$

where

$$\bar{n} = \frac{[H^+] + m_{OH} - [OH^-]}{m_B}$$

m_{OH} and m_B being stoichiometric concentrations of base and boron in solution. The evaluation of the $[OH^-]$ term is derived from emf measurements by iteration solution of Eq. (3) and H^+ from the "dissociation" quotient for water from earlier work of Mesmer et al.[93] The graph in Fig. 25 shows that the equilibrium quotient for Eq. (3) has a small dependence on ionic strength (cf. emf cells determining K_w using K and Na salts as electrolytes), which decreases as the temperature increases. As for the polyborate species, resulting from their

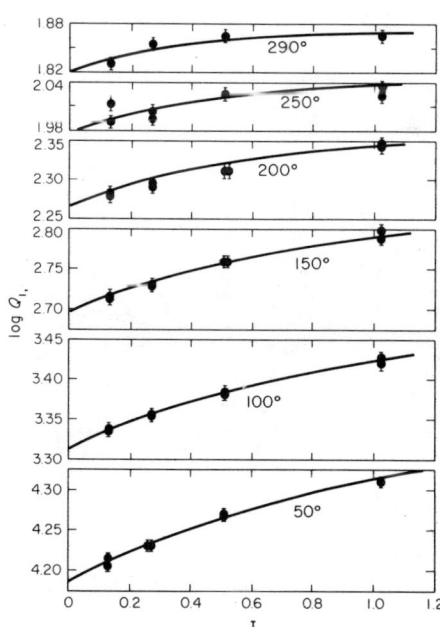

Fig. 25. $\log Q_{1,1}$ as a function of ionic strength. (From Mesmer et al.[11])

analysis of data carried out in concentrated solutions, the various equilibrium quotients Q_{xy} for

$$xB(OH)_3 + y(OH) \rightleftharpoons B_x(OH)_{3x+y}^{y-} \tag{6}$$

furnish the distribution of various species in solution shown in Fig. 26. Mesmer *et al.* point out that existence for these species is clearly evident from Fig. 27 because of the dependence of $\bar{n}$ on log [OH⁻] above about 0.03 m at each of the three temperatures shown. They also cite the work of Ingri *et al.*[94] to support their statements. A minor criticism may be made, however, of the use of relatively simple empirical equations to relate Q with temperature.

Determination of Ionization Constant of Water from EMF Measurements

Interest in the ionization constant of water (K_w) under varying conditions continues to be maintained, presumably because of its involvement in many biological and chemical processes. A number of non-emf studies, mainly as a result of conductivity work, reviews, and recalculations of earlier studies, have appeared in the last decade, notably Hamann *et al.*[95] (Australia), Franck and Holzapfel[96] (Germany), Quist[97] (USA), Bignold *et al.*[98] (UK), Fisher[99] (USA), Ahluwolia and Cobble[100] (USA), Clever[101] (USA) and Perkovets and Kryukov[102] (USSR), published K_w data, late in 1969, between 25 and 150°C

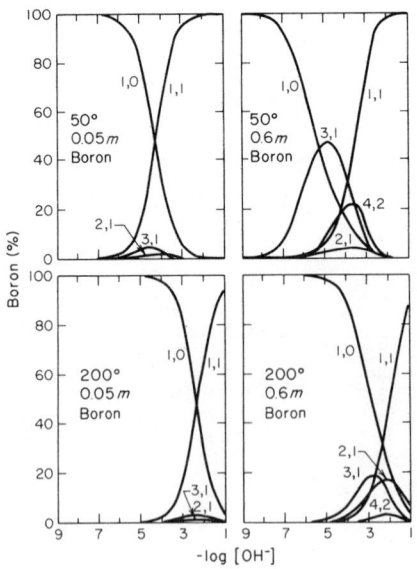

Fig. 26. Distribution of species calculated for solutions containing 0.05 and 0.60 m boron at 50 and 200°C. The species are represented by the notation (x, v) for the formula $B_x(OH)$ $3xtv^{v-}$. (From Mesmer *et al.*[11])

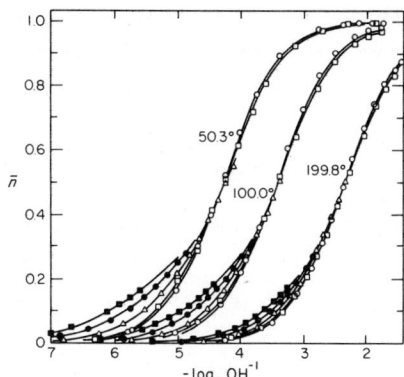

Fig. 27. The effect of boron concentration and temperature on the hydrolysis of boric acid in $1\,M$ KCl. ■, $0.6\,m$; ●, $0.4\,m$; △, $0.2\,m$; □, $0.1\,m$; ○, $0.3\,m$. (From Mesmer et al.[11])

using the Harned-type cell,

$$\text{Pt}, \text{H}_2 \,|\, \text{NaOH NaCl} \,|\, \text{AgCl}, \text{Ag}$$

$$\textbf{(7b)}$$

and the $E°$'s for AgCl, Ag of Lietzke. Dobson and Thirsk[53] (UK) submitted a paper in September of 1969 (but not published until 1971) on K_w to 200°C, using cells (7a) and (7b), discussed in another section of the present chapter (see also Type 2, Fig. 1):

$$\text{Pt}, \text{H}_2 \,|\, \text{HCl NaCl} \,|\, \text{AgCl}, \text{Ag}$$

$$\textbf{(7a)}$$

The work of Dobson et al.[53] removed the objection of using additional standard electrode potential data, because cells (7a) and (7b) were measured simultaneously. Moreover, for the same reason, no correction for partial pressures of hydrogen were necessary, but they were necessary for the work of Perkovets and Kryukov[56,102]; both studies were at SVP. The emf for either cell is

$$E = E°_{\text{AgCl,Ag}} - \frac{RT}{F} \ln a_{\text{HCl}} \tag{7}$$

For cell (7b) the equation may be expanded to

$$E' = E + \frac{RT}{F} \ln \frac{m_{\text{Cl}}}{m_{\text{OH}}} = \left(E° - \frac{RT}{F} \ln K_w \right) - \frac{RT}{F} \frac{\gamma_{\text{Cl}} a_{\text{H}_2\text{O}}}{\gamma_{\text{OH}}} \tag{8}$$

Perkovets used a one-term Debye–Hückel activity expression and used E' ionic strength plots to find pK_w. For cells **(7a)** and **(7b)** combined, Dobson *et al.*, on the other hand, gave

$$\frac{F\Delta E}{RT} = -\ln K_w + \ln a_{H_2O} + \frac{\ln a^2 m}{m - a} + \ln\frac{\gamma_{H}\gamma_{OH}\gamma_{Cl}''}{\gamma_{Cl'}} \tag{9}$$

where the primes on activity coefficients refer to alkaline and acid cells, respectively; a is the molality of either hydroxide or acid and m is the ionic strength. More "correct", Guggenheim[103,104] activity expressions were employed; thus,

$$-\log K_{w''} = \frac{F\Delta E}{2.3RT} - \log\frac{a^2 m}{m - a} + \frac{2Am^{1/2}}{(1 + m)^{1/2}} \tag{10}$$

and also equals

$$-\log K_{w''} = -\log K_w + 0.016\,m + a(B_{HCl} - B_{NaCl})$$
$$+ m(B_{HCl} + B_{NaOH}) \tag{11}$$

The $-\log K_w''$ versus ionic strength plots gave straight lines with intercepts of $\log K_w + a(B_{HCl} - B_{NaCl})$ and a slope of $0.016 + B_{HCl} + B_{NaOH}$. The value of $a(B_{HCl} - B_{NaCl})$ was put for various reasons[53] to equal about 0.003.

Values of pK_w at selected temperatures for Perkovets and Dobson's work, including values from other sources, are shown in Fig. 28. The paper by

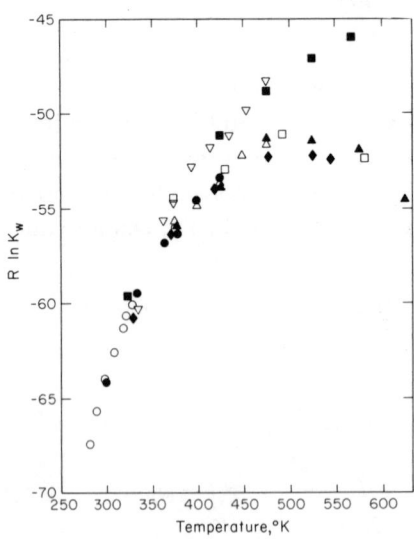

Fig. 28. $R\ln K_w$ versus temperature. $\triangle$, Dobson *et al.*[53]; ∇, Dobson *et al.*[80]; $\bigcirc$, Harned *et al.*[89]; $\square$, Noyes[132]; $\bullet$, Perkovets and Kryukov[102]; $\blacksquare$, Mesmer *et al.*[105]; and $\blacklozenge$, Bignold *et al.*[98]

Dobson and Thirsk[53] goes into depth on the subsequent fitting of K_w data to empirical, or other, expressions in temperature. The problems of extrapolating data to the $K_{w,max}$ temperature range, around 230–250°C, by any empirical equation in temperature were studied closely. One of the quantitative conclusions of the paper was that the most reliable or statistically correct type of equation was one which furnished the observed $K_{w,max}$ temperature. At that time, because reliable and precise experimental data available did not cover a sufficiently high range in temperature, no further progress was made. It is therefore interesting now to compare the values of $K_{w,max}$ temperatures that are produced when the raw data referred to in Refs. 98 and 99 are analyzed. Simple plotting of data would not be sufficiently accurate[53] because of scatter in the data. Therefore, the data are best analyzed by fitting to the most reliable empirical equations in T, those of Clarke and Glew.[88] This was achieved by a least-squares computer program and evaluation of coefficients in the Clarke and Glew equation, whose first four terms are

$$R \ln K_w = -\frac{\Delta g_\Theta^\circ}{\Theta} + \Delta H_\Theta^\circ \left(\frac{1}{\Theta} - \frac{1}{T} \right) + \Delta C_{p\Theta} \left(\frac{\Theta}{T} - 1 + \ln \frac{T}{\Theta} \right)$$
$$+ \frac{\Theta}{2} \left(\frac{d\Delta C_{p\Theta}}{dT} \right)_\Theta \left(\frac{T}{\Theta} - \frac{\Theta}{T} - 2\ln \frac{T}{\Theta} \right) + \cdots \quad (12)$$

and subsequent terms include functions of $(d^2\Delta C_p/dT^2)_\Theta$, $(d^3\Delta C_p/dT^3)_\Theta$. Θ is a chosen reference temperature, which is usually 298°K. The other terms have the usual significance and represent thermodynamic changes during the ionization of water.

In Table 5 are given the values of $K_{w,max}$ temperature produced by fitting two different sets of data, those of Fisher[99] and Bignold et al.,[98] successively to

Table 5. Predicted $K_{w,max}$ Temperature in °C

Data set	Parameter	Number of constants			
		2	3	4	5
Fisher	Variance δ^2 for $R \ln K_w$(observed)	13.8	1.47	3.1×10^{-1}	2.9×10^{-1}
	$K_{w,max}$ temperature	a	223	244	280
Bignold	Variance δ^2 for $R \ln K_w$(observed)	4.75	1.1×10^{-1}	7.7×10^{-2}	1.7×10^{-2}
	$K_{w,max}$ temperature	a	245	237	228

a No max given between 100 and 350°C.

Clarke and Glew equations with two, three, four, and five constants. The dissimilarity between the two sets of results as a whole show that the problem of deciding what is the correct value of $K_{w,max}$ temperature is still unfortunately unanswered. The relatively lower variance[53,54] δ^2 for the fits for the Bignold data when compared with the work of Fisher, perhaps,[53] although not necessarily, indicates that the former data are more precise. However, if this is so, which value does one choose—245, 237, or 228°C?

Mesmer et al.[105] (Oak Ridge, USA) submitted a paper in October of 1969 (published April 1970) concerned with the "apparent" dissociation quotient of water (Q'_w). These workers made emf measurements on the concentration cell (17) up to 292°C,

$$\text{Pt, H}_2 \left| \underset{\text{HCl}(0.00917\,m),\,\text{KCl}(1.0\,m)}{1} \right\| \underset{\text{KOH}(0.008245\,m),\,\text{KCl}(1.0\,m)}{2} \left| \text{H}_2, \text{Pt} \right.$$

(17)

whose cell emf is

$$E = \frac{RT}{F} \ln \frac{m_{\text{H}_1} m_{\text{OH}_2}}{Q'_w} - \sum D_i(m_{i_1} - m_{i_2})$$ (13)

Q'_w is related to K_w by

$$K_w = Q'_w \frac{\gamma_{\text{H}} \gamma_{\text{OH}}}{a_{\text{H}_2\text{O}}}$$ (14)

or

$$\ln Q'_w = (E_1 - E_2)\frac{F}{RT} + \ln m_{\text{H}_1} m_{\text{OH}_2}$$ (15)

The D_i terms are liquid junction potentials derived from the Henderson equation and conductivity data. Although the validity of use of the Henderson equation is in some doubt,[106] the low concentration of ions make any contributions or errors negligible. The values of Q'_w, however, will vary on the ionic strength. Dobson et al.[80] have recently made some measurements on cell (18) as a function of temperature and pressure, up to 250°C and 1.8 kbars:

$$\text{Pd–H}\,(\alpha + \beta)\,|\,\text{NaOH}_{(a)}\text{NaCl}_{(m)}\,|\,\text{AgCl, Ag}$$

(18)

This cell is identical to cell (7b), but here the $E°$ expression also includes the $E°$ for Pd–H $(\alpha + \beta)$, derived from another paper given at the HTHPE:

$E^{\circ\prime} = E^{\circ}_{AgCl,Ag} - E^{\circ}_{Pd-H(\alpha+\beta)}$. Thus, if the values of $0.01\,m$ NaOH and $0.01\,m$ NaCl are used, the relationship given in Ref. 80 rearranges to

$$-\ln K'_w = 2.303 pK'_w \approx 2.303 pK_w \qquad (16)$$

$$= \frac{F(E - E')}{RT} \ln \frac{m_{Cl}}{m_{OH}} \qquad (17)$$

so it would be expected that pK'_w will also vary with ionic strength. However, it is known that the value of pK'_w varies only slightly with ionic strength both for sodium- or potassium-salt-bearing electrolyte cells.[103,104] For example, at $25°C$, values of pK'_w for sodium salts are 14.000 and 13.999 at $0.01\,m$ and $0.1\,m$, respectively; the corresponding values for potassium salts are 14.001 and 14.05.

Similar concentrations of chloride and hydroxide were used in the cell of Mesmer *et al.* but not, of course, with respect to proton concentration or total ionic strength. In compartment (1) of Mesmer's cell, $m_H = m_{Cl}$; a similar equation occurs in the cell of Dobson. Thus, if the two sets of data are consistent with one another, plots of $-R \ln Q'_w$ and $-R \ln K'_w$ with temperature should have similar characteristics. Thus, in Fig. 28 the similarity is shown to be remarkable. This agreement also substantiates the values of $E^{\circ}_{Pd-H(\alpha+\beta)}$ which have been recently derived.[80]

Finally, in this section it is interesting to look at the effects of pressure on K_w from the recent measurements of Whitfield.[5] Whitfield employed the cell

$$Ag,AgCl \left| HCl_{(m_1)} NaCl_{(m_2)} \right| \begin{array}{c} glass \\ membrane \end{array} \left| NaOH_{(m_1)} NaCl_{(m_2)} \right| AgCl, Ag$$

$$(19)$$

whose emf and relationship to K_w is

$$\frac{FE}{RT} = -\ln K_w + \ln \gamma_{H_2O} + \ln M + \ln \frac{\gamma'_H \gamma'_{Cl}}{\gamma_H \gamma_{Cl}} \qquad (18)$$

where the primes indicate NaCl/HCl mixtures.

$$\ln \gamma_{H_2O} = \ln \frac{\gamma_{OH} \gamma_H}{a_{H_2O}} \qquad \text{and} \qquad \ln H = \ln \frac{m'_H m_{OH} m'_{Cl}}{m_{Cl}}$$

Whitfield only made measurements at 0.07 and $0.007\,I$.

Shown on Fig. 29 are values of pK_w derived from the data taken at 0.07 and compared with values produced by Hamann[95] at $25°C$ as a function of pressure of similar ionic strength and the data of Kearns[107] and Millero *et*

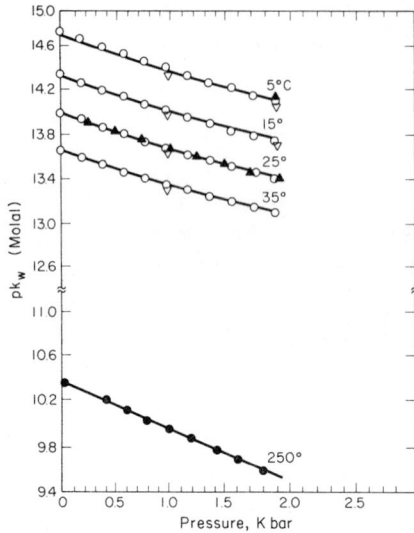

Fig. 29. Effect of pressure on pK_w. $\bigcirc$, Whitfield[5,77]; $\blacktriangle$, Kearns[107]; $\triangledown$, Millero *et al.*[108]; and $\bullet$, Dobson *et al.*[80]

al.[108] at infinite dilution. It is apparent that the effects of pressure on K_w are identical over the pressure range, because of the parallel nature of the curves, and for dissimilar ionic strengths at least from infinite dilution to about 0.1 *m* ionic strength at the lower temperatures. Moreover, if one now considers the data produced by Dobson *et al.*[80] at 250°C, plotted on a lower scale in Fig. 29, the same may be said for the effect of pressure at much higher temperatures.

Measurement of pH from EMF Studies

It is inappropriate in this paper to discuss the pros and cons of the effectiveness or desirability of a pH scale, operational or otherwise. On the practical scale it is usually defined as $pH = -\log a_{H^+} = -\log m_{H^+}\gamma_{H^+} = -\log m_{H^+}\gamma_\pm$. Thus, if the proton concentration and the mean activity coefficients of the particular solution are known or estimated, the pH may be calculated. When activity coefficient data are not known experimentally, Debye–Hückel extended or Guggenheim empirical relationships are used. The temperature characteristics of the empirical parameters used in these activity expressions are required for each type of electrolyte; unfortunately, only a very few electrolytes have been studied thoroughly. A later section, concerned with activity coefficients, discusses this aspect in more detail.

For wide pH ranges the use of common electrolytes (HCl, NaOH, etc.) are not possible because of their sensitivity to dilution, trace gases, and attack on the components of the system. Buffer solutions are considered not to have these disadvantages and to have wide application to elevated-temperature work. Although a number of well-known buffers have been studied at elevated temperatures, many have yet to have their characteristics documented. Thus, the study of cells employing pH-sensitive electrodes, generally glass electrodes, in order to calibrate these buffers, are the major preoccupation of this aspect of the use of emf measurements. A short list is given in Table 6 at selected pH values of some buffers at various temperatures that are of particular interest.

The problem of chemical attack is not as severe for pressure studies as for temperature in excess of 100°C; however, most of the well-known buffers are very complicated from the composition aspect. Thus, while providing a solution of constant pH, the effects of the various ions on the solvent/solute

Table 6. pH of Reference Solutions between 25 and 200°C

Reference number	Solutions (moles/kg H_2O)	Temperature (°C)							
		25	100	125	150	175	200	225	250
37	HCl 0.1	1.8[a]	1.12	1.13	1.14	1.15	1.17	1.19	1.21
37	HCl 0.01	2.04[a]	2.05	2.05	2.06	2.06	2.07	2.08	2.09
41	$K_2C_3H_4O_6$	3.57	3.69	3.79	3.92				
	(saturated at 25°C)					—	—	—	—
110b	(Tartrate)	3.56	3.68	3.79	3.90				
41	$KHC_8H_4O_4$ 0.05	4.01	4.24	4.37	4.50	—	—	—	—
110b	(Phthalate)	—	4.25	4.37	4.51	—			
41	$KC_7H_5O_2$ 0.01	4.19	4.32	4.43	4.54				
	$C_7H_6O_2$ 0.01					—	—	—	—
110a	(Benzoate)	—	4.30	4.43	4.53				
37	$NaCH_3COO$				5.03	5.15	5.35	5.56	5.80
41	CH_3COOH 0.01	4.72	4.92	5.01	5.13				
110a	(Acetate)	—	4.92	5.01	5.13				
37	KH_2PO_4 0.025	6.86[a]	6.88	6.92	7.04	7.15	7.30	7.45	7.60
41	Na_2HPO_4 0.025	6.86	6.96	6.96	7.08				
110a	(Phosphate)	6.86	6.91	7.02	7.14				
37	$Na_2B_4O_7$ 0.01	9.22[a]	8.22	8.75	8.65	8.60	8.56	8.58	8.50
41	(Borax)	9.16	8.81	8.73	8.66				
102	NaOH 0.01	11.95	10.24	9.86	9.60				
102	NaOH 0.1	12.89	11.17	10.79	10.52				

[a] At 20°C.

relationships as a whole or specifically are still generally unknown. Only a few attempts have been made to study these effects since Disteche and Disteche[74] looked at the effect of pressure on pH, and dissociation constants on buffered and unbuffered glass electrode cells. The early 1960s saw the emf–pH calibration work of Lietzke et al.,[23–29] Fornie et al.,[40] and Hamann.[95] A trivial paper by Ciaccio[109] was published in 1966, who used silver–silver chloride and quinhydrone electrodes to measure the change in pH of KCl solution at 25°C, under pressures up to 68 atm of nitrogen, oxygen, nitrous oxide, and carbon dioxide. The only major work of the later 1960s was carried out by the Russian workers Kryukov, Storastina, Perkovets, and Smolya-kov.[41,110] The first paper from these workers was devoted to pH calibration of tetraoxalate, phthalate, phosphate, and borax buffers, up to 150°C, using a concentration cell [cell (20)] with transport and platinum or glass electrodes:

$$\text{Pt, H}_2 \text{ or} \quad \left| \begin{array}{c} \text{HCl} \\ \text{glass electrodes} \end{array} \right| 0.1\ M \left|\right| \text{buffer} \left| \text{H}_2, \text{Pt} \right.$$

(20)

Comparisons were made in the paper with the earlier work of Bates and Bower[20] and LePeintre[44] where possible. Kryukov and co-workers also looked at the pK's of acetic, benzoic, sulfanillic, salicylic, and tartaric acids over the same temperature range.

In 1970[110] these Russian workers found the pH values of the buffers, but from cells which did not involve liquid junctions. For example, in cell (21) [cf. cells (7a) and (7b)]

$$\text{Pt, H}_2 \left| \begin{array}{c} \text{buffer solution} \\ \text{NaCl} \end{array} \right| \text{AgCl, Ag}$$

(21)

the emf is

$$E = E° - k \log (m_{H^+} \gamma_{H^+} m'_{Cl^-} \gamma_{Cl^-})$$ (19)

where $k = 2.303RT/F$. Using a $p_w H$ term after Guggenheim and Hitchcock,[103,104] we get

$$p_w H = -\log (m_{H^+} \gamma_{H^+} \gamma_{Cl^-}) = \frac{E - E°}{R} \log m_{Cl^-}$$ (20)

At $m_{Cl^-} = 0$,

$$p_wH° = -\log(m_{H^+}\gamma_{H^+}\gamma_{Cl^-})° \qquad (21)$$

or

$$pH = p_wH_{buff.\ soln.} = p_wH° + \log\gamma_{Cl^-}°$$

Values of $\log\gamma_{Cl^-}°$ were estimated in their paper by a Debye–Hückel expression.

The agreement in results, when compared (see Table 6) with earlier work from cells with transport, appeared to be very good. It is to be noted that no mention of time effects are given in any of the Russian papers mentioned so far. Their latest paper, however, is concerned with the construction of pH-glass electrodes using No. 121 SKBAP glass, with either thallium or silver chloride electrodes as inner reference. Apart from constructional details some tests and comparisons are also made up to 150°C. The results of emf measurements using the same glass electrode over a period of months is given as well as a correlation of phthalate, acetate, benzoate, sulfonate, and salicylate buffers between 100 and 150°C.

Very little has appeared in the literature in recent years on attempts to substantially alter the performance of the glass electrode or even to extend its temperature range. The glass electrodes used in the Russian work referred to above still use aqueous electrolyte inner fillings, and have relatively low maximum temperatures for operation. Neither has any work been published on attempts to create a high-temperature pH-ion selective electrode following along the lines of the semi-solid-state electrode that, say, Orion produces. In recent years, Dobson at Newcastle has been interested in these problems and has now created a high-temperature pH electrode which is totally solid state and relatively chemically inert. The constructional details are the subject of a pending patent and cannot be discussed in this paper. However, shown in Fig. 6 are some temperature plots of cell emf stability with time using this new disk electrode at temperatures up to 200°C. They may be compared with some early work of LePeintre[44] (see Fig. 5), who employed conventional glass electrodes with mercury amalgam inner fillings. The disk electrode's ideal pH response is also indicated in a selected pH range at 25°C in Figs. 30 and 31. In Fig. 30 the electrode was used to determine K_2, the second ionization constant of sulfuric acid, and gives a favorable comparison with other cell data using platinum hydrogen electrodes.[114] Further details of this new concept in electrode design and behavior will be published as soon as possible.[65]

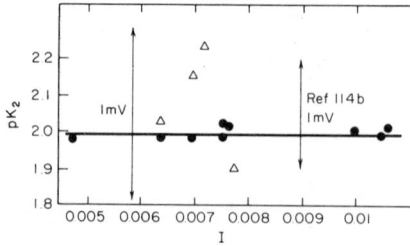

Fig. 30. Experimentally determined $pK_2^{\frac{1}{2}}$ with pH/solid state electrode as a function of ionic strength at 25°C. $\triangle$, pH/solid state electrode from cell, pH/solid state electrode $|^{HCl}_{H_2SO_4}|$ AgCl, Ag; ●, Covington et al.[114] Significance of $\updownarrow$ shows effect on spread of $pK_2^{\frac{1}{2}}$ for I error.

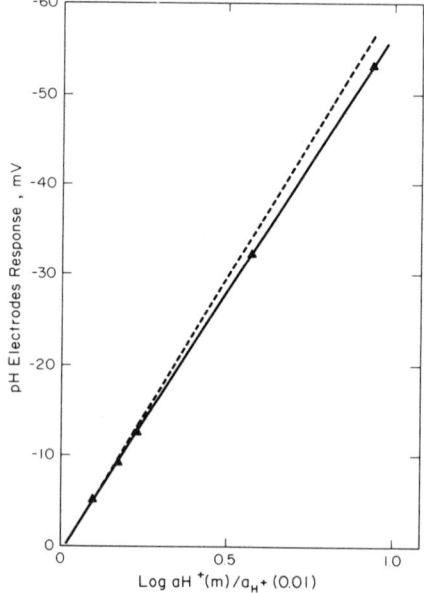

Fig. 31. The pH response of pH electrode derived from pH/solid state/HCl$_m$/AgCl, Ag with changes in concentration at 25°C. (From Covington et al.[114])

Activity Coefficients Data from EMF Measurements

Determinations of activity coefficients from emf measurements are the most precise[111] and have a greater range of applicability, with respect to concentration and temperature, than, say, isopiestic or vapor pressure studies. Unfortunately, a number of the most common electrolytes (HCl, H_2SO_4, NaCl, etc.) have not been studied, by any other method, completely over all ranges of interest in concentration at the elevated temperatures.

Attempts have been made,[112] with varying degrees of success, to extrapolate existing experimental data or use empirical expressions for the activity coefficients. Unfortunately, a serious difficulty is the lack of information on the dependence or nondependence on concentration and temperature of the parameters contained in these equations. As examples to demonstrate the effectiveness of indirect or mathematical attempts to extend activity coefficient data, pure HCl and KCl can be taken.

The Lietzke school has studied directly by emf measurements HCl up to $1\,m$ and 275°C. However, if $\gamma_{\pm,\mathrm{HCl}}$ at say $2\,m$ at 150°C were required, or even KCl at the same concentration but at lower temperatures (say, 70°C), existing published data derived from experimental studies would not be available. Thus, if one takes an activity expression similar to those proposed by Guggenheim,[103,104] for example

$$\log \gamma_\pm = -\frac{z^2 A I^{1/2}}{1 + \rho I^{1/2}} + BI + CI^{3/2} \tag{22}$$

where $\rho = 1.0$, and B and C can be assumed to be collectively related to the specific interaction coefficients of ion–ion and ion–solvent, respectively. It is easy to show that values of B and C may be derived at around 25°C and up to the region of interest in concentration, by rearranging Eq. (22) to

$$\log \gamma_{\pm,\mathrm{exp}} + \frac{z^2 A I^{1/2}}{1 + I^{1/2}} = BI + CI^{3/2} \tag{23}$$

and using experimental activity coefficient values in a least-squares computer program.

However, in order to carry out the calculation at the more elevated temperatures if experimental data are not available, the temperature dependence of A, B, and C must be known beforehand. Values of A at various temperatures are readily found from $\alpha = (2\pi N\rho)^{1/2}(e^2/D\varepsilon_0 kT)^{3/2}$ if $A - \alpha \ln_{10}$[53,113] and a short list of values is given in Table 7.

Evidence[30,53,114] suggests that B and C do depend on temperature. If the $\gamma_{\pm,\mathrm{HCl}}$ data up to $1\,m$ of pure HCl (Lietzke) are analyzed at various temperatures, say up to 150°C, using Eq. (23), the values of B and C are found to vary as shown in Table 8.[116] It is also seen from Table 8 that there is an effect of concentration, in addition, at any particular temperature. Assuming (although it is perhaps not justified) a linear dependence in temperature for B and C, values of dB/dT and dC/dT may be calculated, for the dependence on molality of HCl. Some values are plotted in Fig. 32. Some justification for the linear dependence in temperature of B and C may be found, because smooth lines may be drawn through the points. The points were derived from the

Table 7. Calculated Values of A as a Function of Temperature

Temperature (°K)	A (kg mole^{-1})$^{1/2}$
283.16	0.4990
293.16	0.5061
298.16	0.5107
303.16	0.5145
313.16	0.5240
323.16	0.5345
333.16	0.5460
363.16	0.5858
398.16	0.6422
423.16	0.6897
448.16	0.7449
473.16	0.8098

Lietzke data between 25 and 150°C and high concentration data between 10 and 50°C produced by Harned and Ehlers[89] and Akerlöf and Teare.[115]

It also appears that a little above 1 mole/kg, both dB/dT and dC/dT become invariant with concentration. This observation means that values of B

Table 8. Calculated Values of B and C for HCl between 0.001 and 1.01, and between 25 and 150°C

Temperature (°C)	B (kg mole^{-1})	C [(kg mole^{-1})$^{3/2}$]	Range of ionic strength (I)
25	+0.343581	−0.387667	
60	+0.389839	−0.565814	
90	+0.495956	−0.928533	0.001–0.1
125	+0.552921	−0.938463	
150	+0.644849	−1.318020	
25	+0.261940	−0.119669	
60	−0.266403	−0.127476	
90	−0.273728	−0.154137	0.001–0.5
125	+0.327959	−0.233738	
150	+0.319922	−0.232432	
25	+0.233778	−0.072720	
60	+0.243211	−0.088813	
90	+0.235540	−0.090475	0.001–1.0
125	+0.235540	−0.032963	
150	+0.0260452	−0.038294	

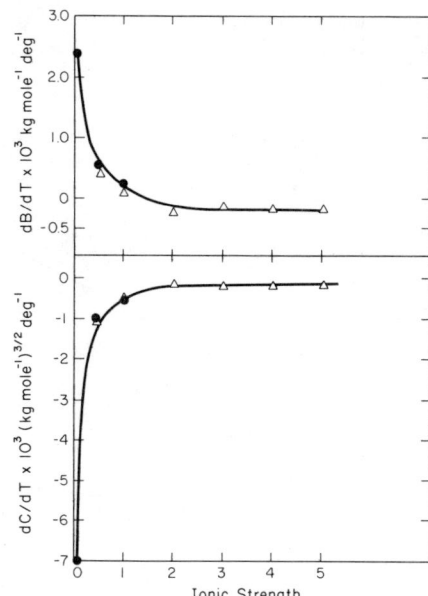

Fig. 32. Rate of change of B and C with temperature. ●, 25–150°C, Lietzke[112]; ▲, 10–50°C, Harned and Ehlers[89] and Akerlöf and Teare.[115]

and C at the high concentrations >1 mole/kg, but derived at lower temperatures, would allow an estimation of B and C at the higher temperatures provided that B and C were assumed to be linear with temperature, or even some known function of temperature. Dobson and Firman,[116] by the same type of analysis discussed above, have produced expressions which may be used to calculate the activity coefficient up to at least $5\,m$ and 150°C for HCl, KCl, CuCl, and so on. For example,

$$(\log \gamma_{\pm,\text{HCl}})_T = -\frac{A_T z^2 I^{1/2}}{1 + I^{1/2}} + (0.232448 - 0.2172T \times 10^{-3})I$$

$$+ (0.40129 - 0.16861T \times 10^{-3})I^{3/2} \tag{24}$$

$$(\log \gamma_{\pm,\text{KCl}})_T = -\frac{A_T z^2 I^{1/2}}{1 + I^{1/2}} + (-0.134964 + 0.599323 \times 10^{-3} \times T)I$$

$$+ (0.052002 - 0.20659 \times 10^{-3}T)I^{3/2} \tag{25}$$

$$\log \gamma_{\pm,\text{CsCl}} = -\frac{A_T z^2 I^{1/2}}{1 + I^{1/2}} + (-0.377567 + 1.19272 \times 10^{-3}T)I$$

$$+ (0.11843 - 0.35995 \times 10^{-3}T)I^{3/2} \tag{26}$$

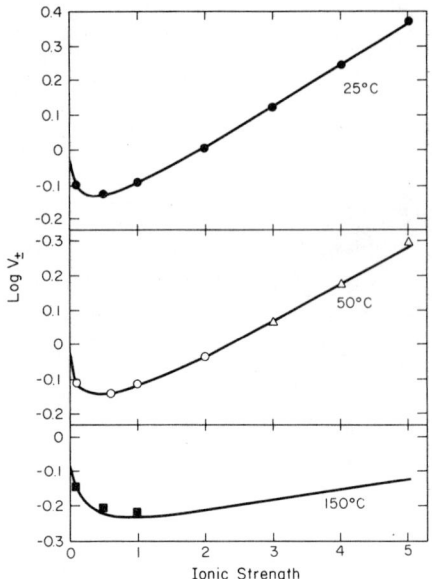

Fig. 33. Comparison of experimental and calculated values using $\log \gamma_\pm$ for HCl at 25, 50, and 150°C. $\triangle$, Akerlöf and Teare[115]; $\bigcirc$, Harned and Ehlers[89]; ■, Lietzke[112]; ●, Dobson and Firman[116]; and ———, calculated at all temperatures.

and shown in Fig. 33 as a comparison of the experimental and calculated values of $\log \gamma_{\pm,\text{HCl}}$ at 25, 50, and 150°C up to 5 m. Estimations[116] of possible errors of these equations can be shown to be at, say, 2 m and 150°C for HCl of about 5%.

Activity Coefficient Data in Mixed Electrolytes from EMF Measurements

Over a period of the last 10 years or so, Lietzke and co-workers at Oak Ridge, Tennessee, have used Harned-type cells [cell (22)]

$$\text{Pt}, \text{H}_2(p=1) \left| \begin{array}{ccc} \text{HCl}_{(m_2)} & \text{MCl}_{x(m_3)} & \text{AgCl}, \text{Ag} \\ \text{or} & \text{or} & \text{or} \\ \text{HBr} & \text{HBr} & \text{AgBr}, \text{Ag} \end{array} \right.$$

(22)

over a wide range of temperature, concentration, and variation in metal (M) chlorides as electrolytes.

From these direct experimental measurements, considerable information has been derived about activity coefficient data and behavior in mixtures in the papers that were published. Their early work of the last decade was concerned with pure HCl,[117] HBr,[118] and later with HBr/KBr,[119] HCl/DCl,[120] and HCl/NaCl[121] mixtures. One of their preoccupations was establishing that the extended Debye–Hückel equation [Eq. (27)] for activity coefficients could be used to express the data, between 25 and 270°C, and at least up to unit molality:

$$\log \gamma_{\pm} = -\frac{s(\rho_0 I^{1/2})}{1 + A(I^{1/2}\rho_0)} + \text{Ext} + BI + \cdots \qquad (27)$$

where s is the Debye–Hückel limiting slope, ρ_0 the density of water, I the ionic strength, $A = 50.29(DT)^{-1/2}a°$, $a°$ the ion size parameter, D the dielectric constant, Ext the extended Gronwell et al.[122] terms, and B the linear coefficient term. The expression being used in cell (**22**) is the general emf equation

$$E = E°\text{Ag},\text{AgX} - \frac{RT}{F}(\ln m_2)(m_2 + m_3) - \frac{2RT}{F}\ln \gamma_{\pm} \qquad (28)$$

The linear B parameters, as well as other higher coefficients, were determined from least-squares treatments of the data using Eqs. (27) and (28).

From this work it was concluded that in general the expression was satisfactory to describe the various solutions studied. A number of interesting observations and conclusions were made as a result of their emf measurements. For example, at constant temperature and ionic strength $\log \gamma_{\pm,\text{HBr}}$ in HBr/KBr mixtures varies linearly with the molality of KBr; moreover, the $\gamma_{\pm,\text{KBr}}$ varies less with changing ionic strength and temperature than does $\gamma_{\pm,\text{HBr}}$ in the same mixtures. The same may be said of the HCl/NaCl mixtures. The first observation is demonstrated in Fig. 34 by plots of $\log \gamma_{\pm,\text{HBr}}$ and HCl against moles of NaBr and NaCl, respectively. A clear comparison can be made between pure HCl and DCl. Figure 35 shows first that the $\gamma_{\pm,\text{DCl}}$ in D_2O is lower than that of $\gamma_{\pm,\text{HCl}}$ in H_2O at all temperatures and concentrations. Second, the difference between the two activity coefficients is greater at 25 and 200°C than at 90°C. Third, the lower the minimum value of the activity coefficient on any curve, the higher the value of m at the minimum. They pointed out that because at 25 and 90° C the DCl curves lie below HCl, such behavior is consistent with a lower dielectric constant for D_2O than for H_2O at each temperature. They also suggested that around 100°C the dielectric constants of D_2O and H_2O should be close together, to be consistent with their data.[130]

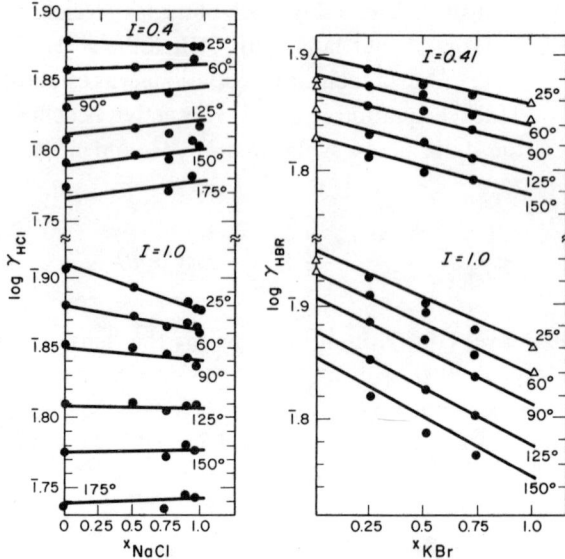

Fig. 34. (Left) $\log \gamma_{HCl}$ versus X_{NaCl} in HCl/NaCl. (From Lietzke *et al.*[121]) (Right) $\log \gamma_{HBr}$ versus X_{KBr} in HBr/NaBr. (From Lietzke and Stoughton![119])

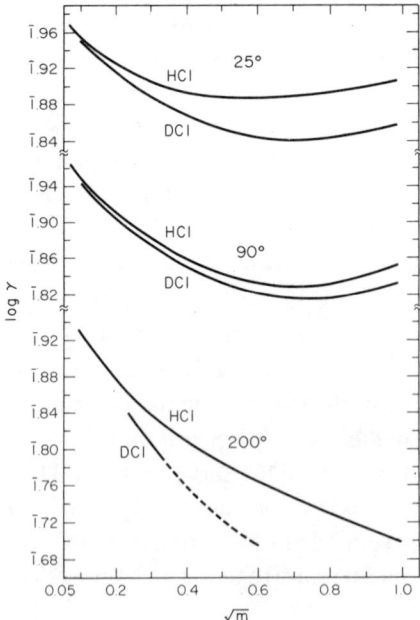

Fig. 35. Plots of $\log \gamma$ versus $m^{1/2}$ for HCl and DCl solutions at 25, 90, and 200° C. (From Lietzke *et al.*[117,120])

More recent emf work by the Lietzke school has exploited cell (**22**) further and has been concerned with $HCl/BaCl_2$,[123] $HCl/LaCl_3$,[124] $HCl/GdCl_3$,[125] and the most recent with HCl/KCl, $RbCl$, $CsCl$, $HgCl$, $CaCl_2$, $SrCl_2$, $AlCl_3$,[126] and $HCl/LiCl$[127] mixtures. As before, they found that at constant temperature and ionic strength, $\log \gamma_{\pm, HCl}$ in mixtures varies linearly with the ionic strength fraction of the salt and in conformity with Harned's rule[128] [see Fig. 36 (left)].

In contrast to the activity coefficients of $BaCl_2$ and $LaCl_3$ [see Fig. 36 (right, top and bottom)], which show a decrease with increase in ionic strength fraction of HCl, $GdCl_3$ shows an increase at comparable ionic strengths. Plots of $\log \gamma_{\pm}$ $GdCl_3$ against X_{HCl} at 25° C resemble quite closely those of NaCl in HCl in that in both systems the activity coefficient of the salt increases more

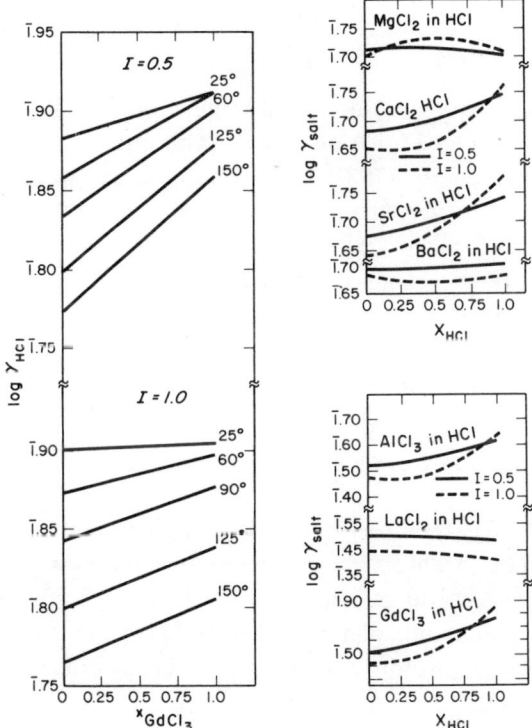

Fig. 36. (Left) $\log \gamma_{HCl}$ versus X_{GdCl_3} in $HCl/GdCl_3$ mixtures. (From Lietzke and Stoughton[125].) (Right, top and bottom) plots of $\log \gamma_{salt}$ versus ionic strength fraction of HCl at 25° C for $HCl/AlCl_3$, $HCl/LaCl_3$, and $HCl/GdCl_3$ mixtures. (From Lietzke et al.[124-126])

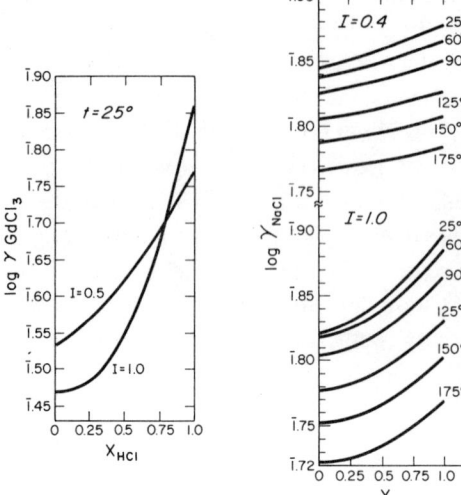

Fig. 37. (Left) $\log \mathrm{GdCl_3}$ versus $\mathrm{XHC/GdCl_3}$ mixtures. (From Lietzke and Stoughton[125].) (Right) $\log \mathrm{NaCl}$ versus XHCl in HCl/NaCl mixtures. (From Lietzke et al.[121])

rapidly in the mixtures at high total ionic strength. These later points are demonstrated in Fig. 37 (left and right). Although emf measurements were generally made up to 175°C in the more recent work by the Lietzke school,[125-127] only derived activity coefficient data at 25°C are discussed in any great detail.

Other workers outside the Lietzke school, for example Körtum and Hausserman in their paper in 1965,[30] have studied activity coefficients by emf measurements. These workers produced activity coefficient data on HI from room temperature up to 200°C.

Transport Numbers from EMF Measurements

Very little has been carried out at all concerning the determination of transport numbers from emf measurements on concentration cells at elevated temperatures and pressures, primarily, it is claimed, because of the difficulties in maintaining liquid junction boundaries. Hills et al.[129] (UK) in 1965 published a paper on proton migration in aqueous solution in which they measured the cell

$$\mathrm{Pt, H_2 \atop Ag\,AgCl} \bigg| HCl_{(m_1)} \bigg| HCl_{(m_2)} \bigg| {H_2, Pt \atop AgCl\,Ag} \qquad {(23a) \atop (23b)}$$

and were able to produce transport data for the hydrogen ion at 25°C and 45°C over the pressure range 1–2000 atm. They compared their work with

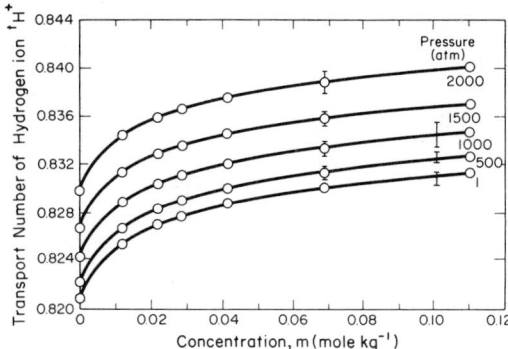

Fig. 38. Isobaric concentration dependence of the transference number of the hydrogen ion at 25°C. I, Wall and Gill[131]; $\bigcirc\Phi$, Hills *et al.*[129]

Wall and Gill[131] (USA), who used the autogenic moving-boundary method. Shown in Fig. 38 are isobar plots of transport numbers and concentrations of HCl at 25°C. The author would now like to present some emf measurements made recently at Newcastle on the concentration cell (**23b**) and Type 4, Fig. 1. A complete presentation of the transport data, including the cations of other chlorides in addition to HCl, will be given elsewhere.[57] However, it is interesting to note here the relatively high precision obtainable in the present series of emf measurements for these cells. In Fig. 38 is shown a typical set of emf data for the $0.01/0.05\,m$ ratio of HCl in the cell as a function of temperature. A best straight line is drawn through the data giving a mean deviation of approximately $\pm 0.5\,\text{mV}$. This type of precision says much about the highly satisfactory nature of the cell design and measuring techniques. Taking the particular composition ratio that is shown in the emf temperature curve on Fig. 39 and assuming that this ratio,

$$\frac{2R}{F}\ln_{10}\log\frac{a_{\text{HCl},0.01}}{a_{\text{HCl},0.05}}$$

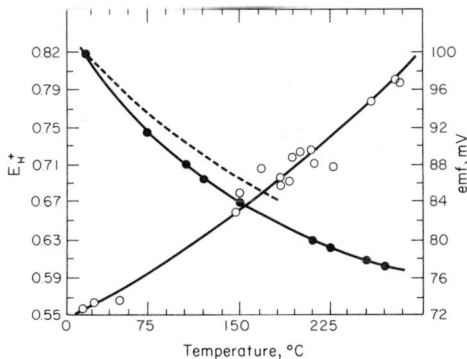

Fig. 39. EMF data and transport number of the proton. $\bigcirc$, EMF of concentration cell; $\bullet$, E_H^+ in HCl. (From Dobson *et al.*[57])

remains approximately the same for the range of temperatures studied, the "apparent" mean transport number of the proton in HCl may be calculated from

$$E = -\frac{2RT}{F}\bar{t}_{H^+} \ln \frac{m_1\gamma_{\pm 1}}{m_2\gamma_{\pm 2}} \tag{29}$$

giving

$$\bar{t}_{H^+} = \frac{E}{T} \times \frac{1}{29.46 \times 10^{-5}} \tag{30}$$

Computed values of $\bar{t}_{H^+}$ are also shown on Fig. 39. The values range from 0.83 at 25°C to 0.59 at 300°C. Shown also on the same figure as a dashed line is the range of value obtained by taking the best crude approximation $(\bar{t}_H)_T(t_H^o)_T(\lambda_H^o/\lambda_{H^+}^o + \lambda_{Cl}^o)_T$ because of lack of data for comparison.

The interesting discrepancies may be due to the nonconsistency of the activity terms in the equation, but it is more likely to be due to the crude approximations necessary in the calculation of the dashed line.

SOME EXAMPLES OF AREAS OF NECESSARY FUTURE WORK

In spite of advances in technology, many problems of a practical nature remain to be solved. While the use of PTFE, for example, has eased problems associated with cell container construction, there is still no proven and generally available material that can be used in excess of 300°C. In this respect, perhaps some of the high-temperature refractory polymers, such as the phosphonitrilic chlorides, or the metallic phosphinates, which have been developed for space research, could be used.

Alternatively, the introduction of the miniature operational amplifier and the solid-state data logging system have certainly aided the acquisition and precision of cell emf data; however, problems still remain specifically connected with the electrodes and electrolytes, the items which furnish the emf of any cell. So, while the environmental biochemist may be concerned about the measurements of sudden detectable changes in concentration of lead or mercury in some river estuary, a control worker in a nuclear plant may need to know continuously the pH of a coolant fluid. In both of these particular cases, as in many others, emf measurements, provided that suitable electrodes are available, can be used to great effect. Unfortunately, there are too few electrodes that are specific for H^+, O_2, alkali metal ions, and metal ions, if wide ranges of temperature and pressure are required.

Further development of old, and the creation of new, electrodes which are also more specific in their reversibility to particular ions are therefore required urgently. This urgency is promulgated by the incipient practical measures, concerning pollution of various sorts, taken by important bodies at government and industrial levels. The practical requirements of general industrial processing plant, anticorrosion, and monitoring facilities are also rapidly increasing and thus generate additional needs. Here also are included the specific needs of, say, those involved in the production of textiles, pharmaceuticals, cosmetics, detergents, printing, and so on. Thus, even a greater exploitation of electrode technology and emf measurements as a whole may be expected within the next decade or so.

Other areas of future use of emf measurements may be indicated. The oceanographer, and those concerned with desalinization, as well as the battery specialist and those involved with energy conversion, continually need more information on the little-known effects of moderate ranges in temperature ($0-100°C$) and pressure ($1-10\,kbar$) on activities, dissociation constants of weak acids and bases, and other thermodynamic quantities. This type of information is generally recognized as being best found by emf study. Finally, a well-proven range of pH buffers suitable between 20 and $300°C$ is still not available, and detailed emf calibrations would be very useful for even a few of the standard buffers.

REFERENCES

1. N. H. Schöön and T. Wannholt, *Sven. Papperstid.*, 431 (August 1972).
2. J. Fiker and J. Santova, *Chem. Prum.* **21**, 237 (1971).
3. P. R. Kryukov *et al.*, Methods of Testing Geothermal Fissures with Regards to Calcification, VINITI, No. 245-68 Dept. USSR.
4. F. W. Smith and G. J. Lauro, 25th Annual-Meeting, Institute of Food Technologists, Kansas City (May 1965).
5. M. J. Whitfield, *Chem. Eng. Data* **17**, 124 (1972); A. Bodanszky and W. Kauzmann, *J. Phys. Chem.* **66**, 177 (1962); L. A. Dunn, R. H. Stokes, and L. G. Hepler, *J. Phys. Chem.* **65**, 2808 (1965); S. D. Hamman, *J. Phys. and Chem.* **67**, 2233 (1963).
6. H. L. Barnes, H. O. Heleson, and A. J. Ellis, *Handbook of Physical Constants*, Geol. Soc. Am. Mem. **97**, 404 (1966).
7. M. G. Gonikberg, *Chemical Equilibria, Reaction Rates at High Pressures*, p. 39, Israel Program for Scientific Translations, Jerusalem (1963).
8. G. J. Hills and P. Ovenden, *Adv. Electrochem.* **4**, 204 (1966).
9. B. B. Owen and S. R. Brinkley, *Chem. Rev.* **29**, 461 (1941).

10. M. E. Indig and D. A. Vermilyea, General Electric Research, Development Center, Schenectady, N.Y. Report 70C269 (1970).
11. R. E. Mesmer, C. F. Baes, and F. H. Sweeton, *Inorg. Chem.* **11**, 537 (1972); see also *J. Phys. Chem.* **74**, 1937 (1970).
12. R. L. Every and W. R. Banks, U.S. Patent 3,462,353 (August 19, 1969).
13. R. W. Staehle and R. L. Cowan, *J. Electrochem. Soc.* **118**, 557 (1971).
14. M. H. Lietzke, *J. Chem. Ed.* **39**, 230 (1962).
15. D. T. Ives and G. Janz, *Reference Electrodes*, Academic Press, New York (1961).
16. G. J. Hills and P. Ovenden, *Adv. Electrochem.* **4**, 180 (1966).
17. D. de G. Jones and H. G. Masterson, *Advances in Corrosion Science and Technology* (M. G. Fontana and R. W. Staehle, eds.), Vol. 1, Plenum Press, New York (1970).
18. G. J. Hills, 2nd Australian Conference on Electrochemistry, Melbourne, pp. 39–51, Royal Australian Chemical Institute (1968).
19. M. H. Lietzke and R. W. Stoughton, *J. Am. Chem. Soc.* **75**, 5226 (1953).
20. R. G. Bates and V. E. Bower, *J. Res. Nat. Bur. Stand.* **53**, 283 (1954).
21. M. H. Lietzke and J. V. Vaughen, *J. Am. Chem. Soc.* **77**, 876 (1955).
22. R. N. Roychoudhury and C. F. Bonilla, *J. Electrochem. Soc.* **103**, 241 (1956); M. H. Lietzke, *J. Am. Chem. Soc.* **77**, 1344 (1955).
23. M. H. Lietzke and J. V. Vaughen, *J. Am. Chem. Soc.* **79**, 4266 (1957).
24. M. H. Lietzke, R. S. Greeley, W. T. Smith, and R. W. Stoughton, *J. Phys. Chem.* **64**, 652 (1960).
25. R. S. Greely, W. T. Smith, R. W. Stoughton, and M. H. Lietzke, *J. Phys. Chem.* **64**, 654 (1960).
26. M. B. Towns, R. S. Greeley, M. H. Lietzke, and R. W. Stoughton, *J. Phys. Chem.* **64**, 1861 (1960).
27. M. H. Lietzke and R. W. Stoughton, *J. Phys. Chem.* **67**, 2573 (1963).
28. M. H. Lietzke and R. W. Stoughton, *J. Phys. Chem.* **68**, 3043 (1964).
29. M. H. Lietzke, H. B. Hupf, and R. W. Stoughton, *J. Phys. Chem.* **69**, 2395 (1965).
30. G. Körtum and W. Hausserman, *Ber. Bunsenges. Phys. Chem.* **69**(7), 594 (1965).
31. M. H. Lietzke, *J. Am. Chem. Soc.* **77**, 1344 (1955).
32. S. Stene, *Rec. Trav. Chim. Pays-Bas* **49**, 1133 (1930).
33. R. N. Roychoudhury and C. F. Bonilla, *J. Electrochem. Soc.* **103**, 241 (1956).
34. O. V. Ingruber, *Pulp. Paper Mag. Can.* **55**, 124 (1956); see also *Ind. Chem.* **32**, 513 (1965); and G. Fiehn, *Chem. Tech.* (*Berlin*) **16**, 369 (1964).
35. H. R. Fricke, *Beiträge zur angewandten Glasforschung* (E. Schott, ed.), Wissenshaftlicher Verlag, Stuttgart (1959).
36. W. Veilstich, *Z. Instrumentenkd.* **6**, 154 (1959).
37. M. LePeintre, *Soc. Ing. Civ. Franc. Bull. Trans.* 3126 **1**, 584 (1960).
38. M. LePeintre, *Soc. Ing. Civ. Franc. Bull. Trans.* 3126 **1**, 584 (1960); see also S. I. Sokolov and A. H. Passynsky, *Z. Phys. Chem.* **160A**, 366 (1932).
39. R. S. Greeley, *Anal. Chem.* **32**, 1717 (1960).
40. R. Fournie, P. Le Clerc, and M. Saint James, *Silic. Ind.* **27**, 33 (1962).
41. P. R. Kryukov, V. D. Perkovets, L. I. Storostina, and P. Smolyak, *Izv. Akad. Nauk. SSR Ser. Khim. Nauk* **7**(2), 29 (1966).
42. W. H. Marburger, J. Anderson, and G. G. Wigle, Argonne National Laboratory, U.S.A., Report ANL 5298 (1954).
43. A. Disteche, *Rev. Sci. Instr.* **30**, 474 (1959).
44. M. LePeintre, *Soc. Ing. Civ. Franc. Bull. Trans.* 3126 **1**, 585 (1960).
45. A. Disteche, *J. Electrochem. Soc.* **109**, 1084 (1962).
46. J. Hamann, *Phys. Chem.* **67**, 2233 (1963).
47. A. Disteche and S. Disteche, *J. Electrochem. Soc.* **112**, 350 (1965).
48. W. R. Hainsworth, H. J. Rowley, and D. A. MacInnes, *J. Am. Chem. Soc.* **116**, 1437 (1924).
49. T. C. Poulter, *Phys. Rev.* **39**, 816 (1932).

50. E. Cohen, I. Fusao, and A. Mozsoveld, *Z. Phys. Chem.* **105**, 155 (1923).

51. J. Casteels, *Neufchatel*, 41 (1920).

52. J. Casteels, *Neufchatel*, 51 (1920).

53. J. V. Dobson and H. R. Thirsk, *Electrochim. Acta* **16**, 315 (1971).

54. J. V. Dobson, R. E. Firman, and H. R. Thirsk, *Electrochim. Acta* **16**, 793 (1971).

55. J. V. Dobson, to be published.

56. P. A. Kryukov, A. G. Kalinina, and E. D. Linov, *Izv. Sib. Otd. Akad. Nauk SSSR Ser. Khim. Nauk.* **3**, 2 (1969).

57. J. V. Dobson, R. E. Firman, and H. R. Thirsk, *Rev. Sci. Instr.* **6**, 24 (1973).

58. J. V. Dobson, to be published.

59. Product Data Sheet, Boron Silicide, Cerac, Inc., Butler, Wis. (1967).

60. J. V. Dobson, *J. Electroanal. Chem. Interfacial Electrochem.* **35**, 128 (1972); J. V. Dobson, M. N. Daglass, and H. R. Thirsk, *Trans. Faraday Soc.* **68**(1), 749, 764 (1972).

61. B. Case and G. J. Bignold, *J. Appl. Electrochem.* **1**, 141 (1971).

62. T. Isaki and K. Arai, *Suiyokaishi.* **16**(7), 367 (1968).

63. J. V. Dobson and R. E. Firman, *J. Electroanal. Chem. Interfacial Electrochem.* (1972–1973).

64. J. V. Dobson, patent no. 1504693, U.K. (March 1978).

65. J. V. Dobson, to be published.

66. M. Lepeintre, C. Malieu, and J. Monjou, French Patent 1,509,928 (1968).

67. R. R. Kelly, Bulletins 476 and 477, U.S. Department of the Interior, Bureau of Mines, U.S. Government Printing Office, Washington, D.C. (1949).

68. J. W. Mellor, *A Comprehensive Treatise on Inorganic Theoretical Chemistry*, Vol. IV, Longmans, Green, London (1952).

69. R. Bates, E. A. Guggenheim, W. F. K. Wynne-Jones, *et al.*, *J. Chem. Phys.* **25**, 361 (1956); **26**, 222 (1957).

70. F. A. Lewis, *The Palladium Hydrogen System*, Academic Press, New York (1967).

71. A. K. Covington, *Ion Selective Electrodes* (R.A. Durst, ed.), Nat. Bur. Stand. Spec. Publ. 314 (1969); see also F. Strafelda and B. Polej, *Chem. Prum.* **7**, 240 (1957).

72. A. R. Covington, C. P. Besboruah, M. Camoes, and J. V. Dobson, *Trans. Faraday Soc.* **69**(1) (1973).

73. A. T. von Wagramjam, M. A. Shamagorzjanz, G. F. Sawtschenkon, L. A. Uwarow, and A. A. Jawitsch, *Z. Phys. Chem.* **238**(4) 167 (1968).

74. A. Disteche and S. Disteche, *J. Electrochem. Soc.* **114**, 330 (1967).

75. S. Ben Yaakov and I. R. Kaplan, *Rev. Sci. Instr.* **39**, 1133 (1968).

76. P. A. Kryukov, A. G. Kalinina, and E. D. Linov, *Izv. Sib. Otd. Akad. Nauk, SSSR Ser. Khim. Nauk* **3**, 3 (1969).

77. M. Whitfield, *Electrochim. Acta* **15**, 83 (1970); see also *J. Electrochem. Soc.* **116**, 1042 (1969); *Rev. Sci. Instr.* **39**, 1053 (1968).

78. K. E. Von Heusler and L. Gaiser, *Ber. Bunsenges. Phys. Chem.* **73**, 1059 (1969).

79. G. J. Hills and D. R. Kinnibrugh, *J. Electrochem. Soc.* **113**, 1111 (1966); see also *Trans. Faraday Soc.* **61**, 326 (1965).

80. J. V. Dobson, B. R. Chapman, and H. R. Thirsk, NACE, HTHPE, Guildford, U.K. (1973).

81. A. J. de Bethune and G. R. Salvi, *J. Electrochem. Soc.* **108**, 672 (1961); **106**, 616 (1959).

82. D. T. Ives and G. Janz, *Reference Electrodes*, p. 75, Academic Press, New York (1961).

83. B. R. Churagulov, Ya. A. Kalashuilov, E. M. Feklichev, and G. I. KhoKhlova, *Russ. J. Phys. Chem.* **41**, 816 (1971).

84. L. F. Kulikova, B. R. Churagulov, and Y. A. Kalashuilov, *Z. Fiz. Khim.* **46**, 530 (1972).

85. L. F. Pogorelova, *Z. Fiz. Chem.* **45**, 818 (1971).

86. A. Seys, T. Fuiju, and A. Van Haute, NACE, HTHPE, Guildford U.K. (1973).

87. G. Biedermann and L. G. Sillen, *Ark. Kemi* **5**, 425 (1953).

88. E. C. W. Clarke and D. N. Glew, *Trans. Faraday Soc.* **62**, 539 (1966); see also P. D. Bolton, *J. Chem. Ed.* **47**, 63 (1970).

89. J. Hamer, *J. Am. Chem. Soc.* **57**(9), 33 (1935); H. S. Harned and R. W. Ehlers, *J. Am. Chem. Soc.* **55**, 2179 (1933); R. G. Bates and V. E. Bower, *J. Res. Nat. Bur. Stand.* **53**, 283 (1954); S. R. Gupta, G. J. Hills, and D. J. G. Ives, *J. Res. Nat. Bur. Stand.* **59**, 1874 (1963).

90. F. Letowski, *J. Rocz. Chem. Pol.* **44**, 1665 (1970).

91. F. Letowski, *J. Rocz. Chem. Pol.* **43**, 1597 (1969).

92. P. Henderson, *Z. Phys. Chem.* **59**, 118 (1907); **63**, 325 (1908).

93. R. E. Mesmer, C. F. Baes, and F. H. Sweeton, *Inorg. Chem.* **11**, 537 (1972); see also *J. Phys. Chem.* **74**, 1937 (1970); see also C. F. Baes, N. J. Meyer, and C. E. Roberts, *Inorg. Chem.* **4**, 518 (1965); R. E. Mesmer and C. F. Baes, *Inorg. Chem.* **6**, 1951 (1967).

94. N. Ingri, G. Lagerstrom, M. Frydman, and L. G. Sillen, *Acta Chem. Scand.* **11**, 1034 (1957); **16**, 439 (1962); **17**, 573, 581 (1963); see also N. Ingri, *Sven. Kem. Tidskr.* **75**, 199 (1963).

95. S. D. Hamann and M. Linton, *Trans. Faraday Soc.* **62**, 2234 (1966); **65**, 2186 (1969); also S. D. Hamann and H. G. David, *Trans. Faraday Soc.* **55**, 72 (1959); **56**, 1043 (1960).

96. E. U. Franck and W. Holzapfel, *Ber. Bunsenges. Phys. Chem.* **70**, 1105 (1966).

97. A. S. Quist, *J. Phys. Chem.* **74**, 3396 (1970).

98. G. J. Bignold, A. D. Brewer, and B. Hearn, *Trans. Faraday Soc.* **67**, 2419 (1971).

99. J. R. Fisher, thesis, Pennsylvania State University (1969).

100. J. C. Ahluwolia and J. W. Cobble, *J. Am. Chem. Soc.* **86**, 5381 (1964).

101. H. L. Clever, *J. Chem. Ed.* **45**, 231 (1968).

102. V. D. Perkovets and P. A. Kryukov, *Izv. Sib. Otd. Akad. Nauk SSSR Ser. Khim. Nauk* **7**, 9 (1969).

103. E. A. Guggenheim, *Thermodynamics*, North-Holland, Amsterdam (1949).

104. E. A. Guggenheim and J. C. Turgeon, *Trans. Faraday Soc.* **51**, 747 (1955).

105. R. E. Mesmer, C. F. Baes, and F. H. Sweeton, *J. Phys. Chem.* **74**, 1937 (1970).

106. G. Bianchi, G. Faita, R. Gulli, and T. Mussini, *Electrochim. Acta* **12**, 439 (1967); P. A. Rock, *Electrochim. Acta* **12**, 1531 (1967).

107. E. R. Kearns, Compressibilities of Some Dilute Aqueous Solutions, Ph.D. thesis, University Microfilms 66-4902, Yale University (1966).

108. F. J. Millero, E. V. Hoff, and L. A. Kahn, *J. Solution Chem.*, submitted 1971.

109. L. L. Ciaccio, *Chem. Ind.* 1525 (1966).

110. (a) V. D. Perkovets and P. A. Kryukov, *Izv. Sib. Otd. Akad. Nauk SSSR Ser. Khim. Nauk*, 622 (1968); (b) P. A. Kryukov and L. I. Storostina, *Izv. Sib. Otd. Akad. Nauk SSSR Ser. Khim. Nauk* **7**(6), 27 (1970).

111. E. J. King, *Acid Base Equilibria*, Pergamon Press, Elmsford, N.Y. (1963).

112. M. H. Lietzke, *Anal. Chem.* **32**, 1717 (1960); also *J. Am. Chem. Soc.* **77**, 1344 (1955).

113. E. A. Guggenheim and J. E. Prue, *Physiochemical Calculations*, North-Holland, Amsterdam (1955).

114. A. K. Covington, J. V. Dobson, and Lord Wynne-Jones, *Electrochim. Acta* **12**, 513, 525 (1967); also *Trans. Faraday Soc.* **56**, 1173 (1960); **61**, 2050, 2058 (1965); **69**, 94 (1973).

115. C. O. Akerlöf and J. W. Teare, *J. Am. Chem. Soc.* **59**, 1853 (1937).

116. J. V. Dobson and R. E. Firman, to be published.

117. M. H. Lietzke, R. S. Greeley, W. T. Smith, and R. W. Stoughton, *J. Phys. Chem.* **64**, 652 (1960); R. S. Greeley, W. T. Smith, R. W. Stoughton, and M. H. Lietzke, *J. Phys. Chem.* **64**, 654 (1960).

118. M. B. Towns, R. S. Greeley, M. H. Lietzke, and R. W. Stoughton, *J. Phys. Chem.* **64**, 1862 (1960).

119. M. H. Lietzke and R. W. Stoughton, *J. Phys. Chem.* **67**, 2577 (1963).

120. M. H. Lietzke and R. W. Stoughton, *J. Phys. Chem.* **68**, 3047 (1964).

121. M. H. Lietzke, H. B. Hupf, and R. W. Stoughton, *J. Phys. Chem.* **69**, 2398 (1965).
122. T. H. Gronwell, V. K. La Mer, and K. Sandved, *Z. Phys.* **29**, 358 (1928).
123. M. H. Lietzke and R. W. Stoughton, *J. Phys. Chem.* **70**, 756 (1966).
124. M. H. Lietzke and R. W. Stoughton, *J. Phys. Chem.* **71**, 662 (1967).
125. M. H. Lietzke and R. W. Stoughton, *J. Phys. Chem.* **72**, 257 (1968).
126. M. H. Lietzke and H. A. O'Brien, *J. Phys. Chem.* **72**, 4408 (1968).
127. M. H. Lietzke and R. W. Stoughton, *J. Tenn. Acad. Sci.* **44**, 66 (1969).
128. H. S. Harned and B. B. Owen, *The Physical Chemistry of Electrolytic Solutions*, Reinhold, New York (1958).
129. G. J. Hills, P. J. Ovenden, and D. R. Whitehouse, Kinetics of Proton Transfer Processes, *Discuss. Faraday Soc.* **39**, 207 (1965).
130. R. Kay, G. A. Vidulich, and K. S. Pribadi, *J. Phys. Chem.* **73**, 445 (1969).
131. G. Wall and T. Gill, *J. Phys. Chem.* **58**, 740 (1954).
132. A. A. Noyes, Carnegie Institute, Washington D.C., Publication 63, (1907); *J. Am. Chem. Soc.* **32**, 159 (1910).

NUCLEATION MECHANISM IN OXIDE FORMATION DURING ANODIC OXIDATION OF ALUMINUM

Paul Csokan

Chemistry and Corrosion Division
General Design Institute for the Engineering Industry
Budapest, Hungary

INTRODUCTION

Numerous research workers have studied the phenomena of oxide formation occurring on the surface of aluminum in various media under the influence of external reaction factors. The physicochemical, mechanical, and structural properties of both the thin oxide membranes and the heavier oxide layers have also been studied. The main variants of the process of oxide formation can be classified according to the character of the influencing factors. These factors are:

1. In atmospheric oxidation an aluminum oxide film of the thickness of several molecular layers is formed extremely rapidly on the clean metallic surface of aluminum under the influence of atmospheric oxygen and atmospheric moisture. Oxidation does not cease after the primary coherent oxide membrane is established; it continues at a decreasing rate and the final thickness of the oxide coating of 0.010–$0.015\,\mu m$ is reached only after a prolonged time (Fig. 1, curve A). According to Herrmann[1] the natural oxide layer with the composition Al_2O_3 (formed under atmospheric influence) always contains a certain amount of aluminum oxihydrate ($AlO \cdot OH$) and water bonded physically or by chemisorption.

2. In thermal oxidation the process is accelerated in the high-temperature atmosphere with its natural or enriched oxygen content; and independently of temperature, the thickness of the oxide layer increases to 0.06–$0.08\,\mu m$ within 12 hr. Subsequently, the process becomes slower and the final thickness of $0.2\,\mu m$ is achieved within several days (Fig. 1, Curve B). According to Cabrera and Mott,[2] Wyon *et al.*,[3] Paganelli,[4] and other

239

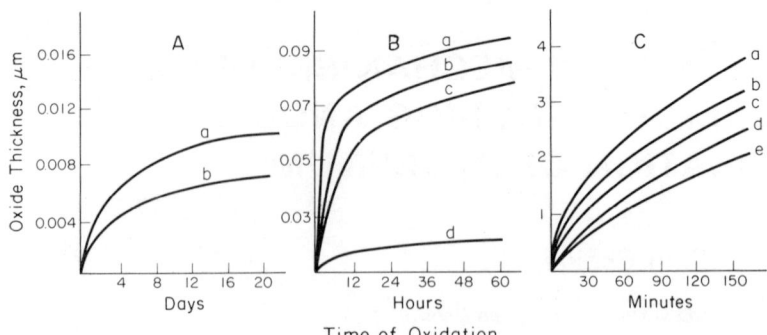

Fig. 1. Change of the rate of oxidation with time. (A) Atmospheric oxidation: (a) Al, 99.6 Al; (b) A, 99.98 Al. (B) Thermal oxidation: (a) 650°C; (b) 500°C; (c) 300°C; (d) 250°C. (C) Chemical oxidation by the MBV process: a: AlCuMg alloy; b: AlMgSi alloy; c: AlMn alloy; d: AlMg alloy; e: 99.5 Al.

authors,[5,6] the thermal oxidation of aluminum involves the initial formation of a barrier oxide film several monolayers thick. If the oxidation temperature does not surpass 500°C, an η-Al_2O_3 layer is formed which is for the most part amorphous; however, at temperatures above 500°C, the layer has predominantly a spinel-type crystal lattice.

In an oxygen-rich atmosphere at 600–650 C, the first few minutes of the thermal treatment produce oxidation nuclei on the surface; these follow the grain boundaries of the crystallites and exhibit an orientation that indicates epitaxy. After prolonged thermal treatment the oxide layer consists of extremely fine and densely arranged grains; under the microscope an

Fig. 2. Oxidation of aluminum at 635°C after 3 hr (200 ×). (Reproduced at 65 %.)

aggregate of rosette-shaped oxide nuclei with microdimensions is observed; the nuclei appear to have grown together with some deformation (Fig. 2).

3. Chemical oxidation is carried out in hot water, in superheated steam, or in solutions containing oxidizing chemicals. According to the modified Bauer–Vogel process (MBV process) chemical oxidation in an alkaline solution of sodium chromate at 90–100°C causes an oxide to form at the metal–electrolyte interface in several phases during which morphologically different formations appear on the metal surface. These formations start with the appearance of oxide nuclei at discrete points; their distribution and size depend on the experimental conditions, while their anisotropic lateral growth and mutual adherence form the heavy oxide coating (Fig. 3). The maximum layer thickness is 1–6 μm, depending on the equilibrium of the oxide formation processes and chemical dissolution of the oxide (Fig. 1, curve C).

4. In electrochemical (anodic) oxidation, the process of oxide formation occurs in an oxidizing solution when an anodic current is applied. Many

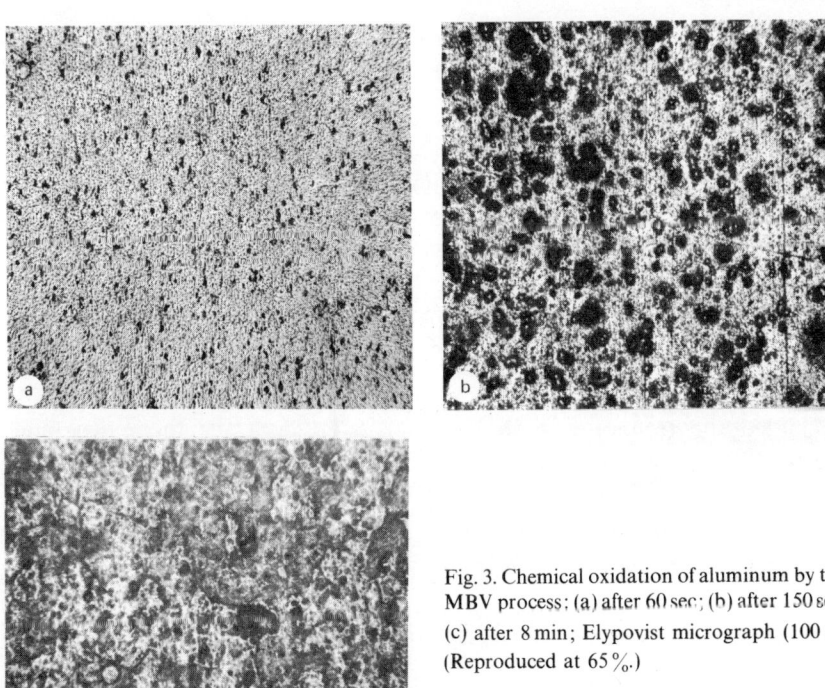

Fig. 3. Chemical oxidation of aluminum by the MBV process: (a) after 60 sec; (b) after 150 sec; (c) after 8 min; Elypovist micrograph (100 ×). (Reproduced at 65%.)

variants of anodic oxidation are known in industrial practice. Oxide coatings can be produced with a variety of properties and thicknesses (anywhere from 1–2 μm to 15–250 μm) on aluminum and its alloys by judicious adjustment of the operating current and solution composition. However, because of the international significance of anodic oxidation for the aluminum processing industries, its technological role, and the importance of the theoretical and practical problems connected with the development of anodizing processes, we shall limit this chapter to the study of anodic oxidation.

We shall describe the results of our research in Hungary on the mechanism of anodic oxide formation and on the structural composition of the oxide layers. In order to promote the evaluation of our conclusions we shall first briefly summarize the conventional theories referring to the explanation of anodic oxide formation processes.

THEORIES EXPLAINING THE PROCESSES OF ANODIC OXIDE FORMATION

The Mechanism of Oxide Formation According to the Model of Keller–Hunter–Robinson

Formerly accepted, but now-obsolete, explanations such as the colloidal theory,[7] which has been popular for a long time, or the theory of "dehydration"[8] will not be considered. Our starting point will be the theory of Keller–Hunter–Robinson,[11] which is based on the assumptions of Setoh and Miyata[9] and Rummel.[10]

Many studies[12–23] in the international literature discuss the Keller–Hunter–Robinson model (KHR model). First among these would be the paper entitled "Anodic Oxidation of Aluminum,"[24] in *Advances in Corrosion Science and Technology*, Vol. 1, in which Sakae Tajima has clearly elucidated the technologies of anodizing from both the theoretical and practical points of view. Since the explanation of the mechanism of oxide formation according to the KHR model is included in the work of Tajima, we shall only stress the main points of the KHR model which are useful for an evaluation of our own research results.

The Processes of Oxide Formation According to the KHR Model

The first stage of anodic oxidation involves the homogeneous formation of a dense and coherent (10–100 Å thick) oxide film throughout the aluminum surface.

The KHR model furnishes no information on the mode of formation, the initial density of the oxide nuclei, or on the fashion in which they form a coherent sealing layer.

The oxide film has a dielectrical nature, and current passes through it only as long as metal ions emerge from the base metal because of the electrical force field between the electrodes, and are able to diffuse into the oxide/electrode boundary. Upon further oxidation, the morphological and structural characteristics of the oxide layer depend on the reactivity of the primary oxide film with the electrolyte and mainly on the chemical solubility of the oxide.

If anodic oxidation is carried out in an electrolyte which is able to chemically dissolve the aluminum oxide produced by the current (i.e., sulfuric acid, oxalic acid, chromic acid, etc.), the oxide will be loosened at the metal surface or at the defect sites, which always occur in oxide films; the current will break down and produce pores in the oxide layer. Aluminum oxide is then re-formed at the bottom of the pores. This process continues until the layer is porous everywhere but practically a uniform structure. It is then covered by an oxide layer whose thickness depends on the length of time the treatment continues. The fundamental part of the aluminum oxide layer which has a structure amorphous under X-ray radiation—the so-called "barrier layer" below the porous covering layer—is usually continuous, coherent, and electrically insulating. According to the KHR model, the formation of pores which are disrupted by the current breakdown and support the continuous conduction of current is a fundamental condition for the formation and thickening of the anodic oxide layer. With the growth of the oxide layer the simple pore openings are extended to pore channels oriented perpendicularly to the base metal.

The pores are surrounded by "oxide cells" which are closely arranged in the shape of hexagonal prisms (Fig. 4). The diameter, volume, mutual distance of the oxide cells and pores, density of distribution, and other geometrical characteristics of the pores depend on the chemistry of the electrolyte, on its concentration and temperature, and on the parameters of the oxidizing current. It must be stressed that according to the KHR model, the closely arranged system of hexagonal oxide cells is not produced by crystallographic laws but exclusively by geometrical (steric) factors.

The stationary equilibrium of the processes of oxide formation and oxide dissolution, which proceed in opposite directions, defines the actual thickness of the double-layer oxide coating. Oxide coatings of a maximum thickness of 30–60 μm can be produced with traditional anodizing processes, and coatings of 100–500 μm with special methods. The former coatings can be dyed

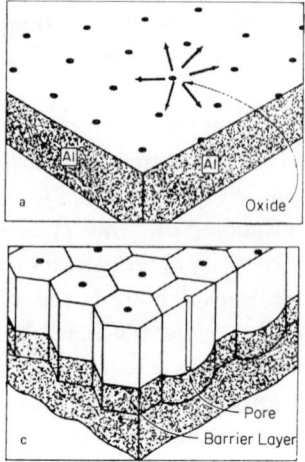

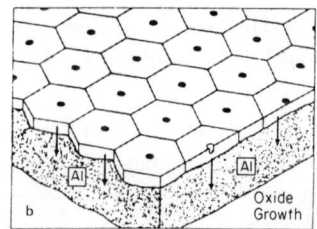

Fig. 4. Keller–Hunter–Robinson model of anodic oxide formation by Barkman (Symp. Anodizing Aluminum; Proc. Univ. Aston in Birmingham, 1967): (a) current breakdowns; (b) formation of a covering layer by the close arrangement of oxide cells; (c) thickening of the covering layer with formation of pore channels.

excellently, and pore-sealing post-treatment results in high corrosion resistance. Therefore, these coatings are used for surface decoration and protection, and because of their special properties, such as high hardness, wear resistance, and electrical insulation, they are also used for special technical purposes.

In electrolytes which dissolve the primarily formed aluminum oxide film very little or not at all (i.e., boric acid, alkali borates, citric acid, tartaric acid, etc.), a dense oxide layer of 0.1–1 μm maximum thickness, with excellent dielectric properties—the barrier-layer type of oxide coating—is formed. To prepare such an oxide layer, a "forming" cell voltage of 100–1000 V is required. To provide a constant current density, the voltage should be adjusted by continuous regulation to the upper voltage value specified. Electrical spark breakdown occurs beyond the maximum specified voltage, usually accompanied by luminous phenomena. We shall return later to the main characteristics of this process (pages 310–311). The thickness of barrier layer-type oxide coatings, which depends on the employed cell voltage, is usually given as 14Å/V.

In electrolytes which extensively dissolve aluminum oxide (i.e., phosphoric acid, a mixture of sulfuric acid and phosphoric acid, hydrochloric acid, etc.) the oxide formed on the aluminum surface during the anodic treatment immediately redissolves almost completely into the oxidizing bath. In solutions of this type an anodic treatment carried out at constant current density causes the cell voltage to increase, and after achieving a maximum, it decreases again. In spite of the redissolution anodizing in such solutions (e.g., in electrolytic polishing), a double oxide layer is produced. It consists of a barrier layer 30–40 Å thick, and a transparent covering layer of less coherent structure, with a thickness of 0.06–0.25 μm.

In addition to the traditional anodizing techniques, oxidative methods have been developed which are carried out in nonaqueous media. Tajima *et al.*[25,26] have prepared very hard, colored oxide coatings in a bath of boric acid dissolved in formamide at 60°C. Methods which specify anodic oxidation in salt melts at high temperatures are of special interest. According to German authors,[27,28] an amorphous aluminum oxide layer of about 100 μm thickness is formed in a melt containing alkali nitrate plus alkali nitrite, at 100–160°C, with a cell voltage of 50–150 V; according to Tajima, such a layer is produced in a melt containing alkali or ammonium hydrosulfite. Because of the high temperature of the medium, this layer is immediately transformed by recrystallization and an extraordinarily hard and wear-resistant layer of α-Al_2O_3 (Carborundum) is obtained.

As Tajima mentioned in *Advances in Corrosion Science and Technology*, Vol. 1, the mechanism of formation of such oxide coatings is essentially similar to the layer formation in aqueous solutions; the difference lies only in the secondary structural transformation provoked by the electrothermal effect.

Further details of the KHR model can be obtained from the literature; only the highlights have been presented here. Next, a few recently discovered examples of the deficiencies of the model will be presented.

Some Deficiencies of the KHR Model

An analysis of the literature reveals a fact that has been disregarded: that most of the theoretical and practical explanations for the mechanism of formation and structure of the sealing and covering layers in anodic oxide coatings depend mainly on conclusions drawn after post anodizing treatment (e.g., rinsing, drying, coloring, pore sealing). These theories involve the morphological parameters and mechanical characteristics of the oxide layer, and to a lesser extent the variation of the cell voltage–current density ratio during anodizing. Surprisingly, very little reference is found in the literature to morphological or other phenomenological investigations carried out *during* the oxidation processes, especially in the initial stage of anodizing.

In studying the operational specifications of the traditional anodizing processes, it is also apparent that the technological parameters are given within very narrow limits. Probably the morphological and structural properties of oxide coatings prepared by these methods served as a basis for creating the KHR model. It seems logical, therefore, that the model so created refers principally to the working area specified by these parameters, and cannot be generally applied to all variants of anodic oxidation of aluminum, which may include some extreme experimental conditions. Although the explanations of the KHR theory have a general validity, its major defect is that it is unable to supply acceptable explanations for phenomena of anodic oxidation recently discovered.

The following examples show the deficiencies of the KHR model. According to the KHR model, the layer formed in the first stage of anodic oxidation and the barrier layers (e.g., the oxide layer obtained in solutions of boric acid or borates) exhibit a practically homogeneous, pore-free dense structure. It has been shown by a special dissolution test that this type of layer is composed of heterogeneous oxide elements with differing solubilities. In addition, Franklin[29] has proved that a small number of scattered pores sometimes form.

The classical theory assumes that the electrolyte solution at the bottom of the pore channels in oxide cells which form the covering layer may be heated by Joule heat to 100–150°C, thus increasing the internal pressure in the pores. At high temperatures and increased pressures, dilated areas should form at the bottom of the pore layer, because of the increased chemical dissolution. Actually, such a phenomenon does not occur, and the pore diameter is always largest at the upper throat, which opens toward the surface. Kaden[30] has already shown the improbability of this superheating of the electrolyte at the bottom of the pore channel by a theoretical calculation of the energy–temperature balance.

According to the KHR model, the structure of the surface at the metal/oxide interface is definitely determined by the current applied at the beginning of the oxidation; later changes of current during the course of anodization have essentialy no influence on the basic structure. Actually, the opposite is the case; the basic structure depends decisively on the current conditions in the final stage of anodic oxidation. Kaden[30] proved the truth of this conclusion by anodizing samples in sulfuric acid and subsequently in oxalic acid. Microscopic observation of polished cross sections of these samples showed that the oxide structure immediately adjacent to the substrate was that characteristic of oxidation in oxalic acid.

It is also impossible to substantiate the assumption of the KHR model that chemical dissolution only acts in pore formation coincidentally with the electrical lines of force (i.e., perpendicularly to the metal base).

According to the geometrical model of KHR the oxide layer is composed of hexagonal column-shaped cells. Electron micrographs of oxides have shown formations which are perpendicular or sloping with respect to the base metal. Straight formations as well as strongly curved or deformed fibers or tubes have also been observed. But so far there is no evidence for the existence of the hexagonal oxide prisms assumed by the KHR model. According to the KHR model, the density of the pores, which are a condition for the formation of the covering layer, depends on the anodizing conditions; they appear in a uniform distribution on the surface. However, many cases have been encountered—both in literature and in our own experiments—where a varying number of pore openings appear in a heavy covering layer, in a completely irregular arrangement, and in varying numbers on different parts of the surface.

According to the KHR theory, the covering layer is electrically insulating, and ionic conduction of the current as well as material transport can only occur via the electolyte solution enclosed within the pore channels. If this assumption is accepted, however, there is no realistic explanation for the increasing thickness of the side walls of the "oxide cells," which also lengthens during anodizing. Kaden has proved by measurements that the insulating ability of the covering layer is negligible and has hardly any influence on the current transport.

According to the KHR theory, the homogeneous or at least quasi-homogeneous base and covering layers are formed continuously. This is true, but only in cases where the oxide coatings are obtained from traditional anodic oxidizing processes. Under other conditions the processes indubitably occur periodically in time and the oxide formations are spatially discontinuous. This is decisively proven by our own experimental results, which will be described later.

Other weak points of the model exist but are too numerous to mention at this point. Only the more glaring deficiencies have been discussed. Our attention now turns to new theories which have been proposed to explain the formation and structure of anodic oxide layers.

Further Theories for Interpreting the Mechanism of Oxide Formation

Murphy and Michelson[31,32] have published a novel theory. We shall omit the details and only mention the salient points of their theory.

These authors propose that after the formation of a dense, barrier Al_2O_3 layer, part of the oxide is transformed into hydroxide and hydrate compounds by the chemical bonding of water, while the parallel formation of a sealing layer progresses toward the substrate. Thus, a covering oxide layer of increasing thickness obtained. It consists of a gel-like matrix formed by the chemical bonding or adsorption of anions and water molecules with the hydroxide and hydrate compounds. Embedded in this gel are finely divided submicrocrystallites of aluminum oxide. Oxidation itself occurs in the interface between the barrier layer and the covering layer; the transformation of aluminum oxide by hydration or peptization continues through the layers in the outward direction. This reaction mechanism produces an oxide coating of individual layers; each layer can have a different chemical composition. Figure 5 shows the Murphy–Michelson model.

Murphy and Michelson assume that pore formation is not a condition of oxide formation; in their view local differences in solubility or electrical breakdown at defect sites play a role in the formation of pores and extended pore channels found in heavier oxide layers. The pores are oriented perpendicularly or at an angle to the surface; under some conditions they produce a fibrous or columnar oxide structure.

The oxide layer formed as described above is permeable to the aluminum ions and to the oxygen-carrying ions. Continuous conduction of current is also promoted by the hydrogen-bridge bonds that form among the aluminum oxide, hydroxide, oxyhydrate particles, the anions, and H_2O molecules. In this polynuclear system of complex composition, the migration of protons and the continuous change of place of hydrogen-bridge bonds (protomery) also promotes current transport.

Ginsberg et al.[33–35] have assumed an intermediate position between the classical theory of KHR and the Murphy–Michelson model. Ginsberg and co-workers did not reject the role of the physical and geometrical factors, but

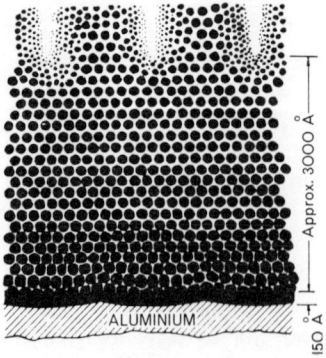

Fig. 5. Anodic oxide formation: Murphy and Michelson model.[31]

instead of a closely spaced oxide cell structure they propose a special fibrous structure. The fibers of the oxide layer are tube-shaped formations; the external part of their wall consists for the most part of amorphous aluminum oxide while the internal part consists of a gel containing hydroxide and bonded anions. The internal cavity of the pore channel is filled with the electrolyte solution. The aluminum hydroxide can also be assumed in the external part of the fibers and in the boundary phase separating the fibers. Since the fibers consist of water-free aluminum oxide which is insoluble in dilute bases, they can be freed by dissolving the material between the fibers in 0.1% sodium hydroxide solution. Ginsberg and co-workers present the convincing results of differential thermoanalysis, IR spectroscopy, electron microscopy, and optical microscopy to prove their assertion.

The results of research carried out by Bogojavlenski and co-workers[36-38] with electrochemical and optical methods are very significant. They have also stated the fact that anodic oxide formation does not occur on the aluminum surface with complete homogeneity; it starts with the appearance of discrete oxide islands, called "monon bodies," which continue to grow, thus producing the oxide layer.

The theory of Akahori is very interesting[39]; he proposes that after the formation of the barrier oxide layer and pores, the electrolyte is vaporized by Joule heat at the bottom of the pores, and simultaneously aluminum metal melts at the base of the pores. The interaction between the molten metal and the electrolyte solution produces a superheated gas film consisting of vapors and oxygen produced by the current; the thickness of this film may attain an order of 500Å if the oxidation voltage is 50 V. Oxygen ions pass from the gas film through the sealing oxide layer toward the liquid aluminum base, and on encountering the aluminum ions, molecules of aluminum oxide are formed. Akahori does not present concrete proof for his mainly speculative explanations, and therefore his conclusions should be accepted only tentatively.

In addition to the recent explanations discussed above we should like to mention the excellent study of Diggle et al.,[40] which not only systematizes the former theses but supplies additional data to explain the mechanism of anodic oxide formation.

The foregoing survey shows that the proven fundamental theses of the classical theories may be accepted; however, further data and investigations are required to explain the formal and structural creation of the oxide layer more realistically and to furnish explanations which can be applied even to extreme technological variants.

In the study of electrolytic oxidation processes, we can also utilize to a certain extent the recent studies of Benard,[41] Rhodin *et al.*,[42,43] Darras and Dominget,[44] Mykura,[45] Doherty,[46] Fischmeister,[47] Paidassi,[48] Franck,[49] and other well-known authors,[50-55] which discuss the kinetics of heterogeneous reactions, mainly in solid/gas systems.

The preceding literature review has presented several explanations for the phenomenon of metal oxidation due to atmospheric, thermal, chemical, or electrochemical factors. The data presented in the literature, together with the aid of a common formal language, can be used to describe oxide formation.

On the following pages we shall summarize our investigations over several years, which have yielded novel results; in our opinion these results may resolve the contradictions of traditional explanations and serve to provide a better understanding of the mechanism of anodic oxide formation.

THE STUDY OF THE ANODIC OXIDATION OF ALUMINUM WITH THE SPECIAL METHODS OF REACTION KINETICS—EXPERIMENTAL PROCEDURE

We have studied the mechanism of anodic oxidation in various electrolytes using conventional methods which have been developed to investigate the electrokinetic processes occurring on the metal surface. The qualitative and quantitative data, the temperature of the oxidizing medium, as well as the electrochemical relations among the current parameters (cell voltage, current density plots, etc.) have been determined by well-established measuring methods.

However, we developed a special method for reaction kinetic testing which has yielded very convincing proofs for our explanations of the mechanism of anodic oxidation; this method consists of taking continuous photographs under the Elypovist microscope from the beginning of anodic oxidation to its end (i.e., during the whole reaction). By preparing such a series of kinematographic moving pictures, or documentation film, we were able to record precisely the morphological and structural characteristics of oxide formation. The theoretical and practical conclusions drawn from these studies are presented in the section on the mechanism of anodic oxide formation.

For a comparative evaluation and for the sake of completeness, the usual physical, mechanical, microscopic, electron optical, and chemical characteristics of the properties of anodix oxide layers are summarized on pages 332–349. The experimental equipment and details developed in our laboratory will be presented in this section.

The photomicrographic films were obtained with the aid of a movie camera **Pentaflex** 16 (VEB Kamera- und Kinowerke, Dresden) which was mounted on an Elypovist microscope (VEB C. Zeiss, Jena). My co-workers, M. Riedl and R. Karacsonyi, have been of great help in constructing this equipment.[56]

The Elypovist microscope (Fig. 6b) is a combination of an optical microscope and an electrolyzing cell; with its aid one can study continuously the morphological and structural changes which occur electrochemically at the surface of the test metal electrode from the beginning to the end of electrolysis. The process can be projected in suitable magnification onto a fixed or moving film. The optical microscope can be fitted out with a $15 \times /0.30$ Apochromat, a $5.5 \times /0.10$ Triplet or a $K15 \times$-Okular optical system; this combination gives a $200 \times$ magnification. Figure 7 shows the sketch of the arrangement.

The electrolyte tank (1a) is suitably filled with the solution selected for the test (about 140 ml), and the pump insert (1b) is fitted between the sealing locks. The electrolyte tank and all other parts of the equipment which contact the electrolyte are made of chemically resistant, high-strength plastic.

Pipes leading out of and back to the electrolyte tank in the lateral direction permit a continuous circulation of the solution. The flow rate of the solution can be regulated by adjusting the pump motor. To avoid any damage to the pump and the electrolyzing cell, the maximum temperature of the electrolyte solution cannot exceed 50°C.

A specimen is prepared from the metal to be tested (aluminum or aluminum alloy); this has a circular and perfectly flat surface about 6 mm in diameter which can be fixed to the diaphragm aperture (2) of the electrolyte

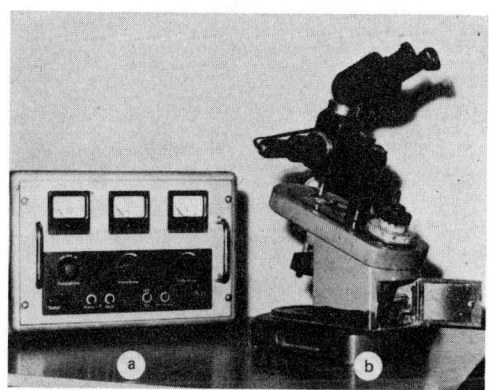

Fig. 6. Elypovist microscope equipment: (a) switching and current-regulating equipment; (b) microscope.

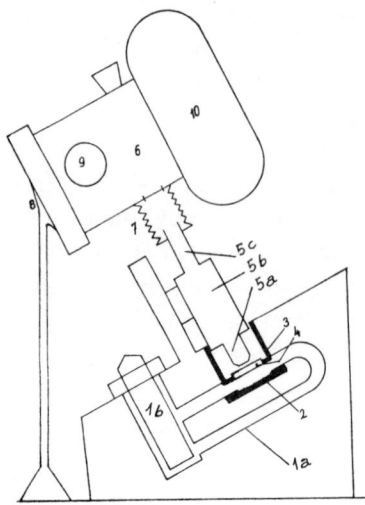

Fig. 7. Sketch of the arrangement of the Elypovist-Pentaflex microcinematographical testing equipment: electrolyte tank with lateral pipe (1a) for passing the oxidizing solution and pump (1b); metal sample pressed against the diaphragm aperture (2); counterelectrode (3); quartz glass window in the tube wall (4); microscope: objective (5a); lens barrel (5b); eyepiece system (5c); Pentaflex film camera (6); adapter (7); tripod (8); auxiliary motor (9); film cassette (10).

pipe with a special clamp. After suitable degreasing and pickling, the metal specimen can be connected to either pole of the current source.

For anodic oxidation or electrolytic polishing, the metal specimen is connected to the anode, and for plating of metal or for the evolution of hydrogen it is connected to the cathode.

The counterelectrode is represented by a cup-shaped element made of chemically resistant stainless steel (3). A window covered by quartz glass is arranged at the bottom of the cup; through this window a suitably lighted electrode surface can be visually inspected or photographed once the photographic equipment is mounted.

The current to the electrolyzing cell is supplied by a switching and regulating apparatus (Fig. 6a) which is connected to a 220-V ac mainline. The transformer and rectifier incorporated in the equipment conduct the current through three connected cables to the electrolyte solution pump, to the illuminating lamp system of the microscope, and to the electrode poles of the electrolyzing cell.

The regulating instruments inserted in this latter cable permit regulation and control of the cell voltage and the current density on a voltmeter and an ammeter, respectively. The DC voltage is variable—from 0 to 15 V or from 0 to 100 V. The current varies from 0 to 10 mA and from 0 to 500 mA. The polarity of the specimen can be changed at will with the aid of a mains-changeover instrument. Recently, a further changeover switch has been included to permit the application of ac current for electrochemical surface treatment.

The Elypovist microscope and the microcinematographic testing method can be used to study not only electrochemical but also purely chemical surface processes. In this latter case the electrolyzing circuit is not applied, but the rest of the equipment is operated as mentioned above.

The arrangement, operation, and maintenance of the microscope are described in the technical publication *Elypovist Elektrolytisches Poliergerät* by the C. Zeiss firm.

For the cinematographic investigations, the Pentaflex movie camera was mounted on the Elypovist microscope with the aid of an adapter made of a metal tube and a bellows which excluded the light. The adapter (7) is joined to the microscope after removing the eyepiece from the microscope; it projects the image of the electrode in the electrolyzing cell directly onto the cinematographic film. The photographic apparatus is mounted on a fixed tripod (8) to eliminate vibrations from the equipment. The sharpness of the image is adjusted directly in the searching mirror of the photographic camera. The image appearing on the 16-mm cinematographic film has a 50–200 times magnification which can be increased when projecting the completed film.

The film speed can be adjusted so that the usual projection rate of the film (24 pictures/sec) gives a good representation of the consecutive changes on the metal surface during electrolysis. This can be generated with the Pentaflex camera because the shooting rate can be varied from 3 to 96 pictures/sec by incorporating an auxiliary motor.

The illuminating equipment of the Elypovist microscope is generally sufficient to illuminate the field of vision. When shooting at high speeds or taking color pictures, however, a Tungsram incandescent lamp and paraboloid mirror can be employed to advantage. In such cases a heat screening sheet should be inserted to provide insulation against the thermal effect of the incandescent lamp.

Optimal conditions for filming (i.e., the film speed), the exposure time, and lighting were generally found by trial and error.

Commercial film with a sensitivity of 17 DIN(=40 ASA)–27 DIN(=400 ASA) and 16 mm (e.g., Forte, Orwo, Agfa, Kodak) proved suitable; the color pictures were taken with commercial color films (e.g., Orwocolor, Agfacolor, Kodacolor). The developing and fixing of the negatives, the preparation of positive film copies, and the synchronization of the documentary sound film were carried out according to the usual method of cinematographic technique.

Finally, we should like to mention that a documentary film has been prepared (time of projection: 25 min) with our microcinematographical equipment, Elypovist-Pentaflex, described above, entitled "The Mechanism of

Nucleation and Electrochemical Transport Processes in Oxide Formation during Anodic Oxidation of Aluminium." It has been shown in the past years at technical and scientific conferences and other events in Budapest, London, Stockholm, Moscow, Hannover, and Tokyo and has received excellent reviews.[57]

We should like to remark that further practical technological details of our research results which have been omitted here can be found in the relevant publications.[58]

To support the explanation which we propose for anodic oxide formation, we shall present a characteristic series of images taken from our moving picture. These pictures, together with our optical and electron photomicrographs and with our results from the physical mechanical and electrochemical investigations, will be used to characterize the process of oxide nucleation and the mechanism of anodic oxidation.

CHARACTERISTICS OF THE MECHANISM OF ANODIC OXIDE FORMATION—NUCLEATION MODEL

The Role of the Technological Factors of Anodic Oxidation in the Morphological and Structural Development of the Oxide Layer

We have studied the effect of the experimental conditions of anodic oxidation on the morphological and structural progress of oxide formation in various technological conditions.

The characteristics of the various stages of anodic oxide formation are determined by the type of electrolyte, its concentration, the temperature, and the time of anodization. The reaction kinetics and equilibrium between the processes of oxide formation and oxide dissolution, as well as the cell voltage and anodizing current density, also play a role in determining oxide characteristics. The composition, microstructure, and surface quality of the base metal contribute to these processes and to the development of the physicochemical and structural properties of the oxide layer.

To obtain a better insight into the relationships among the phenomena of oxide formation, we took films of the experiments in various electrolytes at constant temperature; we varied only the concentration of the electrolyte and the cell voltage of the anodizing current.

In industrial practice the oxide coatings are usually prepared in moderately concentrated oxidizing baths with low or medium cell voltage; only in special cases are high cell voltages used. The parameters used in our experiments are summarized in Table 1.

Table 1. Experimental Values of Oxide Coatings

Solution	Concentration (%)	Temperature (°C)	Voltage (V)	Current density (A/dm²)	Efficiency (%)
Sulfuric acid	10–30	18–20	12–20	0.5–2.5	64–78
Oxalic acid	5–5	18–20	25–60	0.5–1.5	50–70
Chromic acid	1–5	20	5–50	0.3–3.5	46–55

In new processes (e.g., in the so-called autocoloring anodizing process, which has attained worldwide popularity), anodizing is carried out in organic or inorganic electrolyte baths with stepwise regulation of the cell voltage. The oxide coatings obtained with the traditional processes are discussed in the literature. In anodic oxidation carried out in conditions dissimilar to those of the classical processes, the phenomena of oxide formation which cannot be interpreted on the basis of the KHR model appear more pronounced in certain ranges of electrolyte concentrations and cell voltage. To demonstrate this statement, we shall describe the characteristic phenomena of the following series of experiments.

Investigation of the Phenomena of Oxide Formation in Moderately Oxide-Dissolving Environments during Anodizing with DC Current

Specimens for the following tests were prepared from sheet material rolled from 99.9 and 99.5 aluminum and from an AlMgSi alloy. The surface of the sample plates was degreased before anodizing by rinsing in an organic solvent and pickled in a solution containing $HNO + HF + H_2O$.

The oxidizing baths were prepared from reagent-grade solutions of sulfuric acid, oxalic acid, and chromic acid of various concentrations.

The films of oxide formation on the surface of the sample plate anodes in the electrolyzing cell of the Elypovist microscope were taken with a Pentaflex camera at $100–150 \times$ magnification.

Visual observation and the films taken during the anodizing process show that anodic oxide formation does not begin homogeneously at all points of the metal surface—as expected from the statements of the KHR model. Rather, it begins as tiny oxide nuclei dispersed on discrete sites of the metal surface. The number, size, and distribution of the primary oxide nuclei depend on the experimental parameters, especially on the type and concentration of the electrolyte and on the cell voltage of the oxidizing current.

Anodizing in a 20–10% Solution of Sulfuric Acid. The photomicrographs in Fig. 8 show that the first oxide nuclei that appear, at a cell voltage of 12–18 V (GS process, alumilite process), are at definitely discrete sites and heterogeneously distributed. Eventually, the nuclei align themselves along the rolling direction of the metal (micrographs a and b). Their number increases rapidly; in a fraction of a second after switching on the current they cover the whole surface (micrographs c and d). They have a nonuniform distribution with various sizes and shapes. It takes 5–10 sec for the nuclei to impinge; and when this occurs, a porous oxide layer is formed. Its porosity is due to selective oxide dissolution. Such oxide coatings appear frequently in technical publications (Fig. 8e).

Similar phenomena are encountered when anodization is carried out in sulfuric acid of lower concentration, such as 15% or 10%, which is still within the limits of traditional specifications. Upon dilution of the bath, the oxidizing voltage is increased to 15–20 V for a 15% solution and to 20–25 V in a 10% solution.

Anodization in a 5% Sulfuric Acid Solution. When anodization is started at a cell voltage below 22 V, a light-grayish tint appears on the aluminum surface within the first second. This indicates the formation of the basic oxide membrane; subsequently, a great number of small oxide nuclei rapidly appear on the membrane. After the coalescence of this rough, granular layer, pore formation rapidly occurs as a result of the dissolution of the oxide. Finally, the covering oxide layer is formed (Fig. 9).

When anodizing is started with a cell voltage of 26–28 V, the metal surface will darken for a moment—similar to the phenomenon described above—and an intense nucleus formation immediately begins. The first oxidation nuclei grow in three dimensions in a geometrically irregular shape (Fig. 10a–c). The growth of some nuclei stops at the beginning of the process with their size and shape unchanged while other nuclei appear and grow (micrographs d–f). After attaining an upper size limit which characterizes the experimental conditions, the growth of these deformed and calotte-shaped oxide formations is suddenly interrupted; and at the edge of some of these primary oxide nuclei, oxides of a different character—called secondary zones of oxidation—appear and spread laterally with anisotropic growth (Fig. 10g). The layer thickness of the secondary oxide zone is always less than the top part of the calotte of the primary oxide nucleus. While this layer is mostly dark and nontransparent, the secondary oxide is of a much lighter color and often glossy and translucent.

The apparently unoriented, sometimes concentric, but mostly irregular lateral growth of the secondary oxide obeys a certain periodicity. At the reaction front a new oxide product appears in the shape of a narrow band. The boundary between the bands after some time appears in a fine design of lines (Fig. 10h). These periodically repeated reaction waves are the reason why the

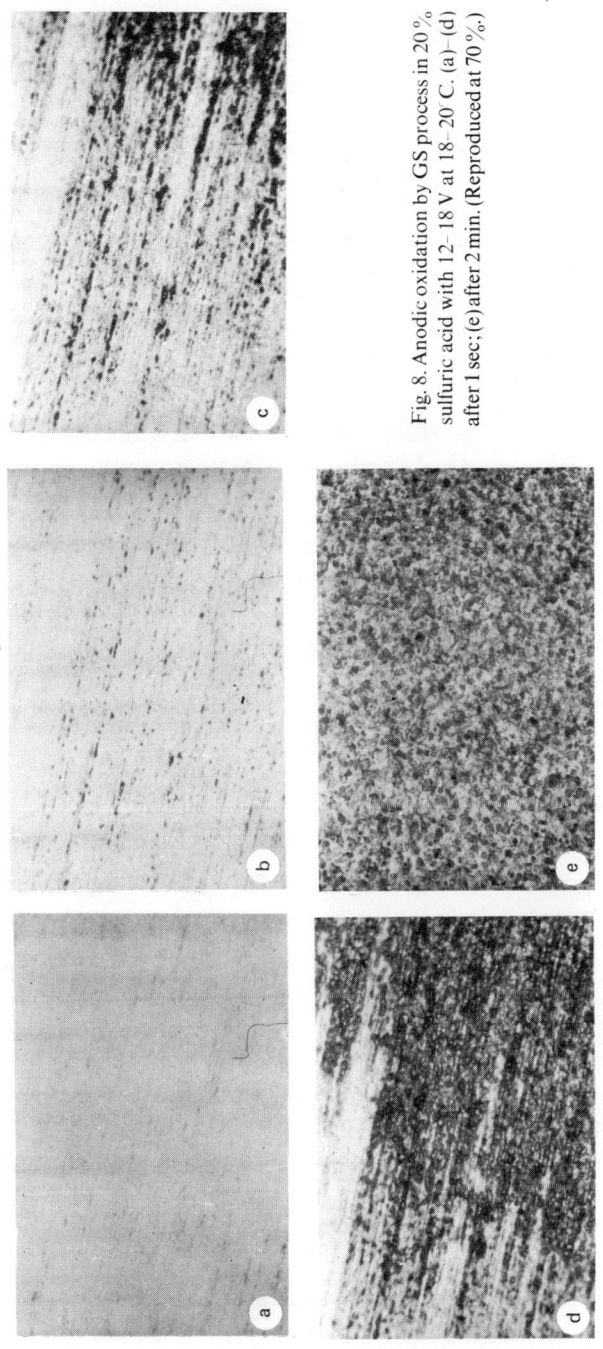

Fig. 8. Anodic oxidation by GS process in 20 % sulfuric acid with 12–18 V at 18–20 C. (a)–(d) after 1 sec; (e) after 2 min. (Reproduced at 70 %.)

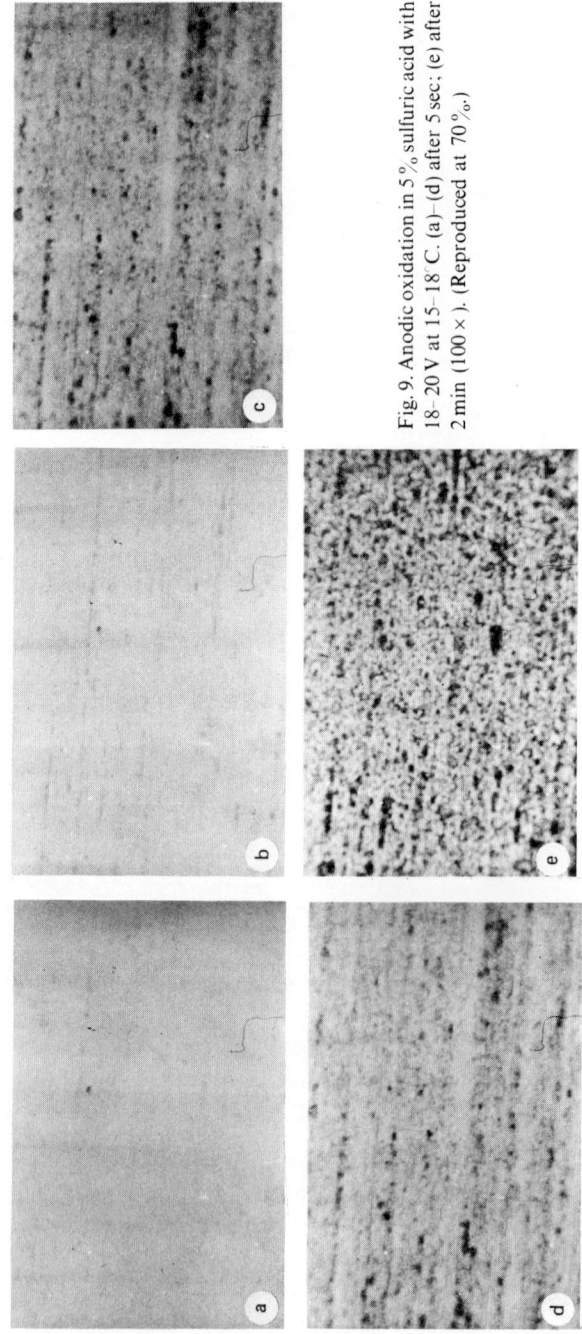

Fig. 9. Anodic oxidation in 5% sulfuric acid with 18–20 V at 15–18° C. (a)–(d) after 5 sec; (e) after 2 min (100 ×). (Reproduced at 70%.)

secondary oxidation form a "terraced" structure in a lobed or rosette-like shape. The lateral extension of the oxidation zones continues until either they touch each other or a primary oxide nucleus which carries no secondary oxide formation at its edge. The fields of secondary oxidation surround primary nuclei but are separated from them temporarily by a depressed area resembling a ditch.

After the total coverage of the metal surface, the processes of formation of primary and secondary oxide nuclei are replaced by another mechanism. The current and material transport continues through the structurally heterogeneous, thickening, oxide layer which separates the metal from the electrolyte. As a consequence of the dissolution of the oxide and of pore formation, the differences in structure and depth between the primary nuclei and the secondary zones of oxidation are slowly obliterated. A prolonged anodization in 5% sulfuric acid yields an oxide coating of more-or-less uniform thickness and pores of various sizes. The oxide elements observed in the early stages of the reaction seldom appear, but when they do, they are observed as pale stains on the surface (Fig. 10i–k).

When anodic oxidation is carried out with a cell voltage double the previous value (i.e., 36–38 V), the primary nuclei appear in the first half-second after switching on the current. These nuclei are similar to the primary nuclei observed in the previous example, but they are less in number (Fig. 11a and b). They grow rapidly and are larger than the previous nuclei. We have actually measured the nuclei size on photomicrographs taken at various times with the Elypovist microscope, and have found that their size distribution can be represented by an asymmetrical Gaussian curve. Since the nuclei continue to grow erratically and new nuclei appear, the oxide islands coalesce. Secondary oxidation zones appear on the sides of the primary nuclei that do not touch other nuclei (Fig. 11h–k). This oxide formation is similar to that previously described (i.e., the propagation is characterized by wave like surges). Their propagation direction is probably directed by random energy or concentration fluctuations and not by the structural elements of the base metal. Each reaction wave produces an oxide field of significant thickness, as can be seen by the shadows on the micrographs; the edges of the zones descend to the level of the first oxide membrane at various angles (Fig. 12).

The reaction waves occur periodically. Each new wave generally starts at irregular intervals and proceeds very quickly, much as in an avalanche. The oxide is formed quite suddenly within these periods in a band parallel with the wave front. The unusually long periods gradually slow down; sometimes they even stop for a moment, only to start again like an avalanche. The fine lines

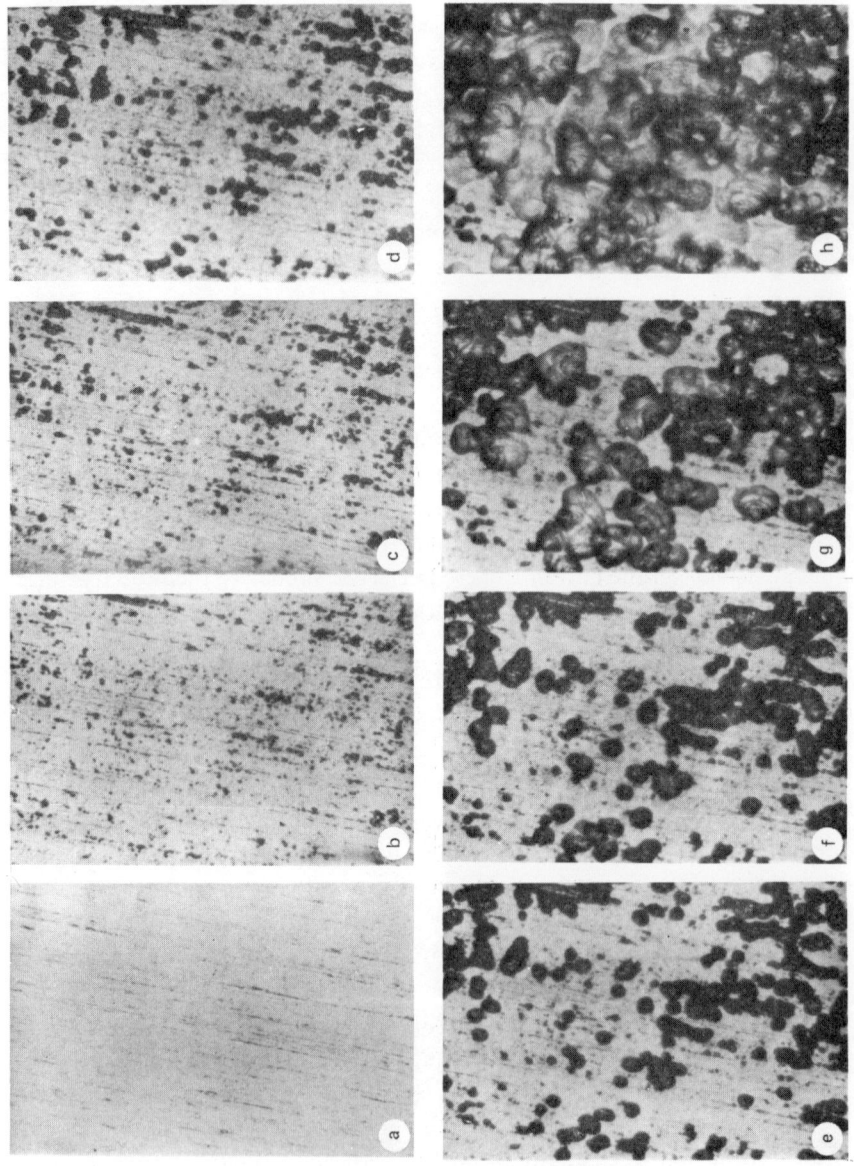

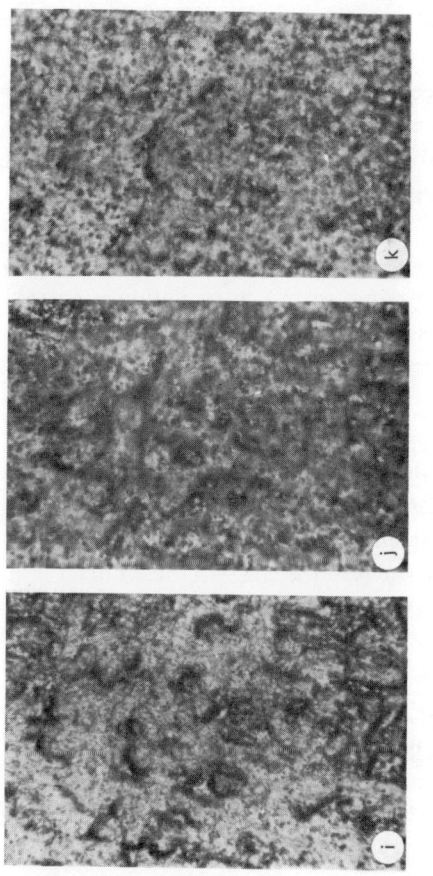

Fig. 10. Anodic oxidation in 5 % sulfuric acid with 26–28 V at ±1°C: (a)–(d) after 10 sec; (e) and (f) after 20 sec; (g) and (h) after 1 min; (i)–(k) after 5 min (100×). (Reproduced at 58%.)

following the wave front are not always visible in the longer periods because of the limited resolution of the Elypovist microscope; when using a microscope with high magnification, however, the lines are definitely visible (Fig. 13). This type of oxide mechanism is caused by local energy fluctuations.

Interestingly, the oxide exhibits a smooth surface, free of pores, in the newly formed bands which are closest to the reaction front. Then a variety of pore openings appear randomly distributed, and the bandlike structure created by the reaction waves is gradually obliterated by the subsequent oxide dissolution.

After the metal surface has been totally covered by primary and secondary oxide elements, and intense pore formation starts by a chemical resolution quite similar to the process above. After the gradual obliteration of the differences between the oxide elements, the usual picture of an oxide coating is observed (Fig. 111–p).

Anodizing in 1% Sulfuric Acid Solution. The nucleation phenomena, which were previously unknown, appear conspicuously when oxidation is carried out in a 1% sulfuric acid solution. It is especially interesting to observe the significant role that the cell voltage plays in forming oxides with different structures.

When oxidation is carried out at a cell voltage of about 50 V, the number of primary oxide nuclei is less than during anodization in a more concentrated solution. These nuclei grow to an average size and are approximately spherical (Fig. 14a–f). Growth stops at some primary oxide nuclei and a laterally spreading zone of secondary oxidation starts at the edges of these nuclei (Fig. 14g–j). This process continues in a fashion similar to that described for oxidation in a 5% sulfuric acid solution (Fig. 14k–n).

These phenomena are even more conspicuous when the cell voltage is reduced. At 36–40 V, only a few oxide nuclei appear initially (Fig. 15a–c). These are then irregularly surrounded by zones of secondary oxidation which may take on a rosette or lobed terrace structure. These formations grow anisotropically until they coalesce and completely cover the metal surface [Fig. 15h–o; this stage is succeeded by the thickening of the layer (Fig. 15p)].

Upon further voltage reduction, another type of oxide formation occurs. When oxidation is started at 32–34 V, a few primary oxide nuclei appear at active sites of the metal surface [i.e., at rolling traces (Fig. 16a–c)], but their growth is rather restricted. Secondary zones of oxidation rapidly form around the primary nuclei (Fig. 16d–i), and after their coalescence, chemical and electrochemical processes induce the formation of a uniform covering layer (Fig. 16j–m).

Anodizing in 0.1% and 0.01% Sulfuric Acid Solutions. In these unusually dilute solutions, aluminum oxide formation occurs in the same fashion as was described above; however, the characteristics of the partial reactions and the morphological and dimensional data of the oxide formations differ from these of the classical models.

When anodizing is carried out in 0.1% sulfuric acid at 55–60 V, primary nucleation is rare; the oxide nuclei which do appear grow to almost twice the size of those on samples anodized in more concentrated oxidizing baths (Fig. 17a–d). The oxide layer which forms by the coalescence of the secondary oxides between the well-developed primary nuclei has a coarser structure than that formed on the earlier samples (Fig. 17e–i).

At 45–50 V the number of primary nuclei is much less when current is switched on (Fig. 18a–c). Their growth soon slows down and is followed by a vigorous formation of secondary oxide zones (Fig. 18d–i). After the surface is covered, the process of layer thickening leads to the formation of a heavy oxide layer with an equilibrium-type structure (Fig. 18k and l).

A further reduction of the cell voltage almost completely prevents the formation of primary oxide nuclei. With a cell voltage of 35 V only a few nuclei are visible (Fig. 19a–c). After some time the metal surface appears gray, which indicates the formation of a thin and coherent oxide membrane; under this membrane the secondary zones begin to form and to propagate laterally with a lobe-shaped structure (Fig. 19d and e). After suitable thickening the coating exhibits the conventional porous surface.

In a 0.01% sulfuric acid solution, anodic oxidation starts only at a relatively high cell voltage of 60–65 V. Few primary oxide nuclei appear, and these rapidly attain large, coarse proportions (Fig. 20). However, the three-dimensional propagation of the nuclei quickly slows down, and finally it stops completely. Surprisingly, the secondary oxidation zones fail to appear after a long time has elapsed and when the cell voltage is increased to 70 75 V. Microscopic investigation at high magnifications of this surface has shown that a thin oxide film forms at the aluminum surface around the primary nuclei; its density and insulating properties prevent the processes of secondary oxidation. When anodization is started and carried at a cell voltage less than 50 V, no nucleation is observed; only a high-density barrier layer with dielectric properties is formed. This is similar to the dielectric layer formed when anodizing in a boric acid solution.

Anodization in 5% and 1% Oxalic Acid Solutions. This type of anodization is similar to the GX process and the chrome-alume process.

Anodic oxidation was studied on aluminum plates immersed in a 5% oxalic acid solution with 45 V applied. As was expected from previous

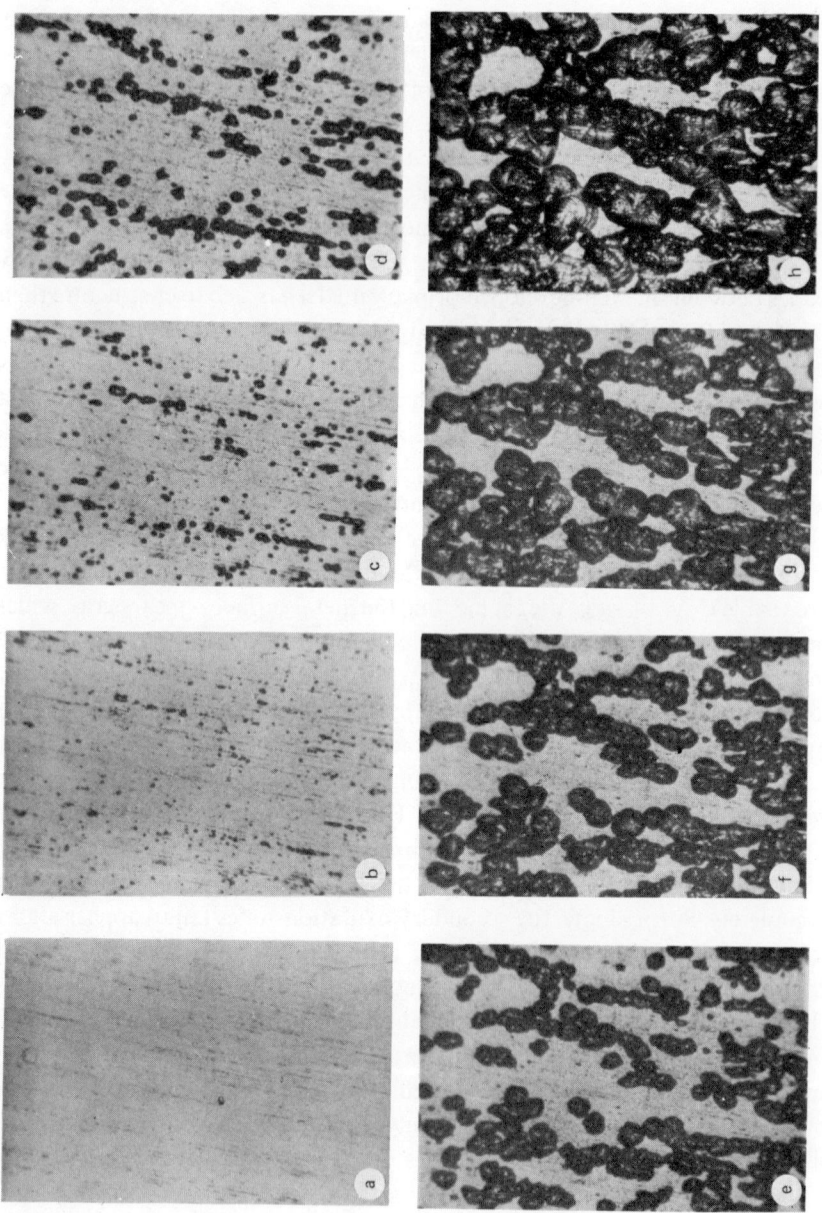

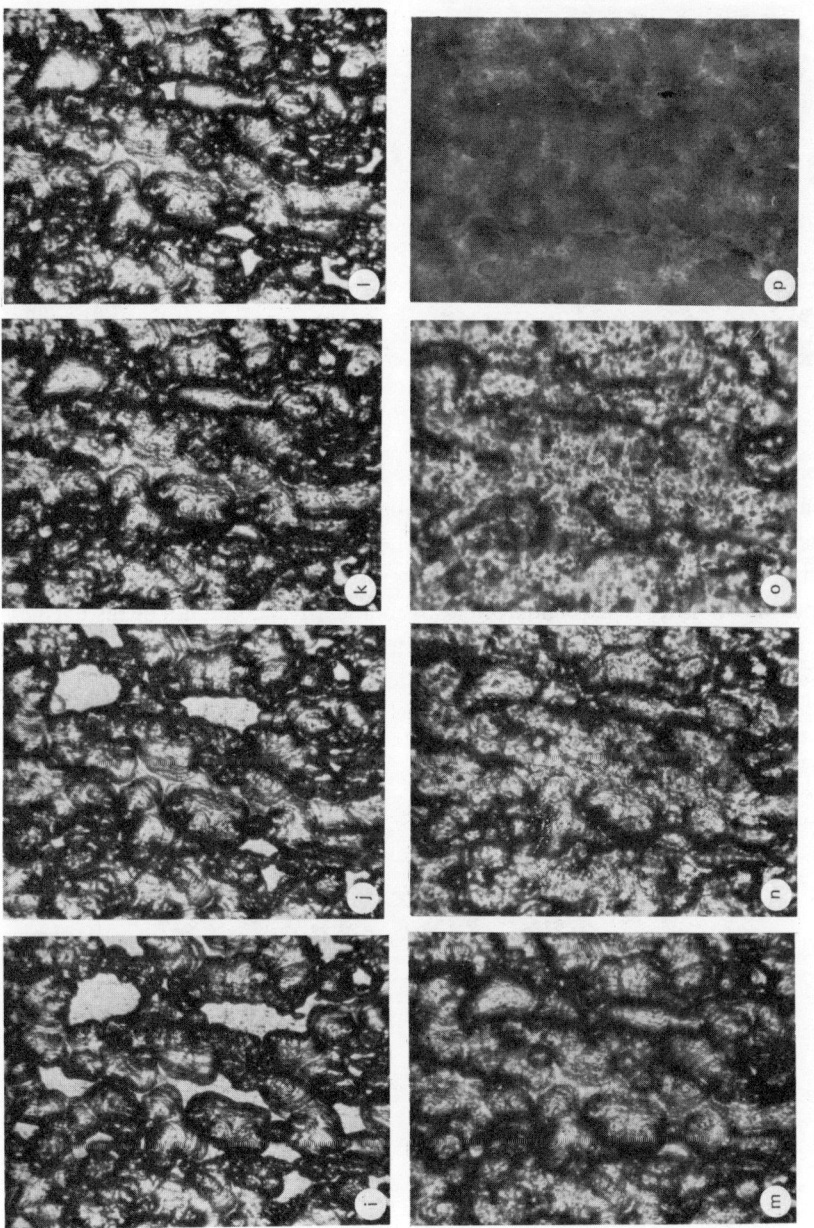

Fig. 11. Anodic oxidation in 5 % sulfuric acid with 36−38 V at ±1°C: (a)−(m) after 10 sec; (n)−(p) after 5 min (100 ×). (Reproduced at 65 %.)

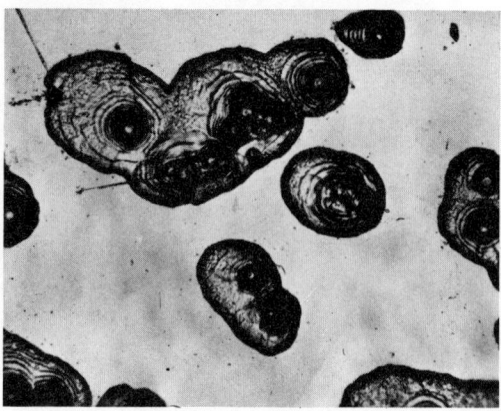

Fig. 12. Formation of primary oxide nuclei and secondary zone of oxidation after 15 sec in 1–5% sulfuric acid with 30–40 V at $\pm 1°C$ (200 ×). (Reproduced at 70%.)

experience with oxide coatings obtained in oxalic acid, the initial oxide membrane was composed of extremely fine grained oxide nuclei. Observable oxide nuclei appear only on large defect sites on the aluminum plate (Fig. 21a). Only the nuclei that appeared on the rolling marks could be seen. If nuclei appeared on other types of microheterogeneities, they could not be resolved by the microscope. Oxide re-solution then induces pore formation, and a thick covering oxide layer with a uniform pore distribution is obtained (Fig. 21b and c).

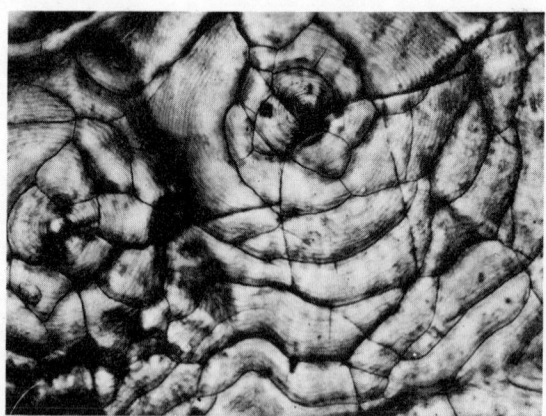

Fig. 13. Coalescence of primary and secondary oxide formations to form a coherent oxide layer after 120 sec in 1–5% sulfuric acid at 45 V at $\pm 1°C$ (200 ×). (Reproduced at 70%.)

When the concentration of oxalic acid is reduced to 1%, an anodizing current of 60 V produces an oriented continuous nucleation similar to anodic oxide formation in 5–10% sulfuric acid. After the first oxide forms (Fig. 22a–c) nucleation occurs haphazardly, leaving bands of bare material (Fig. 22d and e). Subsequently, these bands are populated by finely structured nuclei (Fig. 22f). Prolonged anodixing produces a smooth, uniform coating over the entire surface (Fig. 22g).

It was necessary to investigate the source of the nonuniform bands, so a new set of samples were prepared and a careful microscopic investigation was carried out throughout the anodization treatment. Thus, it was found that the surface condition due to rolling has a greater effect on the initiation of the oxide formation when anodizing in oxalic acid than it does when anodizing in sulfuric acid.

Anodizing in a Mixture of Oxalic Acid and Sulfuric Acid. We mentioned in the introduction to this section that the morphological characteristics of anodic oxide formation depend substantially on the quality of the oxidizing electrolyte. This thesis has been proved by the differences in the superficial oxide formations on samples anodized in sulfuric acid and oxalic acid solutions.

The reaction mechanisms for oxide formation have been found to depend on the type electrolyte, its concentration, and the cell voltage. The interaction among these operating parameters influences the mechanism of oxide formation. We have studied the relationships in many experimental variants, and the following examples are presented to illustrate the results.

A pure aluminum plate was anodized in a mixed oxidizing bath composed of 10% oxalic acid and 1% sulfuric acid at 45 V. The dominant role on coating formation in this case is played by oxalic acid, which is present in a concentration corresponding to the conventional specifications. Results similar to those obtained in a pure 5–10% oxalic acid solution were obtained (as in the previous subsection). An oxide layer is obtained at the end of the process with a uniform density of very small nuclei (Fig. 23).

When the oxidizing treatment is carried out in the same bath with a cell voltage of 25 V (almost half of the former value), partial processes similar to anodizing in dilute sulfuric acid solutions of 1–0.1% with 32–34 V appear instead of the characteristics of anodizing in oxalic acid. After the appearance of a few very small primary nuclei, secondary oxide phases are formed at active sites on the metal surface and spread laterally (Fig. 24a–j), as shown in Figs. 16 and 19 for anodizing in pure sulfuric acid solutions. In the final stage of the process, the structure becomes similar to that obtained when anodizing in oxalic acid at higher voltages (Fig. 24k and l).

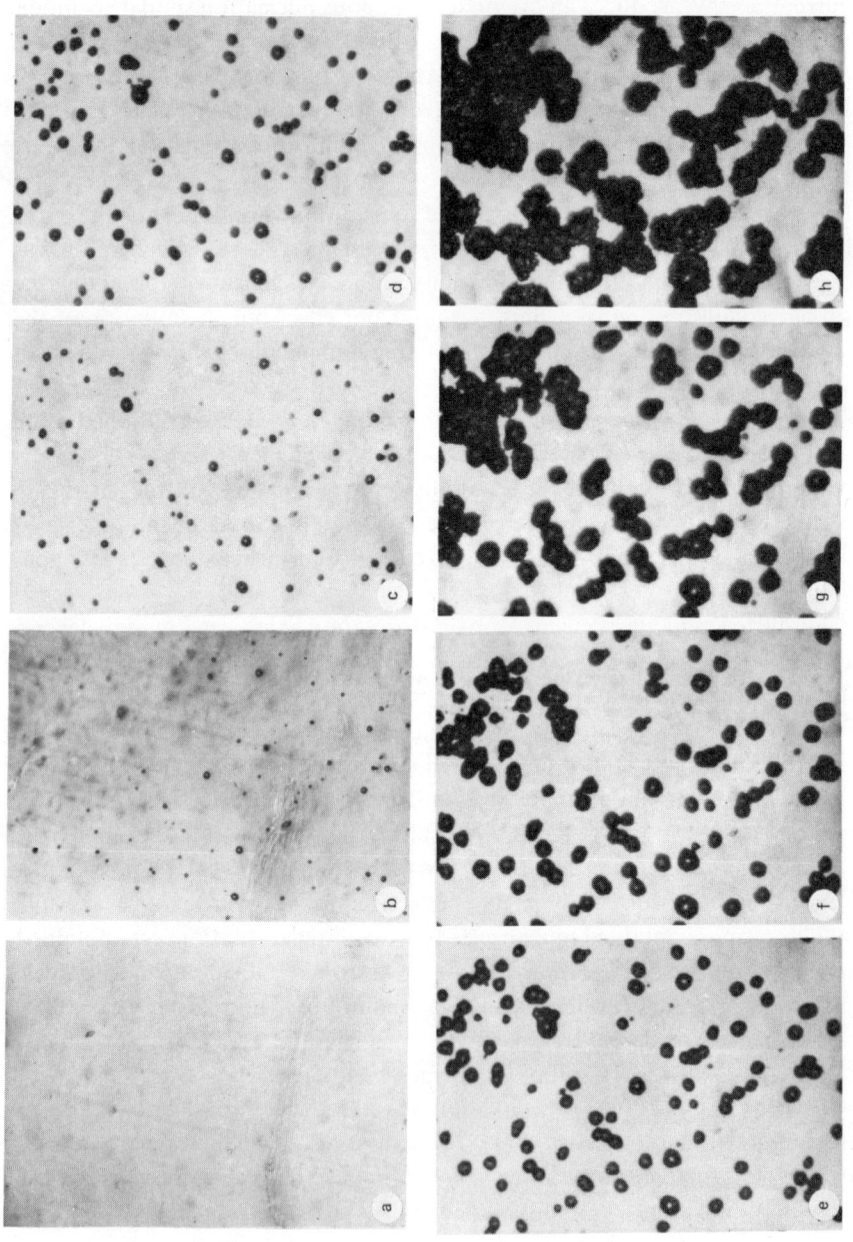

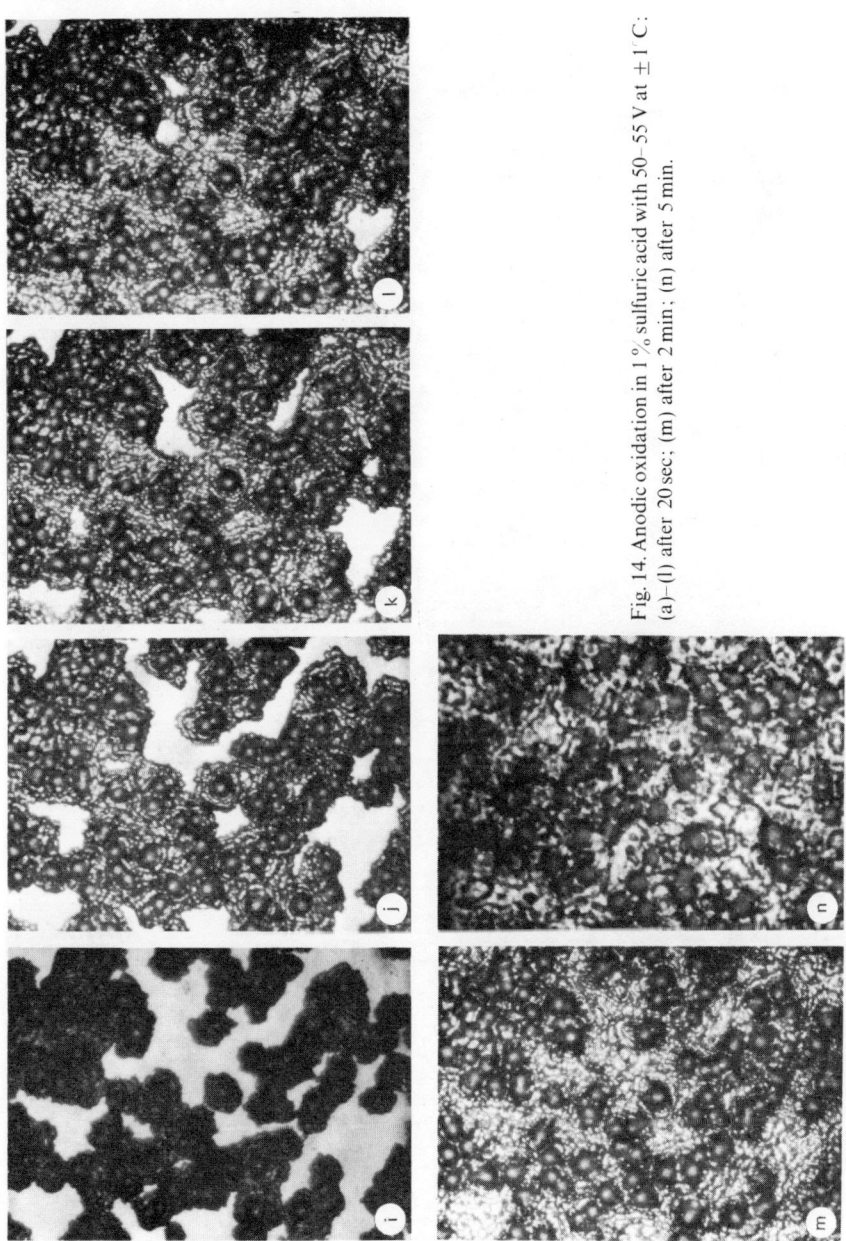

Fig. 14. Anodic oxidation in 1 % sulfuric acid with 50–55 V at ±1 C: (a)–(l) after 20 sec; (m) after 2 min; (n) after 5 min.

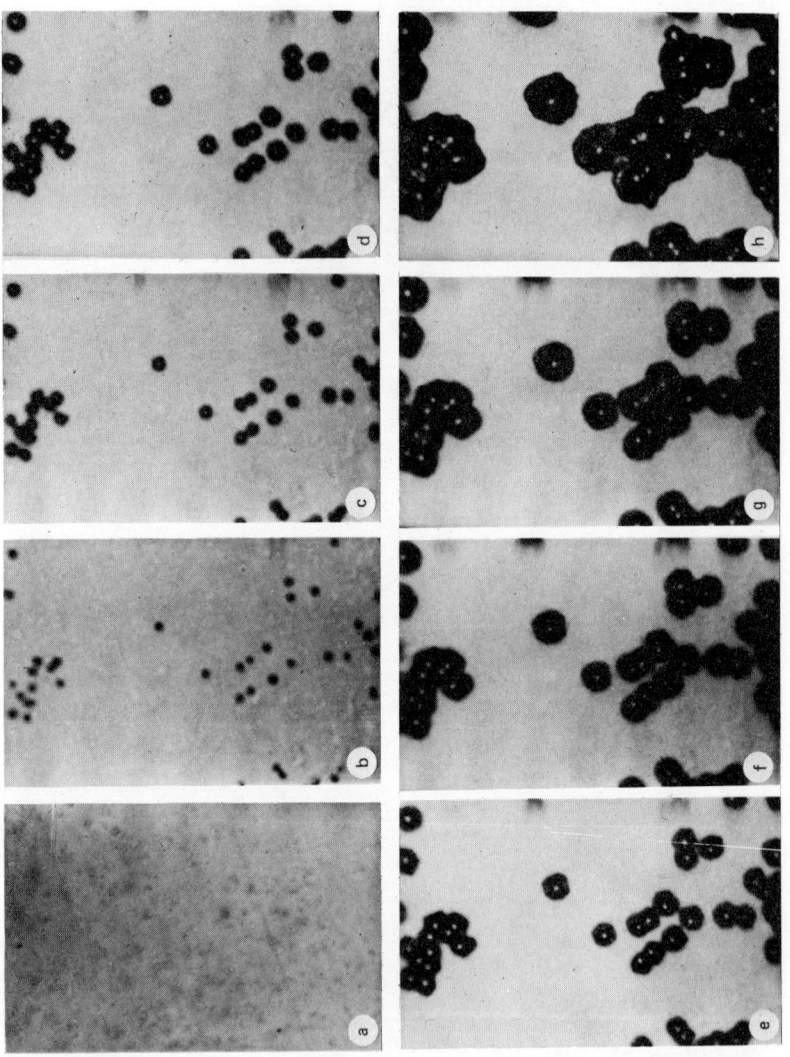

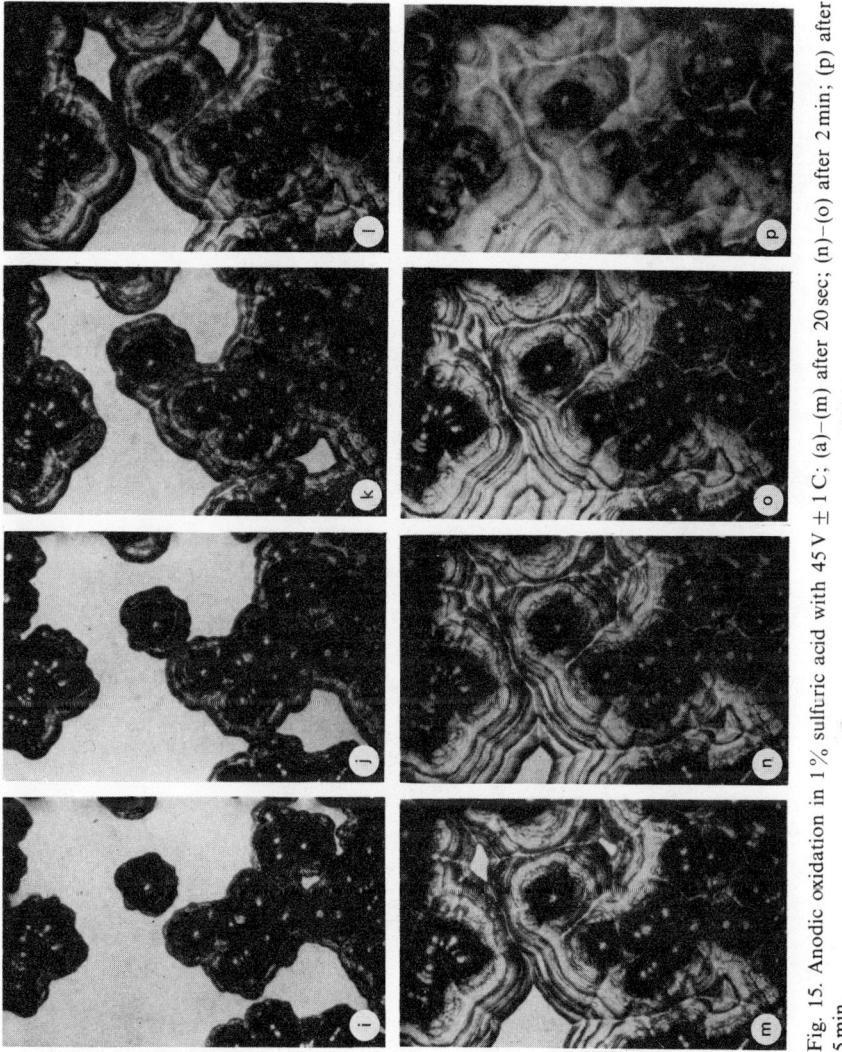

Fig. 15. Anodic oxidation in 1 % sulfuric acid with $45\,V \pm 1\,C$; (a)–(m) after 20 sec; (n)–(o) after 2 min; (p) after 5 min.

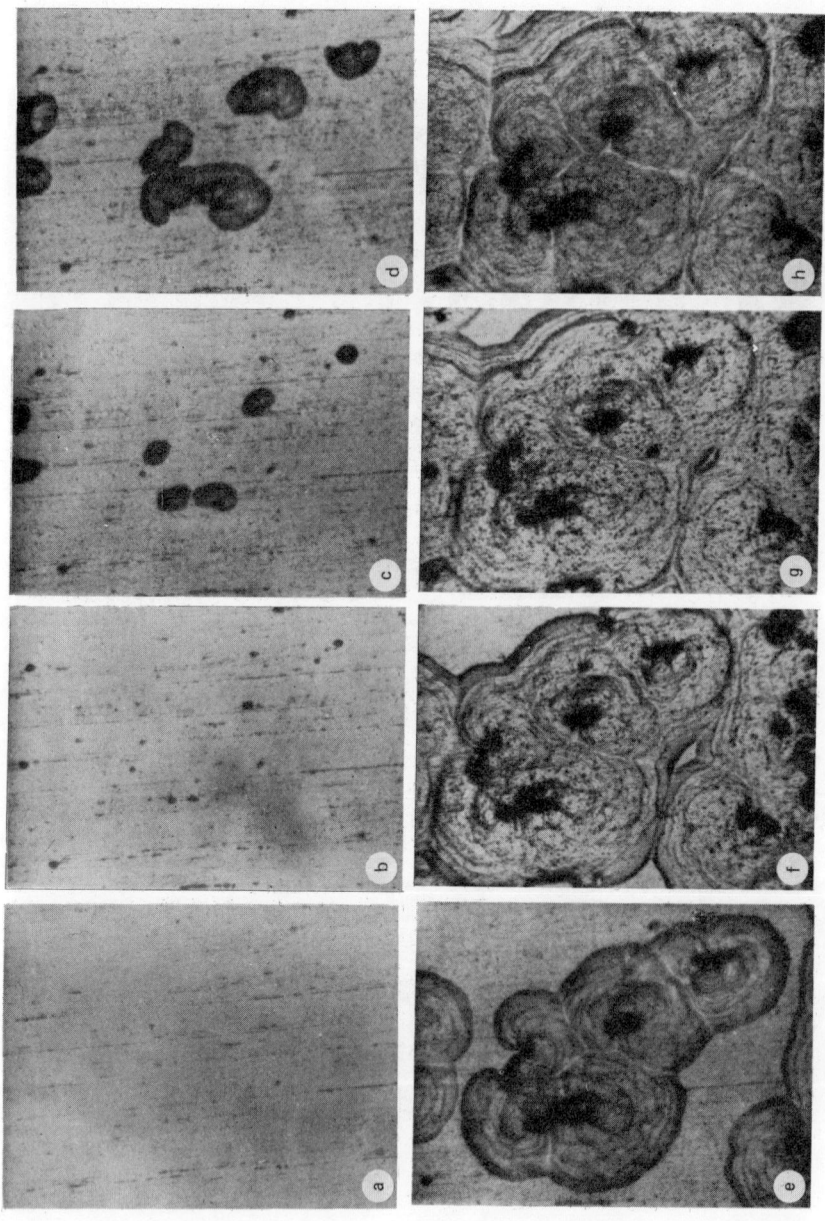

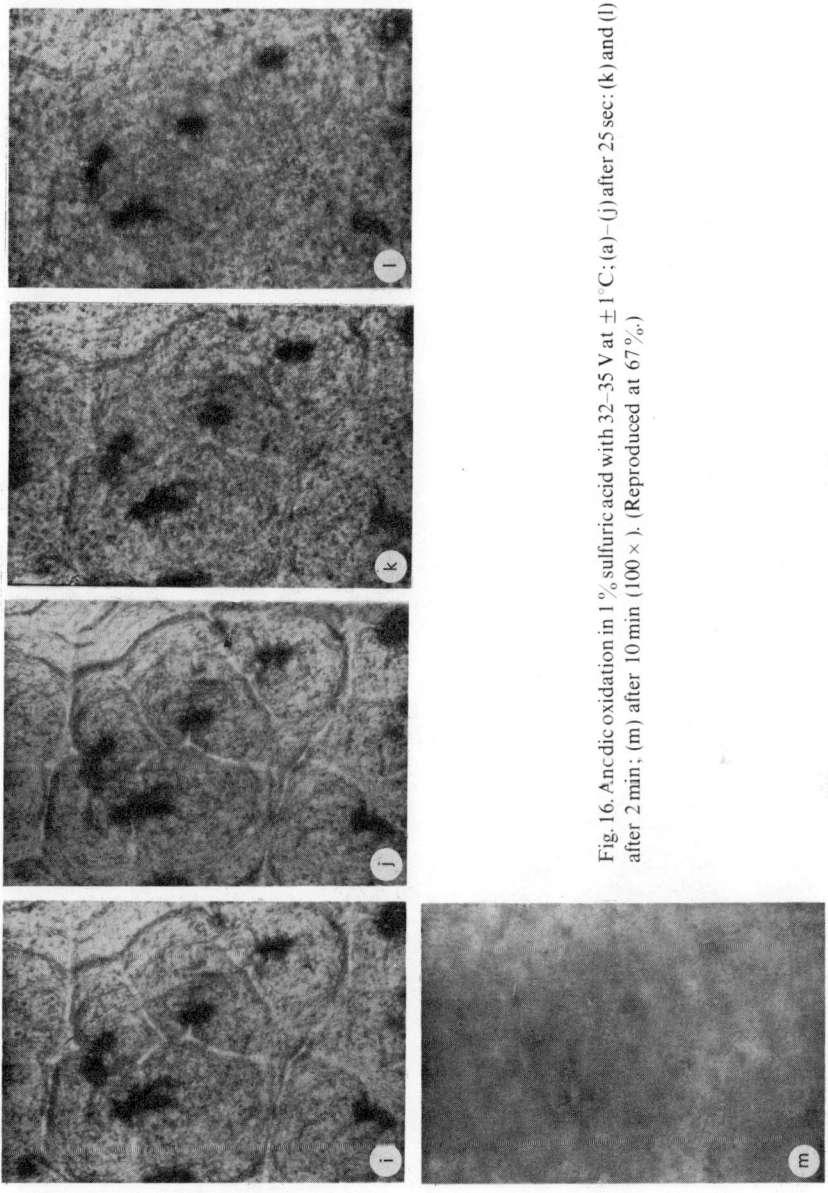

Fig. 16. Anodic oxidation in 1 % sulfuric acid with 32–35 V at ±1°C; (a)–(j) after 25 sec: (k) and (l) after 2 min; (m) after 10 min (100 ×). (Reproduced at 67%.)

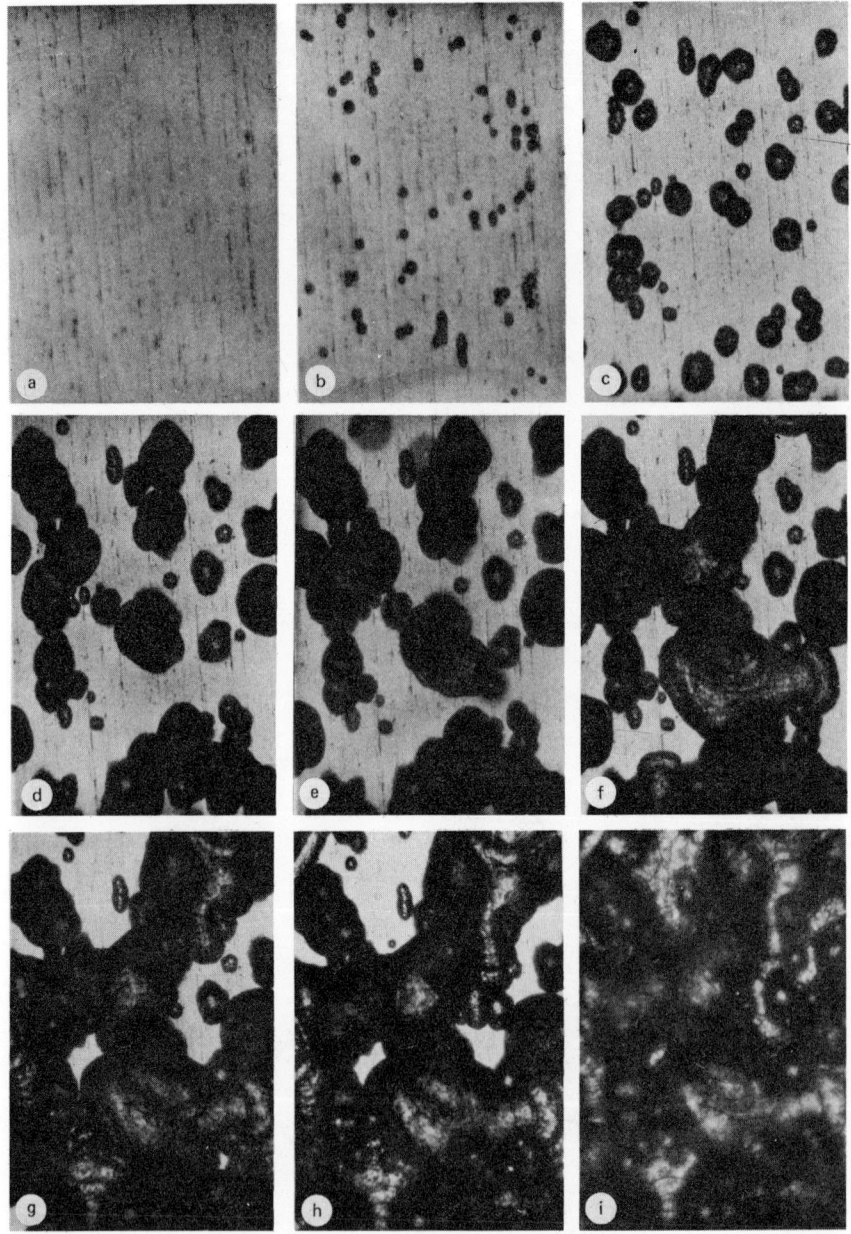

Fig. 17. Anodic oxidation in 0.1 % sulfuric acid with 55–60 V at ± 1 °C: (a)– (h) after 70 sec ; (i) after 2 min.

Anodization in Solutions of Chromic Acid. Anodic oxidation in chromic acid followed the Bengough–Stuart process: 5 % chromic acid at 40°C, with a cell voltage of 10 → 40 → 50 V regulated periodically by a specific time program. The first oxide membrane on the surface of anodized aluminum plate (99.5 % Al) was composed of extremely fine oxide nuclei. At first these nuclei only appear along the rolling marks—but after about 10 sec the whole surface is populated by fine grains (Fig. 25a–e). Oxide re-solution is responsible for the final slightly nonuniform structure. This leaves nuclei of varying size arranged regularly, on a microporous structure (Fig. 25f–h). The total thickness of the oxide layer is no more than a few micrometers, its color is gray, and it is slightly opaque.

Anodization in a Mixture of Organic Acids. For comparison, anodization was also carried out in multicomponent solutions of organic acids (a mixture of sulfosalicyclic acid and formic acid with a total concentration of several percent) with a voltage of 40–80 V. In the present example the cell voltage was 40 V. The appearance, multiplication, and characteristic band-like arrangement of primary nuclei were similar to the phenomenon of anodic oxidation in dilute oxalic acid (Fig. 26a–f). The final oxide coating is fine-grained and dense, but the bandlike structure is visible (Fig. 26g–i). The latter property is probably connected with the epitaxy effect of the base metal.

Investigation of the Phenomena of Oxide Formation in Slightly Oxide-Dissolving Environments with DC Current

In electrolytes with a slight oxide-dissolving ability (solutions of boric acid, borax, tartaric acid, citric acid, etc.) it is known that extremely thin, but dense oxide membranes can be prepared on aluminum; practically no micropores are present in these films.

The highly insulating ability of these membranes requires that the cell voltage be raised stepwise—from the usual starting value of 50 V to several hundred volts to provide the current needed to continue the oxidation process.

In the course of our Elypovist tests we anodized 99.5-grade aluminum plate samples in a 10% boric acid solution at 60–70°C, with cell voltage increased from 50 V to 200 V. A temporary external current source was connected to the electrolyzing cell to assure the higher oxidizing voltage. The increased temperature of the solution was obtained by connecting an electrical heating ring to the liquid circulating pipe of the cell (Fig. 7, 1a). Other parts of the equipment were protected against overheating by a cold-air fan.

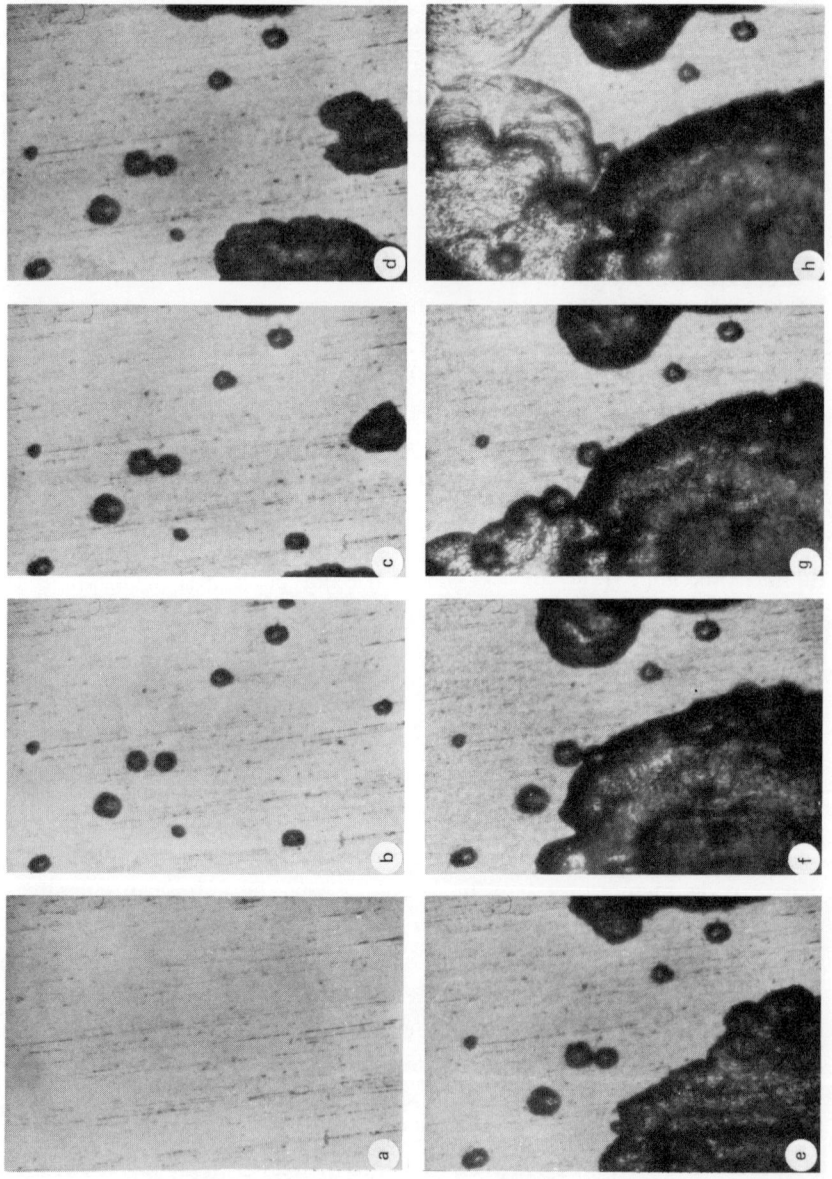

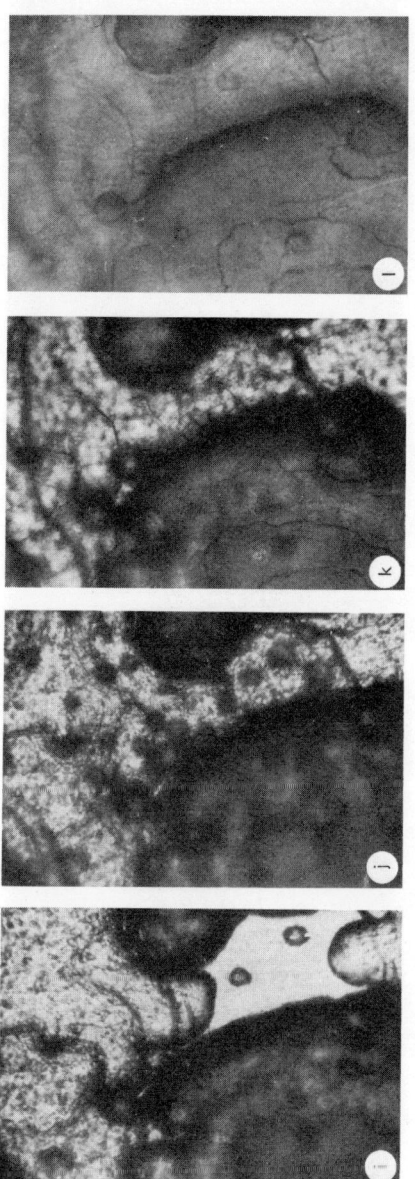

Fig. 18. Anodic oxidation in 0.1 % sulfuric acid with 45–50 V at ±1 C: (a)–(i) after 10 sec; (j) and (k) after 3 min; (l) after 10 min (100 ×). (Reproduced at 67%.)

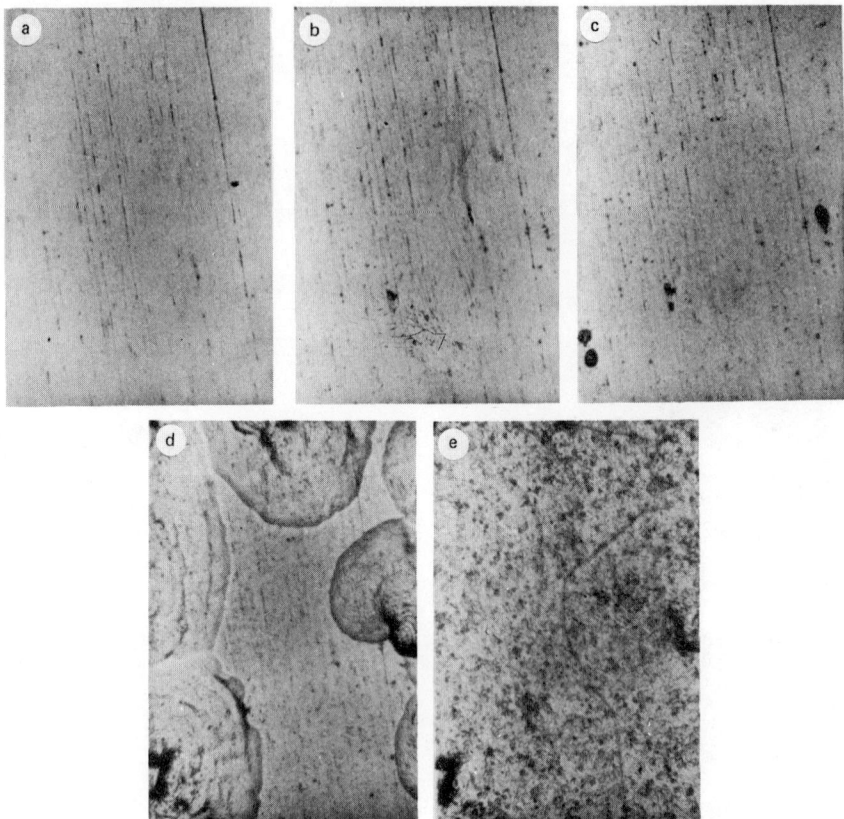

Fig. 19. Anodic oxidation in 0.1 % sulfuric acid with 35–40 V at ± 1°C: (a)–(d) after 150 sec; (e) after 10 min (100 ×). (Reproduced at 65 %.)

Anodization tests were carried out for 5–10 min. In the first second of the experiment, rows of very small oxide nuclei formed on the surface of the plate sample on the sites of the rolling scratches (Fig. 27a); subsequently, a very fine grained oxide coating formed and was accompanied by a slight darkening (Fig. 27b and c). Thickness measurements of the oxide coating gave a depth of the order of 0.1 μm.

The Reaction Periods of Anodic Oxide Formation on the Aluminum Surface

The anodic oxidation of aluminum consists of several consecutive or simultaneous partial processes. This finding is supported by explanations found in the literature for the oxidation of metal surfaces which are in contact

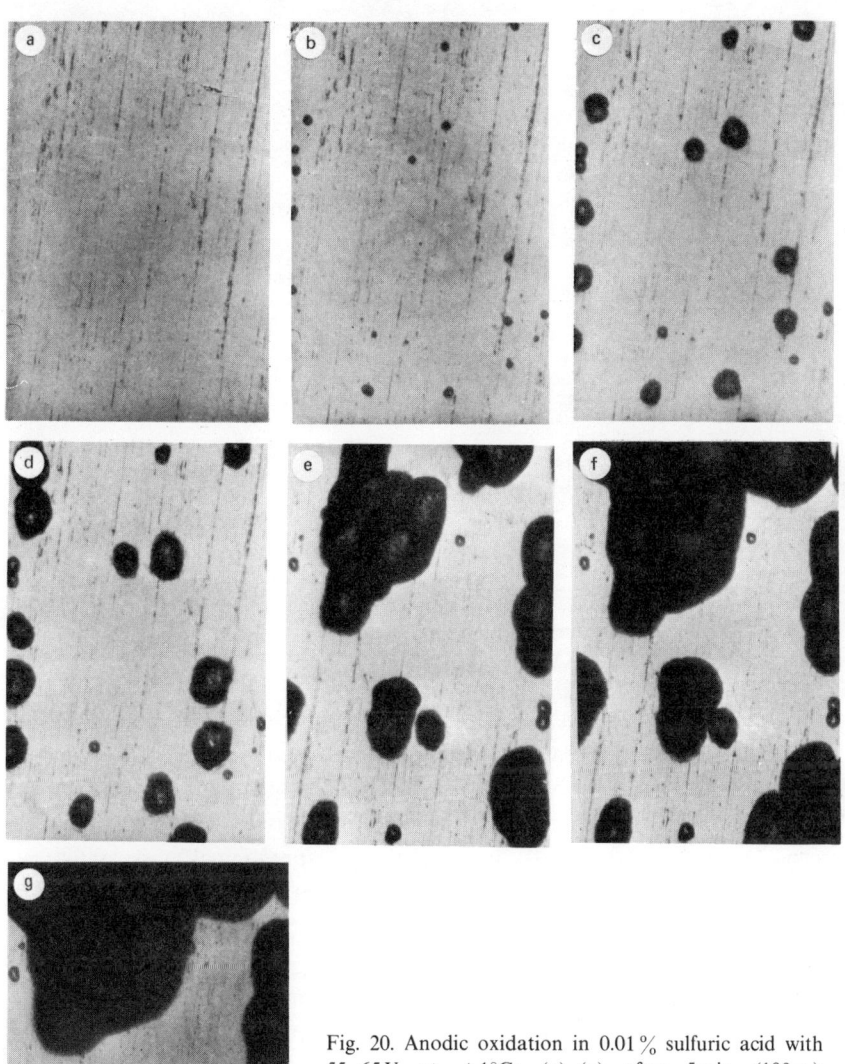

Fig. 20. Anodic oxidation in 0.01 % sulfuric acid with 55–65 V at ±1°C: (a)–(g) after 5 min (100 ×). (Reproduced at 65 %.)

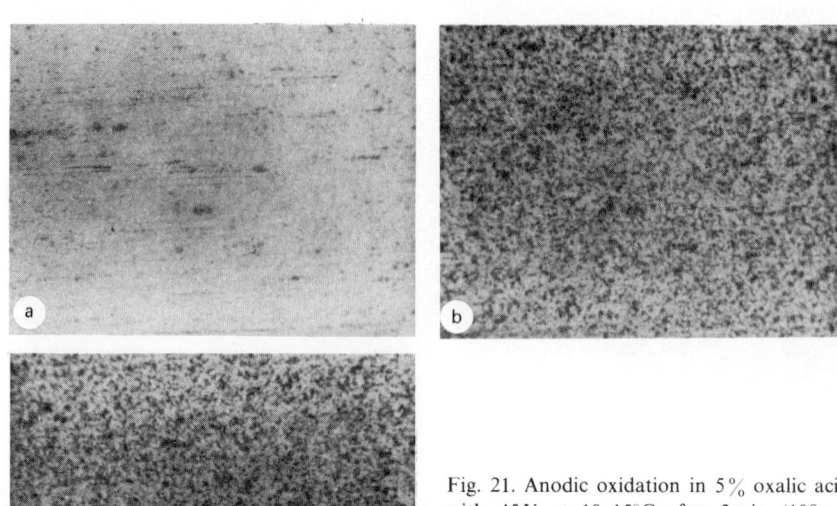

Fig. 21. Anodic oxidation in 5% oxalic acid with 45 V at 10–15°C after 2 min (100 ×). (Reproduced at 69 %.)

with oxygen gas or oxidizing liquids under thermal, chemical, or electrochemical conditions, by our results obtained with physicochemical, electrochemical, microscopic, and electron microscopic methods, and by the reaction kinetic analysis obtained by the films taken with the Elypovist microscope. By taking into account the morphological and structural characteristics and chemical and electrochemical traits of the oxidation process, the partial processes of the reaction mechanism can be grouped in the following order:

1. Initial stage (induced period). Oxygen atoms or oxygen carrying particles (anions with oxidizing effect) are bonded to the metal surface by physical adsorption or chemisorption. In the sorption layer, aggregates of molecules are formed on active sites of the metal surface by diffusion movements. After the state of critical supersaturation is achieved, an oxide membrane (γ-Al_2O_3) of monomolecular or oligomolecular thickness is formed. It starts from the aggregations of molecules on the active sites as reaction nuclei; after a fixed interval this covers the whole metal surface.

2. Formation Stage (formation of the barrier oxide layer). In electrolytes which do not chemically dissolve aluminum oxide, the thickness of the oxide

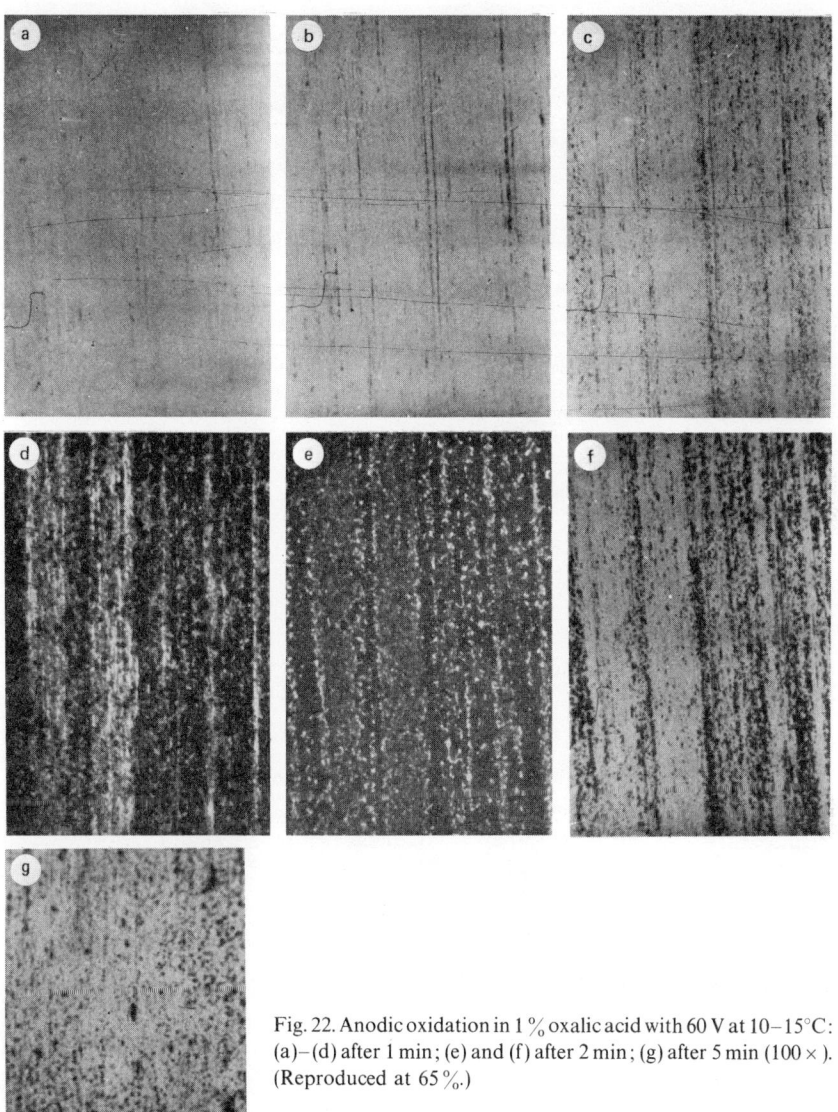

Fig. 22. Anodic oxidation in 1 % oxalic acid with 60 V at 10–15°C: (a)–(d) after 1 min; (e) and (f) after 2 min; (g) after 5 min (100 ×). (Reproduced at 65 %.)

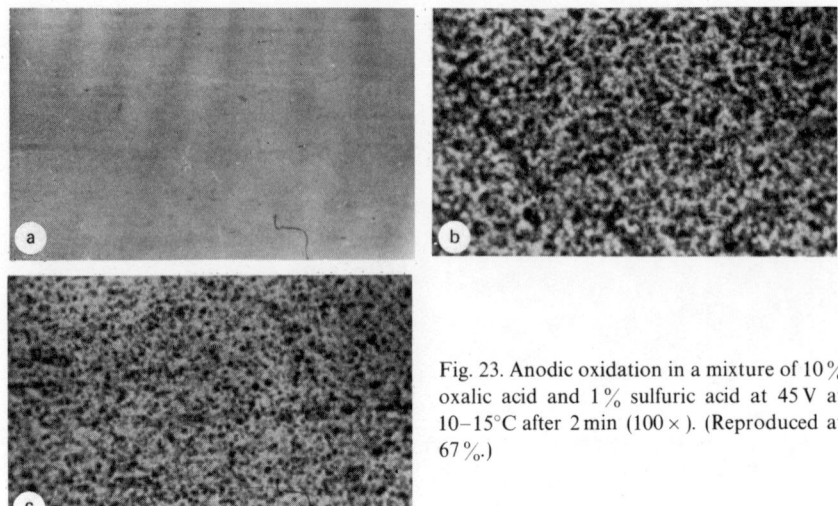

Fig. 23. Anodic oxidation in a mixture of 10%
oxalic acid and 1% sulfuric acid at 45 V at
10–15°C after 2 min (100 ×). (Reproduced at
67%.)

layer increases to a limit defined by the current conduction and the so-called
barrier layer is formed. It is a dense oxide coating with dielectric properties.
The composition and structure of the oxide membrane are generally uniform;
however, microinhomogeneities are always present.

3. Layer formation by processes of nucleation. In electrolytes which
chemically dissolve aluminum oxide, the oxide membrane appearing at the
end of the initial period does not change into a dense barrier oxide layer with a
specific thickness. Its chemical and structural composition change only when
repeated reactions occur, such as a series of chemical dissolution, hydration,
peptization, or incorporation of foreign ions. The thickness then increases
continuously; at the same time, further barrier oxide membranes are always
formed at the metal/oxide boundary.

Primary nuclei appear on the active sites with sufficient current
conductivity to form a heterogeneous structure with a thickness of several
molecular layers. The number, distribution, size, and morphological
characteristics of these nuclei are determined by the properties of the base
metal and by the experimental parameters (the qualitative and quantitative
composition of the electrolyte solution, its temperature, and the oxidizing
current).

Starting from the edge of the nuclei, the so-called secondary zones of
oxidation appear and cover the free surface between the nuclei by lateral
growth; this interval is defined by the experimental parameters.

4. Forming period of the covering oxide layer. After the surface has been completely covered, the reactions continue in various directions on the oxide layer. In the covering layer, pores of various sizes are formed, depending on the experimental parameters and the local differences in the reactivity of the oxide product (chemical solubility, state of energy as a consequence of structural deformations, etc.). Because of energy differences between discrete points of the metal/oxide boundary phase, continuous or periodically repeated reaction waves progress vertically in the oxide layer. In this way a fine laminar oxide structure forms (tertiary oxide layers). The thickness of the covering oxide layer and its final stationary structure depend on the equilibrium between the processes of oxide formation and chemical dissolution. The microstructure of the oxide layer and in some cases also its macrostructure contain geometrical inhomogeneities and inequalities (pore size, pore distribution, pore wall thickness, layer thickness, etc.).

The Processes in the Initial Period

The study of the short-term changes occurring in the initial period of the oxidation of aluminum is extremely difficult because the initial phenomena can barely be observed on the metal surface. Further, during anodization the interface is inaccessible to the usual physicochemical or optical test methods. The reactions in the initial period of anodic oxidation can be interpreted with the aid of reaction kinetics of interface processes in solid/gas systems and with the morphological changes observed through the Elypovist microscope during further stages of anodization.

The Formation of an Oxygen Film by Adsorption or Chemisorption. As has been stated in the KHR theory, the oxygen produced by the anodic current in the first moments of anodic oxidation is adsorbed on the aluminum surface. Some authors speak of the adsorption of elementary oxygen, others (e.g., Glayman[59]) of the adsorption of hydroxyl radicals or other oxygen carriers; for the sake of simplicity we shall employ the expression "oxygen adsorption."

Opinions diverge concerning the nature of the adsorption. The earlier explanations only assumed a physical adsorption due to van der Waals forces; Baumann[60] and other authors assumed chemical bonding (i.e., chemisorption). According to Bénard,[41] there is no sense in aggravating the discussion concerning adsorption or chemisorption, since evidently in the initial stage of oxidative processes (depending on temperature, gas pressure, affinity between the adsorbent and oxygen, etc.) a van der Waals bond or chemical bonding accompanied by an exchange of electrons can occur with

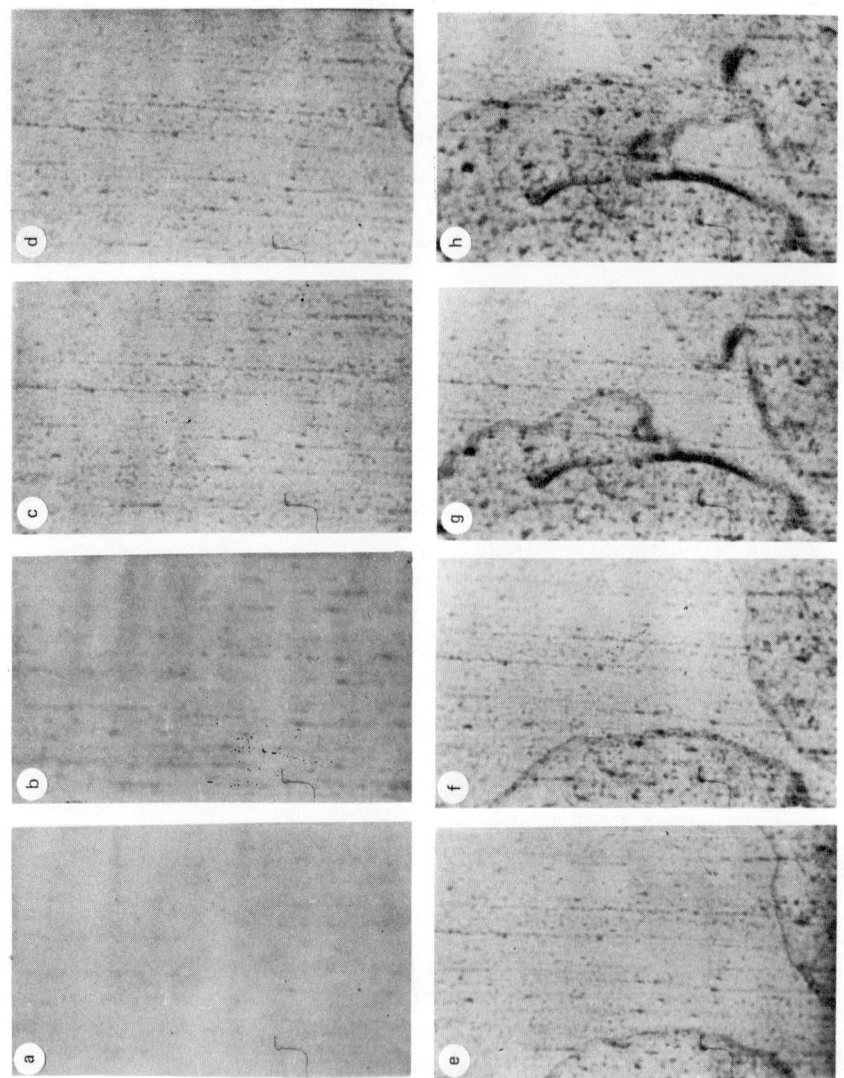

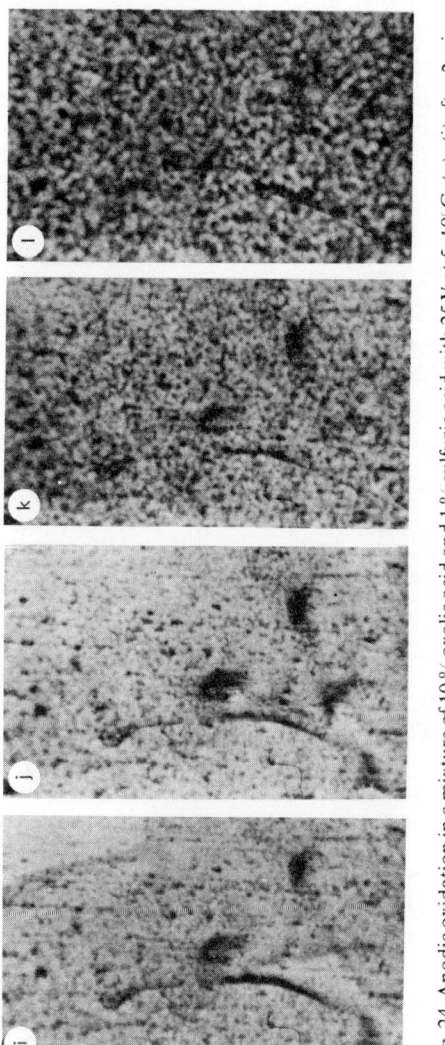

Fig. 24. Anodic oxidation in a mixture of 10 % oxalic acid and 1 % sulfuric acid with 25 V at 5–10°C: (a)–(i) after 2 min; (j)–(k) after 3 min; (l) after 6 min (100 ×). (Reproduced at 67%.)

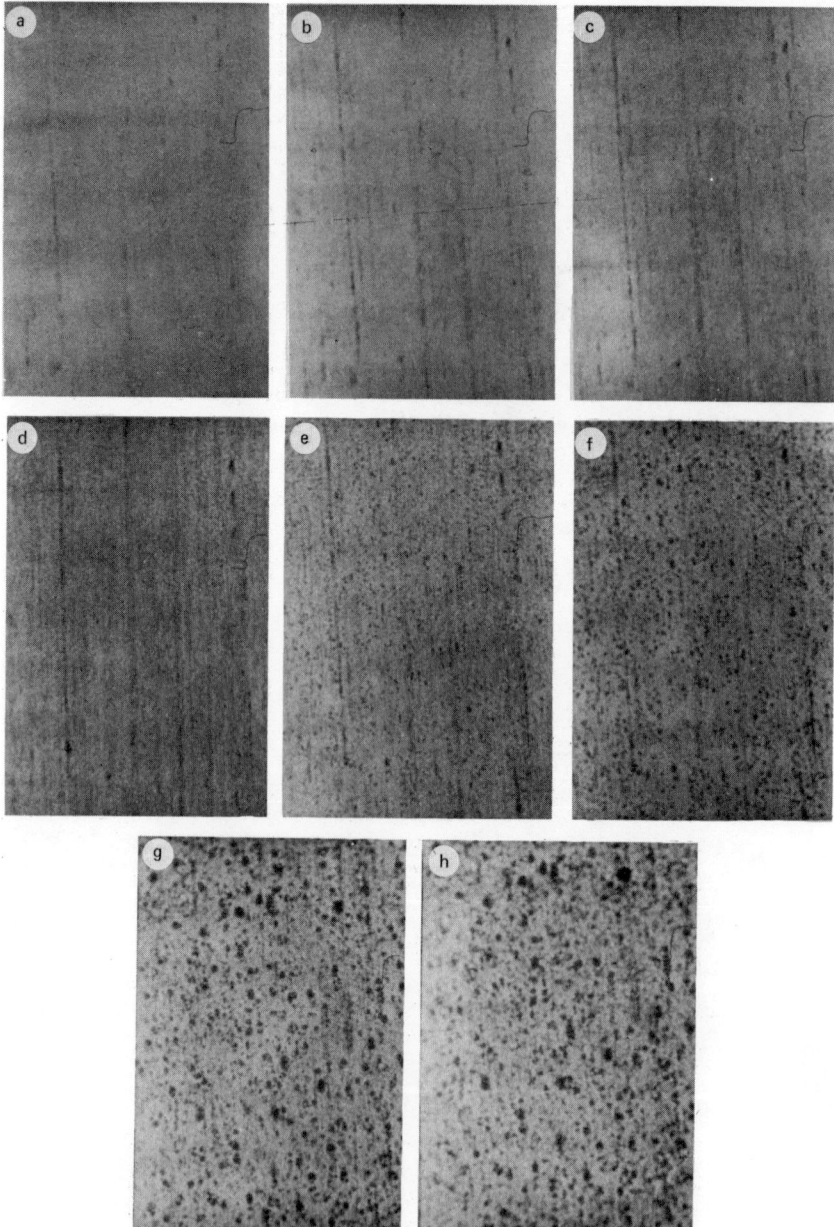

Fig. 25. Anodic oxidation in 5 % chromic acid with the cell voltage regulated continuously from 10 to 40 to 50 V at 10–15°C: (a)–(d) after 2 min; (e)–(g) after 6 min; (h) after 8 min (100 ×). (Reproduced at 65%.)

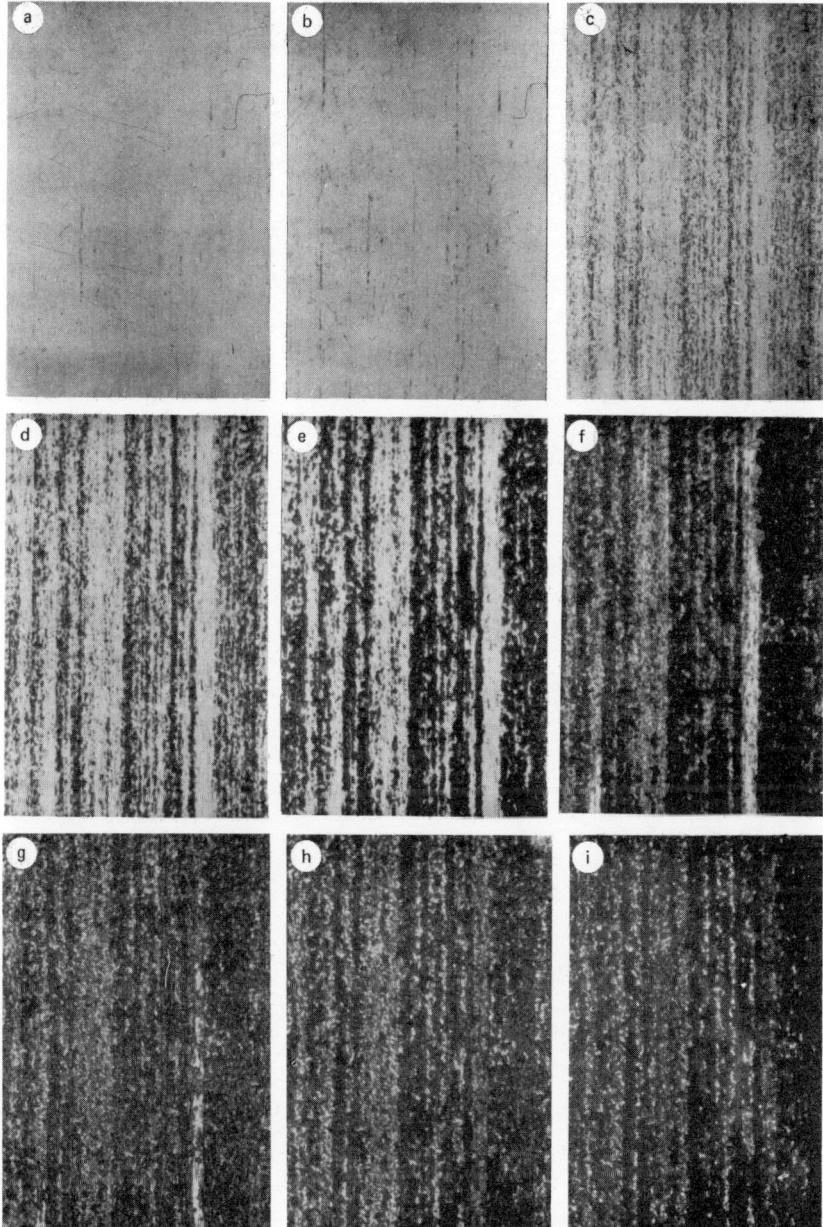

Fig. 26. Anodic oxidation in a mixture of 5% sulfosalicylic acid and 3% formic acid at 40 V at 18–20°C: (a)–(e) after 90 sec; (f)–(i) after 2 min (100 ×). (Reproduced at 65%.)

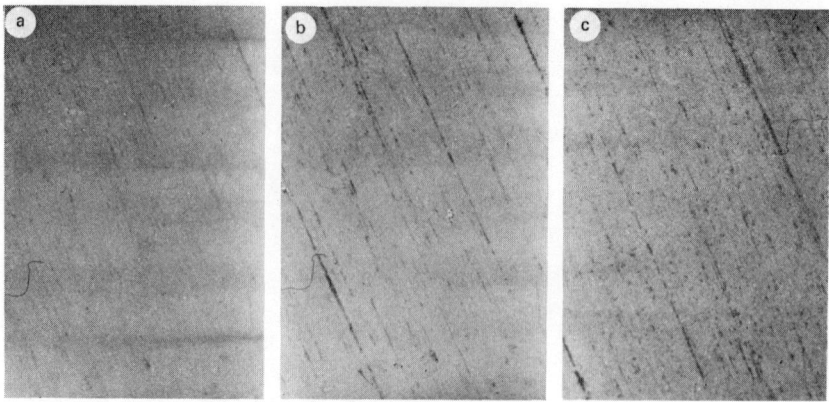

Fig. 27. Anodic oxidation in 10% boric acid with the cell voltage regulated continuously from 50 to 200 V at 60–70°C after 10 min (100 ×). (Reproduced at 65%.)

equal probability. Seith[61] suggests that there is no sharp distinction between the two types of adsorption, because in a given system physical adsorption bonding can be transformed to chemisorption bonding in the course of the chemical processes, as indicated by a change of experimental conditions during the reaction.

The surface concentration of adsorbed particles (Γ) in the solid/gas system obeys the Langmuir isotherm and depends on temperature and pressure. In the solid/liquid system it follows the Helmholtz theorem and depends on the solution concentration:

$$\Gamma = \frac{A}{RT}\,\frac{p}{1/(b + c)} \quad \text{(solid/gas system)} \tag{1}$$

$$\Gamma = \frac{A}{RT}\,\frac{c}{1/(B + c)} \quad \text{(solid/liquid system)} \tag{2}$$

where A, B, and b are material constants; p the gas pressure; and c the concentration of the solution.

The sorption layer is usually monomolecular, but in certain conditions an oligomolecular layer thickness may be expected. In the case of chemisorption, first a metal/oxygen phase with monovalent bonding is formed which tends to completion as an oxide compound bonded with full valency. The tendency to chemisorption is sometimes so strong that diatomic oxygen molecules split into atoms which are strongly bonded to the metal after the establishment of joint electron paths; their separation from the metal (i.e., their desorption) is possible only at high temperatures with a high energy of activation.

Data are found in literature which state that particles can adhere to the "compound film" formed by chemisorption through physical adsorption if the gas pressure or concentration of the solution is sufficiently high. The bonding strength of these particles decreases rapidly with increasing distance from the metal surface.

We know from the literature on the Langmuir isotherms that the adsorption layer is not always closely adherent to the metal surface, even when the concentration approaches the saturation level. Rhodin et al.[43] conclude that adsorption does not occur continuously; rather, it occurs at the high-energy sites of the surface. Active sites are usually produced by orientation characteristics of the crystal structure of the base metal or by a defective metal surface (deformed microstructure, foreign inclusions, heterogeneities caused by local enrichment of alloying elements, etc.). At the defect sites of the microstructurally inhomogeneous metal surface ("imperfection structure" according to Rhodin), the energy levels are usually dissimilar and the adsorption of oxygen occurs first at surface sites with greater activity and subsequently on sites with medium activity; sites with low activity are filled only when saturation has been achieved. As a result of this mechanism, the material adsorbed on the metal surface occupies periodic positions, and the adsorption can be geometrically heterogeneous.

We now refer to the interesting experiments of Vlasakova[62] to study the oxygen and oxide films bonded by physical and chemical sorption on aluminum. Vlasakova directed polarized light at the metal surface and found from the elliptical change of the index of reflection that the effective thickness of an adsorbed oxygen layer on a mirrorlike surface changes in the saturated state from 30 to 35Å.

Exoelectron emission studies were conducted by Vlasakova on these samples and on aluminum surfaces oxidized by the MBV process in an aqueous alkaline solution. These tests have shown that the points of adhesion of the sorbed oxygen film are at the sites which have a high surface activity. Vlasakova postulates that van der Waals forces are mainly responsible for the bonding, but there is an equal chance for a chemisorbed oxygen film. Its composition can be expressed by the formula

$$Al-O-Al=O \leftrightarrow Al-O-Al\overset{\displaystyle O}{\underset{\displaystyle O}{\diagup\diagdown}} \leftrightarrow Al-O\cdots O-Al\overset{\displaystyle O}{\underset{\displaystyle O}{\diagup\diagdown}} \leftrightarrow$$

$$\leftrightarrow O-Al-O-Al=O$$

which depends on the concentration of the environment and on the experimental conditions. The oxygen film contains covalent bonds and has an amorphous structure.

Diffusion Processes in the Metal Surface Phase. It was shown above that the supersaturation concentration of adsorbed or chemisorbed species is different (it depends on the momentary energy state) at different sites on the metal surface. Haruyama[64] has proved experimentally that the energy of sorption can vary from point to point, depending on the quality of the surface and on the experimental conditions; in the area around zones of high activity, a concentration gradient appears corresponding to the energy conditions.

The activities at different points of uniform high-purity metals are only slightly different, and on such surfaces no substantial concentration differences occur in the sorption film. However, the surface of technical-grade metals is always strongly heterogeneous, and more significant differences in reactivity and concentration gradients are possible.

According to Rhodin *et al.*[43] the oxygen atoms colliding with the surface in the active zones and "caught" in it can migrate along a concentration gradient (which corresponds to differences in surface energy) by surface diffusion or migration; at active sites they produce aggregates of molecules or "clusters." At the nodes, the aggregates are more-or-less oriented and possess significant stability.

Seith[61] has also shown by radioactive isotope tests that foreign particles colliding with the surface and adsorbing on it ("ad-atoms") can diffuse on the solid metal in all energetically permitted directions; they aggregate at points and form large agglomerations. Their final distribution is dependent on the experimental parameters.

This process continues until a certain supersaturation is achieved, and then at discrete points the covalent bonds are replaced by ionic bonds; this results in the irreversible formation of aluminum oxide.[63] The electron micrograph in Fig. 28 shows the oriented oxide formations appearing on subgrain surfaces of a polycrystalline base metal.

After achieving critical supersaturation, if the energetic conditions are favorable, γ-Al_2O_3 in stoichiometric composition can form. It starts from a molecule (or an aggregate which is one to three molecular layers thick) and spreads to cover the entire metal surface.

Mykura[45] postulates that lateral surface diffusion can occur not only in the monomolecular sorption film but also in the oligomolecular oxide layers, and at the same time aluminum ions or oxygen ions can also diffuse perpendicularly to the substrate outward or inward through the oxide layer. The perpendicular to the surface determines whether oxide formation serves to

Fig. 28. Formation of oxide agglomerations at the surface of active subgrains during the first few seconds of anodic oxidation in 1 % sulfuric acid at $\pm 1°C$ with 50 V. Electron micrograph (20,000 ×). (Reproduced at 80 %.)

increase the thickness of the first barrier oxide film, or whether further oxygen agglomerations or nuclei appear on the active sites (i.e., whether nucleation processes appear).

The Elypovist films made in polarized light show the changes in the processes of adsorption, diffusion, and formation of oxygen or oxide membranes, all of which depend on the reaction time of the initial period. The pictures from one film sequence (Fig. 29) show a change from light to dark on areas which are high and low in surface energy; these areas are determined by the crystallographic or microstructural conditions of the base metal. These areas indicate the local development of oxygen and oxide membranes representing phase boundaries.

Processes in the Nucleation Period

Formation of Barrier Oxide Layers. It was mentioned earlier (in the section on the processes of oxide formation according to the KHR model) that very densely arranged and relatively small oxide nuclei appear on the oxygen–oxide membrane initially formed when a suitably oxidizing electrolyte and high cell voltage is employed. This combination only slightly dissolves aluminum oxide, if at all.

These nuclei form a very dense layer. The oxidizing voltage ("forming" voltage) is adjusted within 100 to 500 or even 1000 V to obtain the constant

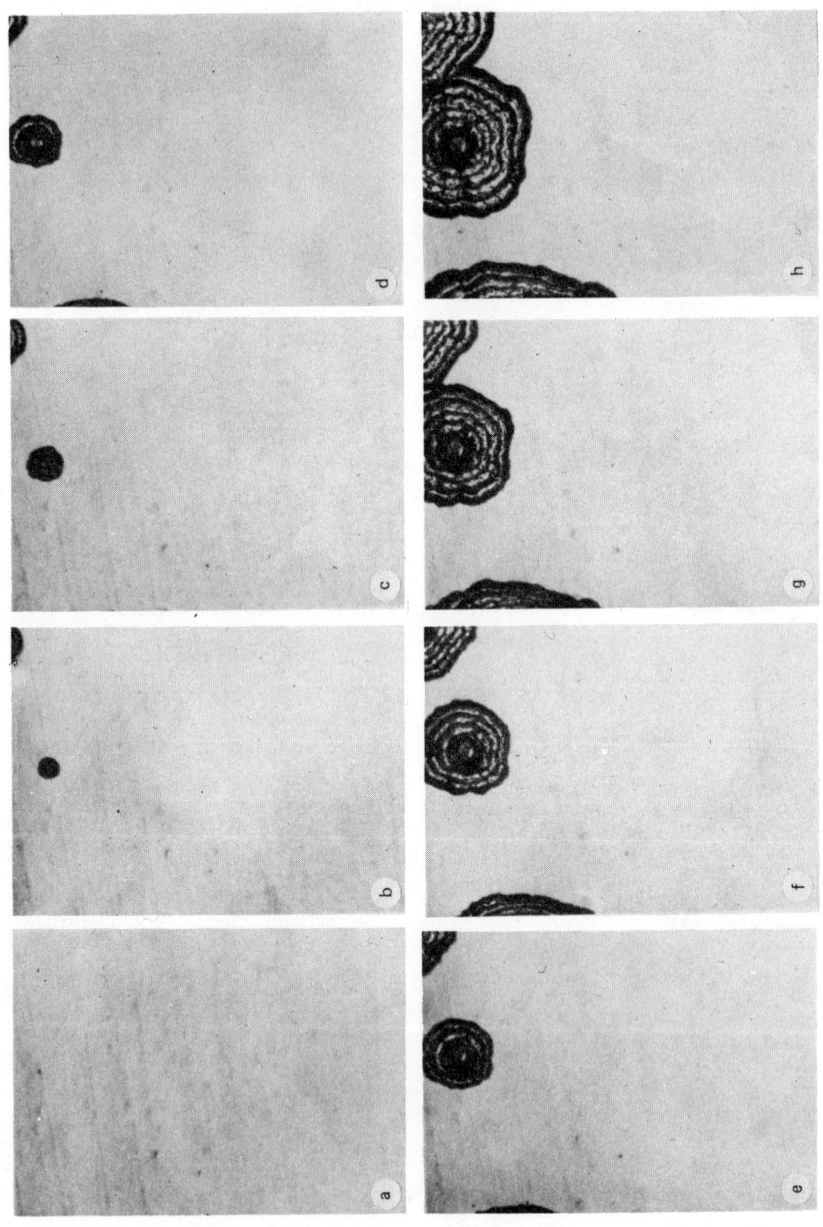

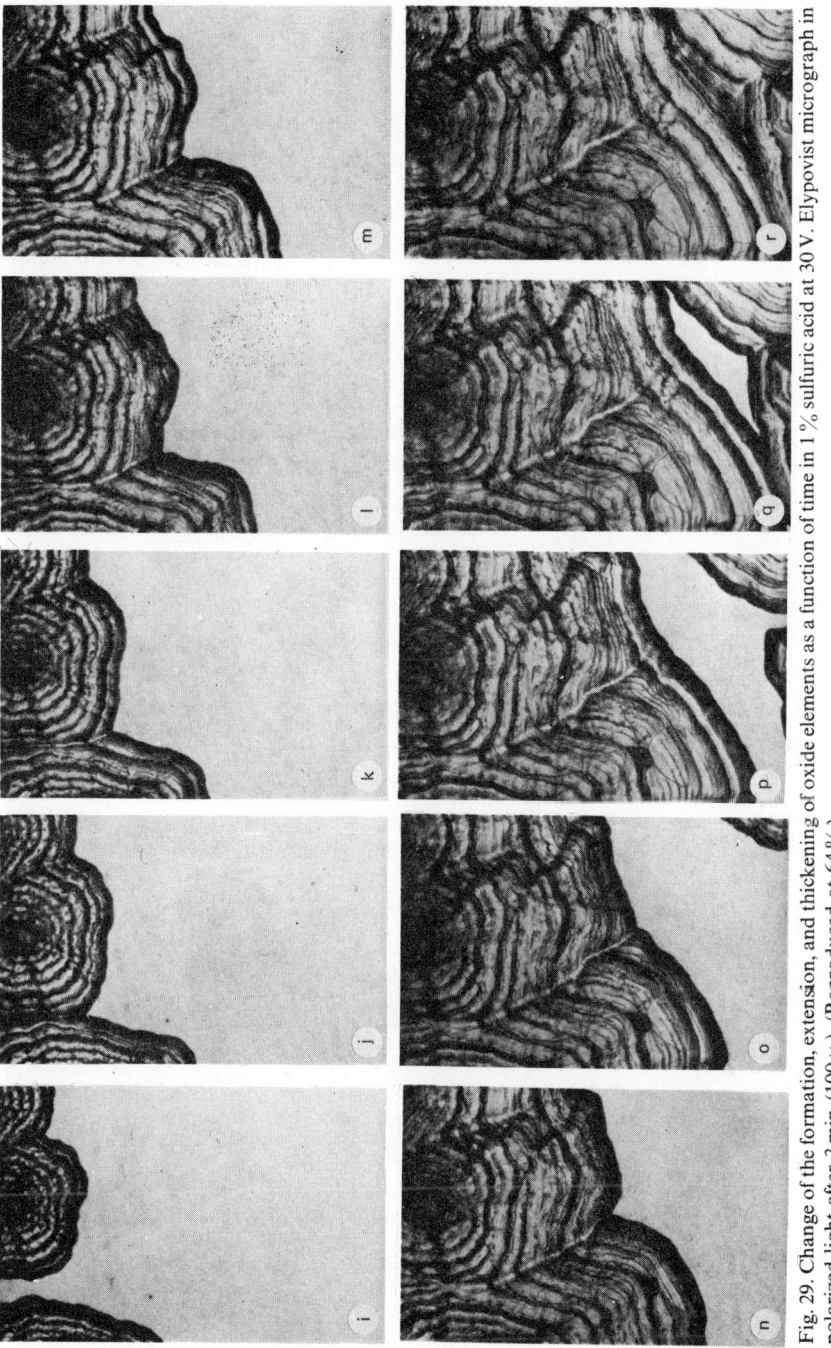

Fig. 29. Change of the formation, extension, and thickening of oxide elements as a function of time in 1 % sulfuric acid at 30 V. Elypovist micrograph in polarized light after 3 min (100×). (Reproduced at 64 %.)

current density required for oxide formation. These barrier oxide layers are usually thin, not more than 0.1–1 μm thick.

However, the microscopic study of the barrier oxide layers in polarized light shows the fine-grained, dense structure of the oxide, although the surface of these oxide layers is smooth translucent or slightly clouded (see Fig. 27). In such coatings the gaps formed by current breakdown are closed by the progressive oxidation; porosity is practically negligible in the final oxide layer.

Layer Formation by the Growth of Primary Oxide Nuclei and Secondary Zones of Oxidation. In electrolytes which chemically dissolve aluminum oxide and oxidation is continued after the initial period, primary oxide nuclei appear at the energetically active sites of the base oxide membrane which possess sufficient current conductivity. The number, size, morphological characteristics, and distribution density of the primary oxide nuclei are determined by the parameters of oxidation, but the properties of the base metal also play an important part.

The growth of the oxide nuclei generally proceeds in three dimensions: longitudinally and transversally in the plane parallel to the metal surface and perpendicularly to the surface.

The perpendicular growth direction is favored over the other two directions since growth can occur both outward and inward toward the metal surface since metallic aluminum can be transformed to aluminum oxide. This growth observed by Keller–Hunter–Robinson and other authors is shown in Fig. 30, which presents the micrograph of the polished section of oxide nuclei formed in the first stage of anodization.

Oxidation studies with various experimental conditions showed how the number of primary oxide nuclei depends on the experimental parameters. By

Fig. 30. Three-dimensional growth of oxide nuclei on the surface of the base metal (300 ×). (Reproduced at 80%.)

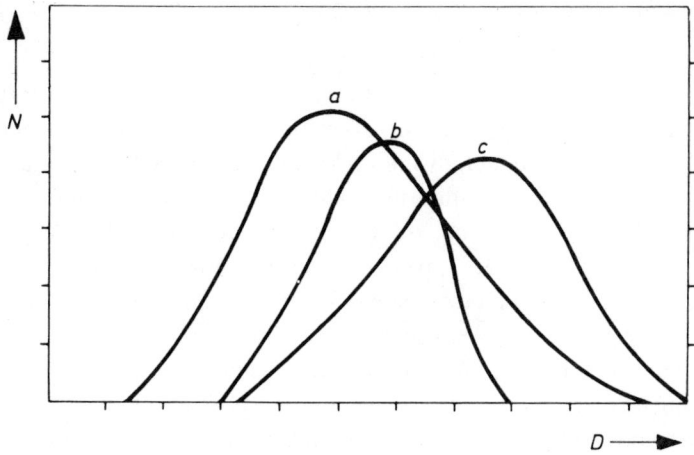

Fig. 31. Distribution of the number (N) of oxide nuclei of various dimensions (D) during thermal and anodic oxidation: (a) Thermal oxidation in high-pressure gas; (b) anodic oxidation in 20 % sulfuric acid at 15–18 V; (c) anodic oxidation in 1 % sulfuric at 50 V.

comparing the size (D) and number (N) of nuclei formed in a 20%, 5%, 1%, 0.1% and 0.01% sulfuric acid solution, a definite dependence of the maximum values and curve symmetry (represented in the $D–N$ distribution diagram proposed by Fischmeister[47]) was found (Fig. 31). The width, height, and summetry of the maximum band depend on the electrolyte concentration and on the cell voltage.

The main factors in nuclei growth can then be summarized as follows:

1. The lateral surface diffusion from the edge of a nuclei conserves the critical oxygen supersaturation indefinitely. Accordingly, the lateral propagation decreases rapidly. Thus, the process becomes more and more anisotropic.

2. Energetically distinguishable sites (microstructural or crystallographic defects) are the points where primary nuclei exclusively appear when oxidation occurs at low temperatures with a low cell voltage. At defect-free sites, the oxidation starts and progresses more slowly.

3. When oxidation is started at a higher temperature with higher cell voltage, the number of nuclei is greater and their distribution is more uniform than the formations appearing at low energy levels. They appear not only on crystallographic or microstructural defects but also randomly.

Processes in the Formation of a Covering Layer

Chemical and Electrochemical Aspects of Oxide Formation. In the initial stages of anodizing, a large number of very small primary nuclei cover the base metal in a fraction of a second; this event occurs in moderately concentrated electrolytes where a moderately high cell voltage is applied. The surface of the final oxide coating appears uniform and fine grained when magnified (Fig. 32).

Anodizing in more dilute solutions with a low to medium cell voltage causes secondary oxides to appear between the smaller number of growing primary nuclei. The secondary oxides cover the base metal by lateral extension with a coherent oxide layer which carries on its surface the traces of the nucleation structure. The final oxide layer is heterogeneous (Fig. 33a and b), but during further oxidation the thickening process smooths out the layer and the final heavy oxide develops a surface similar to the former picture and easily recognized in traditional anodizing literature (Fig. 33c).

The heavy covering layer is formed by material transport through ion and electron diffusion across the first coherent oxide layer and across the thickening layer. This statement is supported not only by the chemical reactions by Murphy and Michelson but also by the conclusions of Ginsberg and co-workers about the current conductivity of the oxide layer.

Accordingly,[31-35] the actual processes of oxidation proceed in the intermediate portion (transition region) of the oxide layer instead of on the metal/oxide or oxide/electrolyte interfaces; the oxide product is always in

Fig. 32. Surface of oxide layer prepared by the GS method in 20% sulfuric acid with 15 V after 15 min (250 ×). (Reproduced at 75%.)

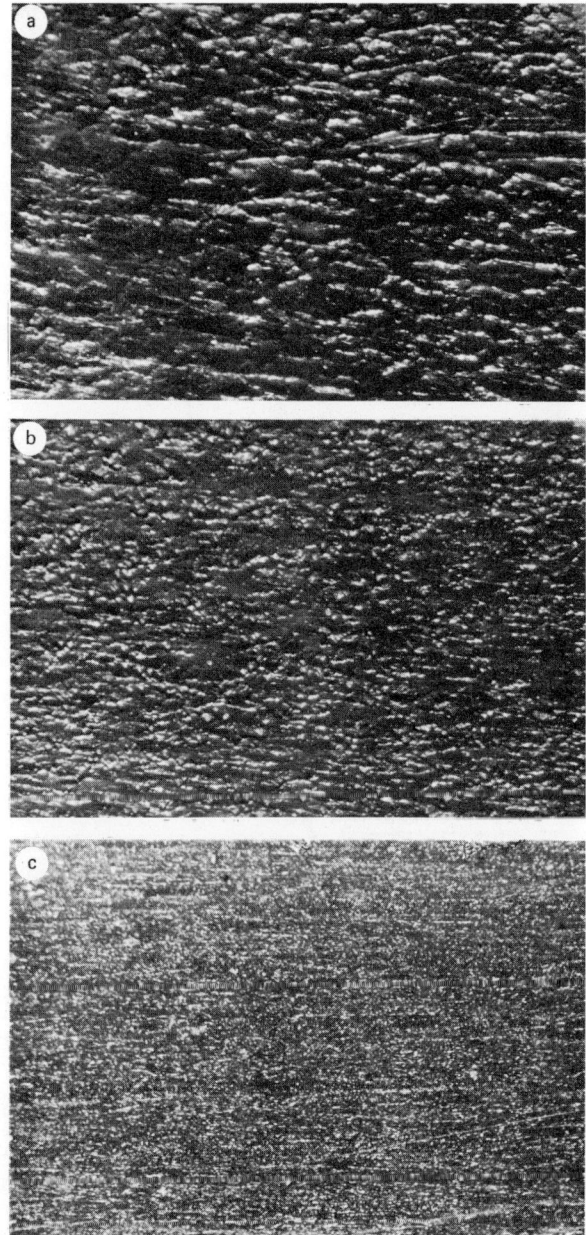

Fig. 33. Oxide layer prepared in 5 % sulfuric acid with 30 V: (a) after 5 min; (b) after 10 min; (c) after 30 min (250 ×). (Reproduced at 70 %.)

laminar form. Therefore, the conclusion that current breakdown across the layer opens pores, as stated by the KHR model, cannot be regarded as a condition of anodic oxide formation. Instead, pore formation depends decisively on the dissolving power of the oxidizing electrolyte, and on the anodizing parameters, while current breakdown is a phenomenon which accompanies pore formation.

In the forming stage other chemical transformations occur concurrently with the solution of the oxide. Some of the first oxide products are transformed by hydration or peptization to hydroxide or hydrate compounds in chainlike coupling, forming a dispersed (colloidal) system with the unchanged aluminum oxide particles.[31] This layer possesses a permeability for material and current transport which further supports oxidation.

Dorsey[65] models the structure of the oxide layer due to chemical transformations as follows:

Primer oxide   Secondary oxide   Porous covering oxide layer

Pore Formation by Chemical Re-solution of the Oxide. In addition to thickening and smoothing of the oxide layer, pore formation is important to the forming stage of the final covering layer. It is the porosity of the final coating that influences the physicochemical, mechanical, and corrosion properties of the oxide layer.

Chemical re-solution is more intense at energetically favored sites of the layer; naturally this depends on the operational parameters, especially on the concentration and temperature of the anodizing solution. Therefore, pore formation starts by a selective dissolution. Data can be found in the literature on the number, size, surface distribution, density, and other properties of various types of oxide coatings used in industrial practice. Oxide layers obtained with traditional anodizing processes are supposedly homogeneously porous, but, in opposition to the regular hexagonal geometric pattern assumed by the KHR model, they are actually heterogenous, as can be shown by the electron micrographs shown in Fig. 34 and 35. At higher magnification the significant differences between the geometry, size, the distance between pores, and the fibrous structure of pore channels formed in heavier oxide layers are even more apparent (Fig. 36).

In oxides prepared in more extreme experimental condition, the pore formation depends on the morphological and structural characteristics of the nucleation. In oxide films prepared in 1% and 0.1% sulfuric acid solutions with high and medium applied voltage, the pores are irregular in size and distribution. An oxide formed at a high cell voltage (Fig. 37) in 1% sulfuric acid shows pores arranged irregularly around a primary nucleus. In the moderate applied voltage range (Fig. 38) the pores are arranged in rows in a concentrically propagating zone of secondary oxidation around a primary oxide nucleus. The pattern of rows following the reaction waves of the secondary zones of oxidation is readily apparent in the pore pattern (Fig. 39). At higher magnification the differences in shape and size of the pores are visible (Fig. 40).

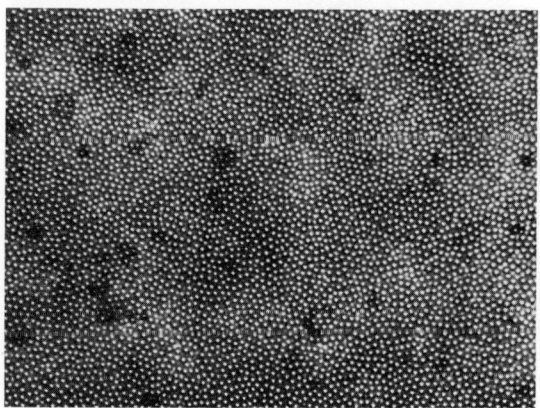

Fig. 34. Porosity of an oxide layer prepared in 20% sulfuric acid with 12 V at 18°C for 2 min. Electron micrograph (64,000 ×). (Reproduced at 65%.)

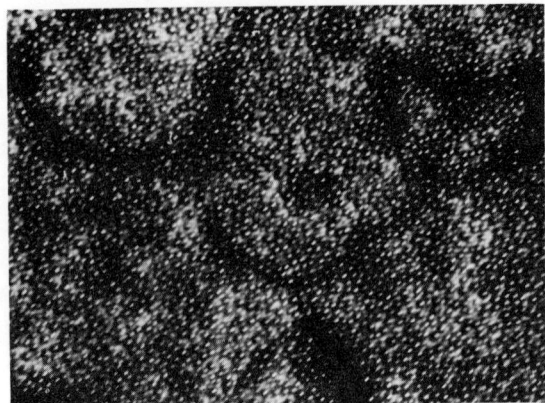

Fig. 35. Porosity of oxide layer prepared in 1 % sulfuric acid with 40 V at $\pm 1°C$ for 5 min. Electron micrograph (62,000 ×). (Reproduced at 65 %.)

The electron micrograph of a fractured oxide shows the fibrous structure of closely coalesced pore channels (Fig. 41). The arrangement of the oxide fibers in irregularly bent or otherwise deformed rows can be connected with mechanical effects which accompany the irregular lateral extension of the oxidation. Figure 42 (right-hand side) shows the surface of an oxide coating prepared in 1 % sulfuric acid with pore openings arranged in rows; at the left a fractured oxide layer is shown with pore channel fibers arranged in irregularly

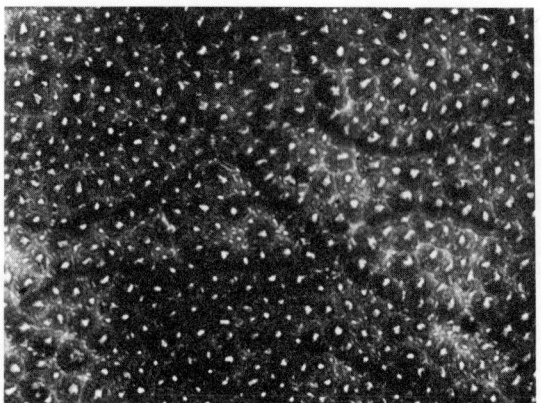

Fig. 36. Porosity of oxide layer prepared in 5 % sulfuric acid with 35 V at $\pm 1°C$ for 10 min. Electron micrograph (67,000 ×). (Reproduced at 65 %.)

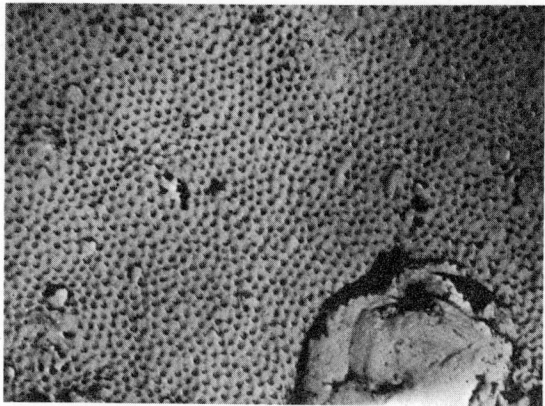

Fig. 37. Irregular pore distribution in an oxide layer prepared in 1 % sulfuric acid at 50 V at ±1°C at around the primary oxide nuclei. Electron micrograph (13,500 ×). (Reproduced at 65 %.)

deformed bundles. The oxide fibers, which carry traces of nucleation processes, the blocks of nuclei wedged together, and the differentiation into layers in parallel with the base surface are not particularly apparent in ordinary light, but they are easily visible in polarized light (Fig. 43). The pore channels do not penetrate directly to the metal, and the barrier layer (the boundary zone between the metal and the covering oxide layer) can be regarded as an actual

Fig. 38. Concentric arrangement of pores around the primary oxide nucleus in an oxide layer prepared in 1 % sulfuric acid with 35 V at ±1°C. Electron micrograph (10,000 ×). (Reproduced at 65 %.)

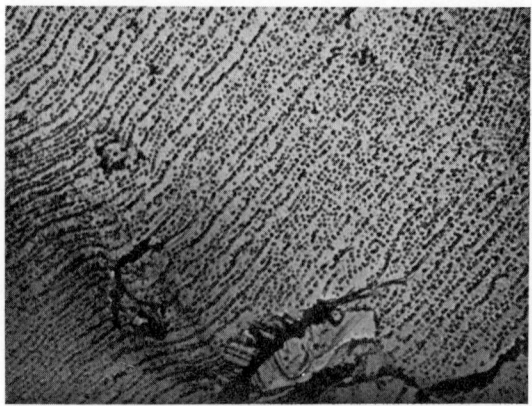

Fig. 39. Rows of pores along the reaction waves of the secondary zones of oxidation in an oxide layer prepared in 1% sulfuric acid at 35 V. Electron micrograph (10,000 ×). (Reproduced at 65%.)

part of the oxide coating. The results of the extensive studies of Domony *et al.*[66,67] and the data from Tajima[24] show that the oxide sealing layer attains a depth which depends on the composition and temperature of the electrolyte and on the voltage applied.

Some details of oxide coatings in natural or in artificial environments by chemical or electrochemical methods are given in Table 2; they permit a

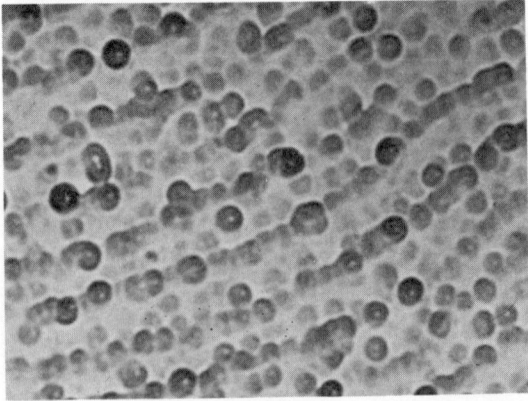

Fig. 40. Pore structure with irregular rows, various sizes, and occasional deformed pores. Electron micrograph (147,000 ×). (Reproduced at 65%.)

Fig. 41. Fibrous structure of pore channels on the fracture of a heavy oxide layer prepared in 1% sulfuric acid at 50 V. Electron micrograph (31,000 ×). (Reproduced at 65%.)

comparison of the structural characteristics which can be expressed numerically.

Since the density or porosity of the oxide coatings influence the application of anodized aluminum, the data mentioned above should be considered when choosing an anodizing treatment.

Kinetic Characteristics of Material Transport and Current Conduction Occurring in the Metal/Oxygen/Electrolyte and Metal/Oxide/Electrolyte Systems. The start of processes of oxide formation depends on several reaction kinetic conditions:

1. The concentration of the reacting components should attain or exceed a definite critical limit of saturation.
2. Chemical or electrochemical energy of activation should be available

   for the migration of the reacting components toward each other and for the production of collisions.
3. The chemical process leading to a stable new phase between components in metastable concentration of supersaturation should be irreversible.

The rate of oxide formation (v) depends mainly on those physical processes occurring in the metal/electrolyte boundary phase or the

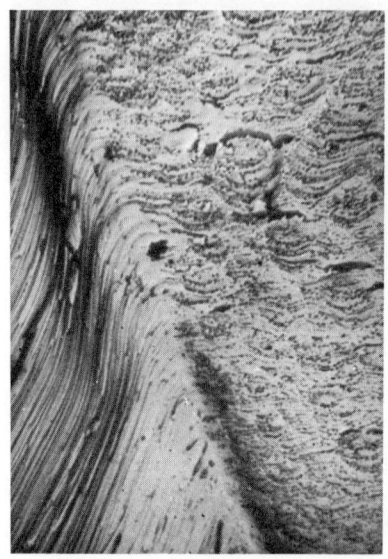

Fig. 42. Oriented arrangement of pore apertures on the surface of a heavy hard oxide layer (right) prepared in 1 % sulfuric acid at 45 V and curved deformation of the pore channel bundles on the lateral fracture (left). Electron micrograph (20,000 ×). (Reproduced at 65 %.)

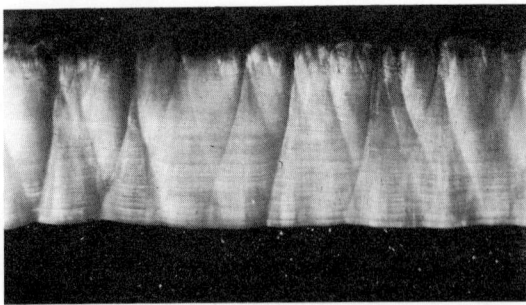

Fig. 43. Micrograph of the heavy hard oxide layer prepared in 1% sulfuric acid at 50 V, in polarized light. (150 ×). (Reproduced at 65%.)

Table 2. Total Oxide Thickness and Pore Distribution Produced under Various Conditions

Medium (technology)	Temperature (°C)	Barrier thickness (μm)	Total thickness (μm)
Dry air atmosphere	20	0.001–0.002	0.001–0.002
Dry oxygen atmosphere	20	0.001–0.002	0.001–0.002
Dry air atmosphere	500	0.002–0.004	0.040–0.060
Dry oxygen atmosphere	500	0.010–0.016	0.030–0.050
Humid atmosphere	20	0.0004–0.001	0.050–0.100
Humid atmosphere	300	0.0008–0.001	0.100–0.200
Boiling water	100	0.0002–0.0015	0.500–2.00
High-temperature and -pressure water	150	ca. 0.001	0.10–5.00
Chemical oxidation (conversion)	70–100	0.0002–0.0008	1.00–5.00
Chemical polishing	50–100	ca. 0.0005	0.01–0.10
Electropolishing	50–60	0.0050–0.0100	0.10–0.20

Anodic oxidation					Pore	
Electrolyte (g/liter)	Voltage (V)				Diameter D (A)	Number (N/μm^2)
200 H_2SO_4	10–12	18–20	0.010–0.015	5.0–35.0	130	1190
150 H_2SO_4	10–12	10–15	0.010–0.013	5.0–35.0	120	1140
150 H_2SO_4	10	15	0.015–0.018	5.0–40.0	120	772
10 H_2SO_4 (hard anodizing)	50–60	+1 to −1	0.015–0.320	100–250	90–250 (300)	100
30 CrO_3	40	42	0.010–0.013	3.0–25.0	430	48
30 CrO_3	40	33–38	0.010–0.013	3.0–20.0	240	12
56 H_3PO_4	30	25	0.010–0.012	2.5–20.0	400	70
400 H_3PO_4	20	25	0.010–0.012	2.5–22.0	330	188
40 H_3PO_4	120	20	0.010–0.011	2.0–15.0	330	90
50 H_3BO_3	50–500	50–100	0.005–0.010	1.0–1.2	110	85

metal/oxide/electrolyte phase system, which bring the reacting components close together to produce a collision. Thus, the momentary concentration of the aluminum and oxygen ions (c_{Al} and c_O) at a given time and place are established:

$$v = kc_{Al}c_O \qquad (3)$$

where k is the rate constant of the formation of aluminum oxide.

Material Transport in the Boundary Phases in the Oxide Layer

From reaction kinetics, diffusion in anodic processes can be separated into two different groups:

1. diffusion in the solution layer immediately adjacent to the solid body;
2. diffusion of ions in or through the metal oxide electrolyte system.

As has been mentioned already, aggregates of molecules are formed by diffusion in the adsorbed or chemisorbed oxygen layer at the metal surface at discrete sites depending on the surface energy distribution. This distribution can be traced through subsequent stages of oxidation. Sites which are more reactive than their surroundings are found both in the barrier oxide layer (composed of several molecular layers) and in the layers formed over the barrier oxide.

Material Transport in the Boundary Phases. The adsorption and/or chemisorption of oxygen and the formation of a stoichiometric oxide membrane in the sorption layer occur as a consequence of the immediate collision between oxygen and metal ions. After the appearance of the oxide membrane, the reacting species meet during nucleation periods.

Nucleation induced by thermal, chemical, or electrochemical effects at the metal surface are interpreted by the model proposed by Franck,[49] shown in Fig. 44. Isolated oxide nuclei develop from the first oxide formations appearing at discrete surface sites; these grow independently and anisotropically either parallel or perpendicular to the metal surface. Point C in the figure is where the oxide nucleus has first appeared and growth has started; this is the nucleus center. The oxide nucleus grows spherically, resembling a calotte, but only to a certain size; then lateral growth predominates. The higher propagation rate in a direction parallel to the metal surface is explained

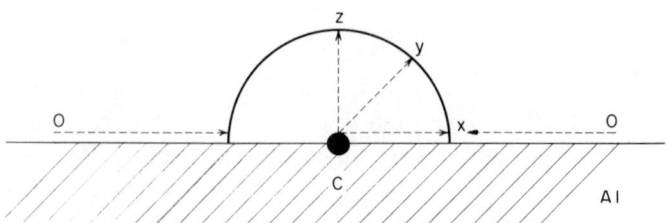

Fig. 44. Model of the growth of oxide nuclei: C, center of the nucleus; x, y, z, directions of metal ion and electron conduction; O, direction of oxygen diffusion on the metal surface. (From Franck.[49])

by the favorable situation at the edge of the nucleus for the transport of reaction components for the passage of current (point x in the figure). Here the metal ions leaving the metal and the anions of the electrolyte encounter the least resistance and the reaction proceeds with the greatest speed. In addition to the favorable transport conditions, the presence of the precipitated solid material at this point promotes the adhesion of further reaction products onto the nucleus.

At other points on the nucleus where the distance from the metal surface is too great (points y and z in the figure), the reaction is slowed substantially because it is difficult for the transport process to occur. The anisotropy of the growth causes the nucleus to resemble a flattened half-sphere.

If the new oxide formation possesses ionic conductivity, diffusion is always more favorable (naturally this depends on the degree of conductivity) perpendicular to the metal surface. Accordingly, oxide is also formed at the nucleus cap, but the reaction there is much slower than the reaction rate at the sides of the sphere. The height of the nucleus calotte then depends on conductivity of the nucleus and the diffusion rate.

The fact that concentrations at the edge of the nucleus differ from those at the top of nucleus can be explained by a disturbance of the equilibrium concentration due to the exhaustion of the reacting components. Additional oxygen and metal ions are transported by diffusion from more distant parts of the metal surface to supplement the reserves of reacting components. The rate of surface diffusion naturally depends on the chemical and electrochemical parameters of the system.

Surface diffusion causes the local concentration to retain the critical supersaturation limit; then the oxide formation starts again and proceeds further. The concentration gradient which then develops in the electrolyte adjacent to the nucleus depends on the amount of current passing through the system. The concentration of the saturation limit has been characterized by Franck with the aid of the diagram shown in Fig. 45. At high current densities saturation is achieved rapidly in the boundary zone, but it decreases rapidly at greater distances from the edge of the nucleus. At low current densities the saturation in the boundary zone is slower to develop and the maximum appears at lower concentrations. However, at these lower current densities, the concentration gradient decreases more slowly with increasing distance from the edge than in oxidation at high current densities. Since the rate of oxide formation basically depends on the concentration, lateral growth increases more or less rapidly at the edge. These variances cause the nuclei to appear in a variety of shapes with a range of boundary angles. The coalescence of nuclei

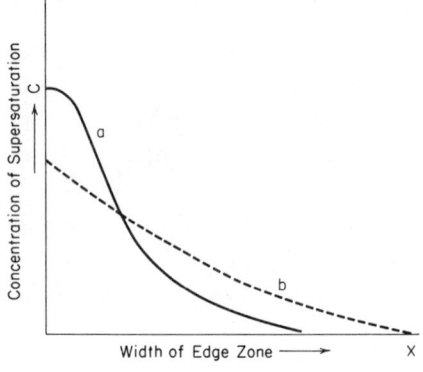

Fig. 45. Concentration profile for the limit of
saturation in the edge zone of the oxide nuclei:
a and b, concentration profiles at high and low
current densities; c, concentration of supersaturation. (From Franck.[49])

and the formation of a coherent oxide layer occurs with different speeds.

Since surface diffusion depends on the reaction time and on the supply of
the reactants, the extension of the edge depends on the experimental
conditions. Therefore, the "reaction waves" start after material supply has
occurred. During the reaction wave the material available at the edge is
consumed by the oxide formation; and when the concentration falls below the
critical limit, the reaction stops again. After the completion of a further process
of material delivery, it starts again when the upper limit of the critical point is
attained. This periodicity is indicated in the tests with the Elypovist
microscope by the wave or rosette shape of the oxide formations.

At low cell voltages or current densities, oxides form slowly; the material
supply by diffusion is also slower due to the low oxygen concentration. The
oxidation front then proceeds more slowly laterally from the edge zone of the
nucleus. Under the Elypovist microscope the periodicity of the reaction and
the terrace-like oxide form is readily visible (Fig. 13).

At the moment when the nucleation period is completed and the oxide
layer is coherent, the layer morphology is rather nonuniform. At higher
magnifications the heterogeneous structure, containing secondary zones of
oxidation and depressions (similar to cracks or deformed channels through the
coalesced oxide formations), can be seen with great clarity (Fig. 46). This
irregularity is gradually obliterated in the next period of an anodization, with
the aid of transport processes occurring laterally and through the oxide layer
(Figs. 47 and 48).

Material Transport through the Oxide Layer. After the formation of the
barrier layer, the material and current transport for the processes which form
the covering oxide layer are supplied by bulk diffusion through the oxide
membrane (lattice or interstitial diffusion). Material to complement oxide

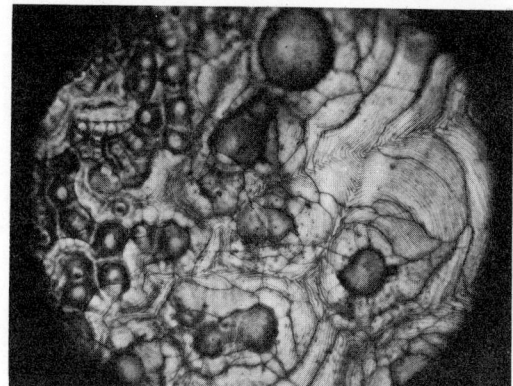

Fig. 46. Coalescence of primary and secondary oxides in 1% sulfuric acid with 45 V at ± 1°C after 100 sec (250 ×). (Reproduced at 65%.)

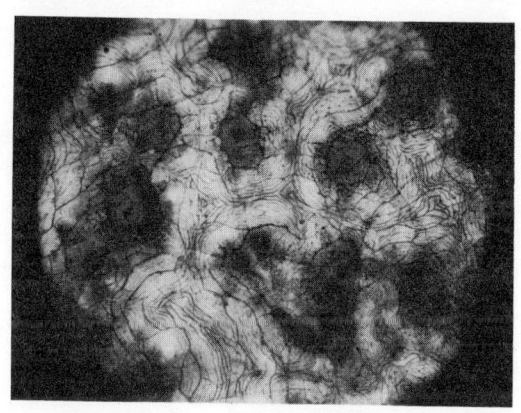

Fig. 47. Same as Fig. 46. Gradual coalescence of primary and secondary oxides formation period of the covering oxide layer after 5 min (250 ×). (Reproduced at 65%.)

Fig. 48. Same as Fig. 46. Surface of the oxide coating due to chemical re-solution and pore formation after 30 min (250 ×). (Reproduced at 65%.)

formation in the lateral direction is supplied by surface diffusion parallel to the metal surface and induced by differences in the chemical or electrochemical potential between the high sites of the metal surface (concentration or potential gradients). Therefore, the diffusion of ions and electrons plays an important part in defining the reaction rate.

From the electrochemical measurements carried out by Murphy and Michelson[31,32] as well as those reported by other workers,[33-38,58,63] it is possible to conclude that apart from current transport by proton migration, the current can also be carried by other ions (metal ions and oxygen-containing anions) within the highly dispersed aluminum oxide, aluminum oxyhydrate, or aluminum hydroxide lattice. This conclusion is in agreement with the earlier results obtained by Rayola and Davidson[68]; they showed that not only Al^{3+} ions but also Al^{2+} and Al^+ ions are formed during anodizing and the trivalent dissociation product which has the smallest ion diameter can move with much greater velocity than either the bi- or monovalent aluminum ions.

The metal cations diffuse outward toward the oxide/electrolyte boundary, while the oxygen anions migrate inward toward the metal/oxide interface. The oxidation process occurs in the oxide region adjacent to the substrate.

The diffusion of material in the oxide layer is summarized and evaluated by Tajima.[24] Therefore, we shall omit detailed literature references.

Current Transport in the Boundary Phases and in the Oxide Layer

Current Conduction in the Boundary Phases. In electrochemical oxidation the concentration of oxygen-carrying ions is rapidly reduced to a minimum in the liquid layer adjacent to the anode (the Helmholtz double layer with a thickness of 10^{-4}–10^{-2} cm) when the current passes. If the same number of ions as were precipitated on the anode are transported by diffusion into the boundary layer fom the solution to supplement the loss of material due to loss of charge, then the intensity of the current passing through the boundary layer is determined by the stationary diffusion rate; this is similar to Fick's[61,121] first law. The diffusion current (i_d) in the liquid boundary layer can be expressed[69] as

$$i_d = zFD \, (dc/dx) \tag{4}$$

where z is the valence, D the diffusion constant, F Faraday's constant, and dc/dx the concentration gradient in the distance x.

In the initial stage of anodic oxidation the concentration gradient in the liquid layer varies from point to point and from moment to moment; therefore, diffusion is non-steady-state. It is because of this continuous variation of current and concentration that the diffusion current cannot be measured. In later phases of anodizing, as in the forming period of the covering layer, the local changes of concentration at the oxide/electrolyte phase boundary parallel with the surface can be regarded as being approximately uniform; however, perpendicular to the metal surface, nonstationary processes should be encountered. By neglecting these latter processes, the diffusion current[69] can be expressed by

$$i_d = \frac{zFD^{1/2}}{(t\pi)^{1/2}} (c_0 - c_f) \tag{5}$$

where c_0 and c_f express the concentration in the solution and in the superficial liquid boundary layer, respectively; t is anodizing time; and the other symbols have the previously defined values. However, practice has shown that this expression has its shortcomings.

Current Conduction in the Oxide Layer. The details of current conduction in the oxide layer depend on the quality of the base metal and on the anodization conditions. The variants of current conduction in different electrolytes can be represented by the changes in oxidizing cell voltage (U) as a function of the time (t) at a constant current density (Fig. 49).

Current Conduction on the Barrier Oxide Layers: Electroluminescence. According to Günterschultze and Betz[70] and Ulrich,[71] the structure of sealing layers produced in electrolytes which do not dissolve aluminum oxide is not

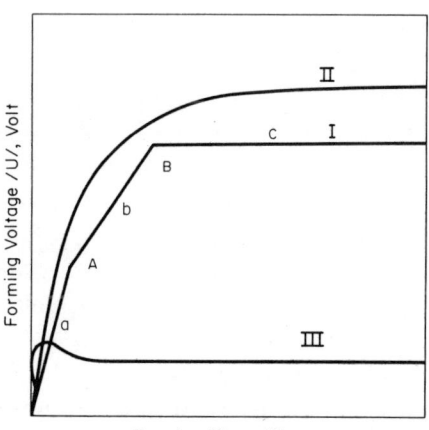

Fig. 49. Dependence of the forming cell voltage on anodization time with a constant current density: I, ideal valve anode; II, transition anode; III, porous oxide forming anode; A, spark voltage; B, maximal cell voltage; a, luminescent zone; b, spark discharge zone; c, zone of maximum spark voltage.

entirely uniform. Excess aluminum is present in the area adjacent to the base metal: in the area near the electrolyte the ratio of aluminum to oxygen corresponds to the stoichiometric Al_2O_3 composition. Two Schottky cation vacancy sites are present in the elementary lattice of the face-centered-cubic γ-Al_2O_3, as demonstrated by Ulrich,[71] Verwey,[72] and Kordes.[73] Therefore, the possibility of ionic diffusion and current conduction in the barrier layer can be interpreted partly by the excess of aluminum and partly by the presence of vacancy sites in the external zone.

The fact that a certain amount of anions or water molecules can be incorporated in the barrier film from the anodizing electrolyte argues against the total insulating capacity of the barrier layer. These water molecules are present either as hydrate water[74,75] ($Al_2O_3 \cdot H_2O, Al_2O_3 \cdot 3H_2O$), bonded in the compound (amorphous $Al(OH)_3$, $AlO \cdot OH$), or bonded by chemisorption.[76] These components can introduce a certain amount of conductivity to the dense barrier layer.

Kaden and Ginsberg[77] state that the current density due to ionic conduction in the layer (i) depends primarily on the applied cell voltage (U):

$$i = \alpha \exp\left(\frac{\beta U}{d}\right) \tag{6}$$

where d is the layer thickness, $\alpha \sim 10^{-22} A/cm^2$, and $\beta \sim 4 \times 10^{-6} V$. The value of the force field (F) is found by the expression

$$F = U/d \tag{7}$$

As can be seen from diagram I in Fig. 49, for an ideal barrier layer, the increase in cell voltage is initially proportional to the time of anodization. At the limiting value A (i.e., at the boundary of the so-called "spark voltage") spark discharges occur at the surface of the anode accompanied by luminescence. After passing A, the forming voltage continues to increase; on attaining the turning point B, sparking is increased to such an extent that constant luminescence at certain surface sites results.

Although several aspects of sparking are still under discussion, Ulrich[71] has satisfactorily explained some aspects of the phenomenon. His results are based on tests conducted in boric acid and borax solutions, which gave a time-dependent change of the forming voltage [i.e., $U = f(t)$].

At the onset of anodizing (Fig. 49, I, a), ionic conduction through the oxide layer dominates because of the outward diffusion of aluminum ions. At this stage electronic current conduction is negligible. The first inflection point (A) of the curve, $U = f(t)$, is a result of the electric force field, which increases as the layer thickens to a depth which exceeds the energy level of the Helmholtz double layer formed at the electrolyte/oxide interface. It causes a sudden metal ionization and a loss of electrons by the anions. After the "electron avalanche" a high charge density appears at certain sites of the metal boundary; this current pulse results in local superheating and in spark discharges acompanied by luminescence.

When sparking occurs, the electronic current represents a substantial portion of the current, and the ionic conduction is negligible (Fig. 49, I, b).

An express condition of sparking is that the anions of the electrolyte should be able to simultaneously emit a great number of electrons when a suitable voltage is applied. Our own experiments in dilute, low-temperature chromic acid solutions have supported this conclusion.

Aluminum plates (99.5 and 99.9% Al) were anodized in a 3–5% chromic acid solution at temperatures between $+1°$ to $-1°C$. This solution has an artificially reduced oxide dissolving ability. The applied voltage was gradually increased from 10 to 100 V. Between 60 and 70 V, a very small current density was detected. At 65 V the measurable current density was still only $0.5 \, A/dm^2$ (Fig. 50). At 73–75 V a slight luminescence is observed at the edges of the anode (Fig. 51a). When the voltage is increased further, the degree and intensity of the luminous phenomena rapidly increase (Fig. 51b and c); at 85 V the current density increases to $5 \, A/dm^2$ (Fig. 50). When this voltage is held for

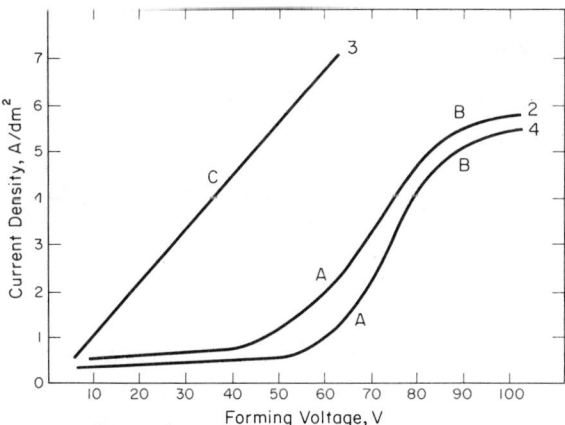

Fig. 50. Relationship between current density and cell voltage for aluminum anodes with different compositions: 1, 99.5Al; 2, AlMg5; 3, AlSi5. A, luminescence; B, spark breakdown; C, anode dissolution.

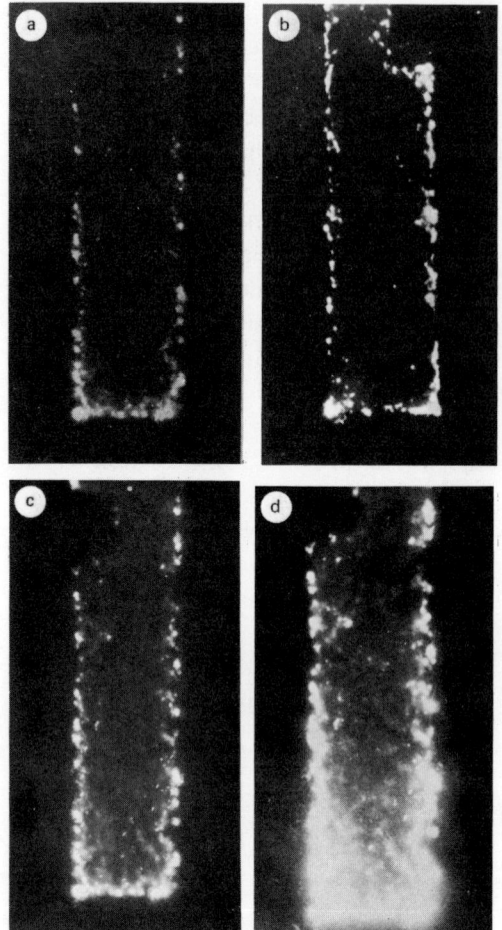

Fig. 51. Phenomena of luminescence and spark discharge on a
99.5Al anode in 3–5 % chromic acid
at $\pm 1°C$: (a) 75 V; (b) 80 V; (c)
85 V; (d) 90–100 V (image dimen-
sion ratio 1 : 1).

a time, the current density eventually decreases. After 30 min it drops to
$2 \, A/dm^2$ and continues to gradually decrease.

At 90 V or more the anode starts to glow from the bright sparks that are
then produced (Fig. 51d). Some of the sparks zigzag across the anode surface
and disappear, only to be replaced by new sparks. As the oxide layer thickens,
the number and intensity of the luminous points decrease. The final result is a
slight luminescence of a larger area of the anode surface. The anodic film
obtained in this manner is iridescent, with bright interference colors; the
zigzag motion of the sparks is also traced out on the anodized surface as in
Fig. 52.

Increasing the cell voltage to 100 V results in localized erosion at areas

Fig. 52. Surface of the 99.5Al anode plate shown in Fig. 51 after anodizing 10 min.

where the sparks formed on the metal surface shown in Fig. 53. Sometimes a green precipitate of solid crystalline trivalent chromium oxide is found in these eroded spots, indicating that the localized electron loss of the anions of the electrolyte occurred at a very high rate, due to the high cell voltage, by the reaction

$$2CrO_4^{2-} \rightarrow Cr_2O_3 + \tfrac{5}{2}O_2 + 4e^-$$

The increase in the local force field which causes sparking resulting in luminescence depends on the quality of the electrolyte, the current, and also to a substantial degree on the quality of the anode surface (i.e., on the composition of the base metal, its microstructural irregularities, its geometrical roughness, on edge and corner effects, etc.). Compositional effects can be demonstrated by the fact that spark discharges on $AlMg_5$ samples appear earlier than on pure aluminum, while in $AlSi_5$ samples no sparks form, even when the voltage is increased. However, in this latter case, low voltages (20–30 V) causes pitting, which results in a rapid increase in the current density.

Conduction in the Covering Oxide Layer. No sparking is observed in electrolytes which readily dissolve aluminum oxide; the anodizing voltage in these solutions is lower than that required for sparking (Fig. 49, III).

In these solutions the oxide at the external surface of the barrier layer is completely dissolved during anodization. The remaining portion is transformed, incorporating water as hydroxides and hydrates; the area close to the base metal has a continuous barrier layer. A chemical aggregate is

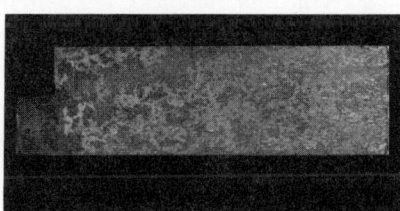

Fig. 53. Surface of the 99.5Al anode plate shown in Fig. 51 after anodizing 40 min.

formed on the continually thickening covering oxide layer. In this structure the unchanged submicrocrystallites of aluminum oxide are colloidally embedded in the gel-like hydroxide and hydrate compounds.

Aluminum ions and oxygen carriers (OH ions, H_2O molecules) diffuse through the oxide layer; thus, this layer supports a current. In addition to the diffusion processes, Murphy and Michelson[31,32] also assumed that in the covering oxide layer, H_2O molecules are joined by "hydrogen bridge" bonds to the aluminum oxide molecules and to the occluded anions. In this polynuclear system the covalent bonds and "hydrogen bridge" bonds are able to change places, or move in a certain direction (protomery). The changing of bonds and the simultaneous exchange of electrical charges between particles occur along adjacent atom groups very rapidly in the direction of the electrical force lines. This rapid migration of charges continuously supports the current transport due to ionic and electronic diffusion. Eigen[78] states that the bond-changing frequency can be as high as 10^{13}–10^{14}/sec. If this is the case, then the current transport is lower than electronic conduction in pure metals by only one or two orders of magnitude. However, only oxyanions can form hydrogen bonds and conduct a high ionic current.

Ionic and electronic conduction proceeds through the oxide normal to the substrate. Interpreting this phenomenon, particularly the lateral growth of the interlaminar oxide, parallel to the substrate, involves an assumption that the electrical force field within the oxide coating is not uniform throughout but that potential differences arise (probably due to epitaxial influence). This allows conduction and material transport parallel to the surface until the potential differences are eliminated. References to ionic conduction in the metal/oxide/electrolyte system can be found in the work of Ginsberg *et al.*,[33,35] Diggle *et al.*,[40] Dorsey,[65] Charlesby,[79] Hoar and Yahalom,[80] Winkel *et al.*,[81] Brace,[82] Neufeld and Ali,[83] and other authors.

Hoar and Yahalom[80] give the current density (i) by the equation

$$i = A \exp (B E_{F/y}) \tag{8}$$

in which A and B are system constants and $E_{F/y}$ represents the potential difference (forming voltage, in volts) between the metal and the electrolyte and y (cm) is the film thickness. The exponential ratio is due to the nature of the ion transport process and regulates the total current density of current. This relation results from the nature of the ionic transport process and regulates the total current density through the system regardless of whether it signified transport at the metal/oxide boundary, within the oxide, or at the oxide/electrolyte interface.

Different values of the constants A and B can be found in the literature, but these differences have no influence on the theoretical considerations. The problem is that i may vary at irregular rates and in irregular directions during nucleation. The oxide layer with ionic conduction has a higher electrical resistance than the electrolyte; and therefore during nucleation when the coverage of the metal surface is incomplete, the current passing through the system is concentrated at points which are uncovered. Then, since the area of bare metal is gradually reduced because of nuclei growth, anodization proceeds at a constantly changing current density instead of at a constant current density, which would depend on the total area to be anodized. Thus, the current value in the preceding equation is at best an average current value assuming that all points on the surface are equal. It does not provide a definite current value.

After total coverage has been attained, current conduction proceeds by ionic or electronic diffusion through the layer and eventually only by

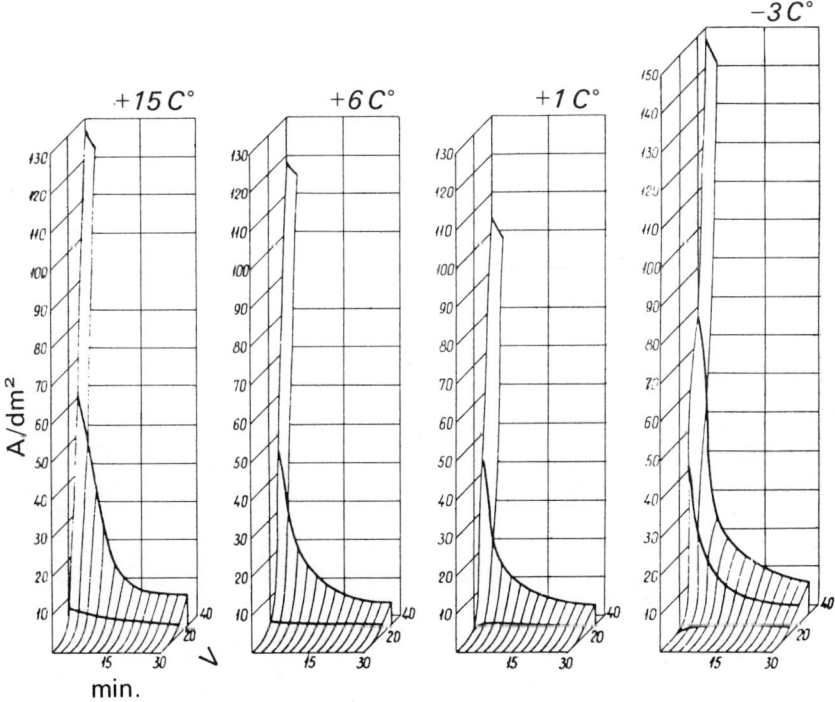

Fig. 54. The relationship between cell voltage and current density as a function of time in 20% sulfuric acid at various temperatures.

protomeric migration of bonds. Then the current density (i) decreases, at a uniform rate proportional to the time of anodization or to the thickening of the layer until a stationary equilibrium value is achieved. Here it is possible to characterize the current transport to a good approximation with the aid of Eq. (8). Figures 54 to 58 are space diagrams showing the relationship i–V–t for anodic oxidation in sulfuric acid at different temperatures and concentrations.

For reaction kinetics the Müller rule[84] is of interest. This rule characterizes layer formation by the time required for forming an oxide consisting of laterally propagating and coalescing nuclei and secondary oxidation. The "time of coverage" (τ_p) depends on the material properties of the system and on the intensity of the chemical effect on the current density (I):

$$\tau_p = kI^{-m} \qquad (9)$$

where k and m are material constants of the system.

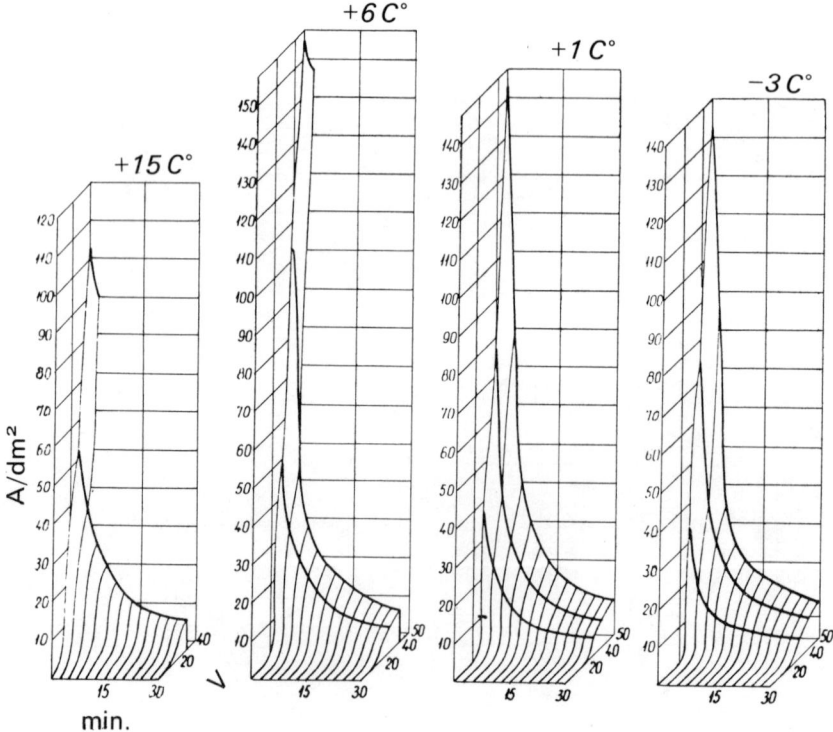

Fig. 55. The relationship between cell voltage and current density as a function of time in 10% sulfuric acid at various temperatures.

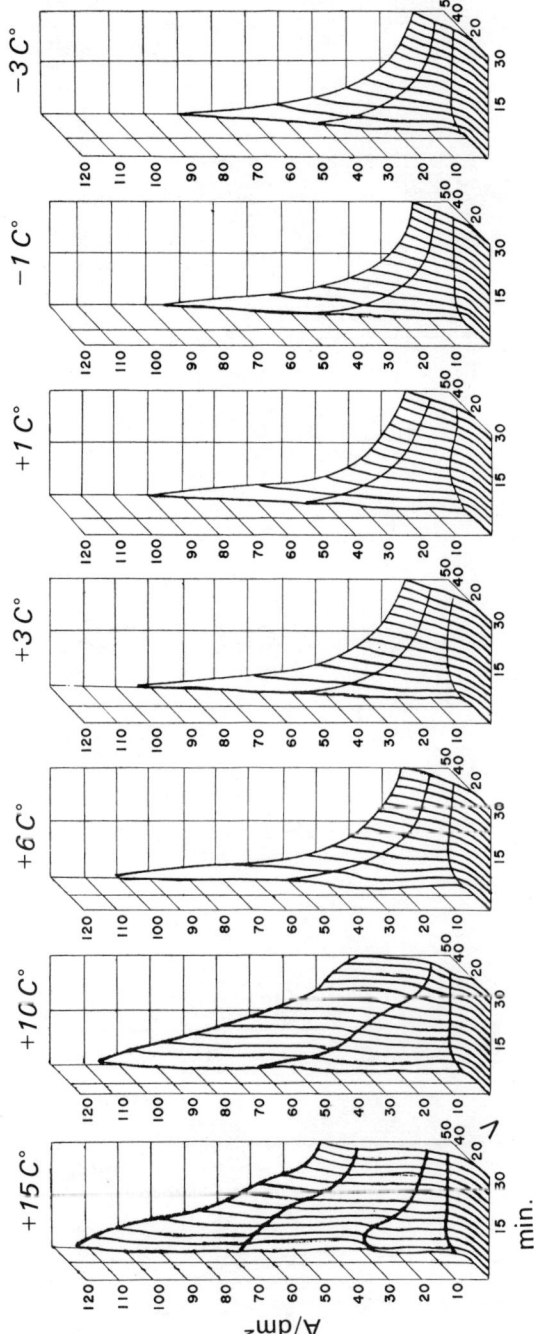

Fig. 56. The relationship between cell voltage and current density as a function of time in 5% sulfuric acid at various temperatures.

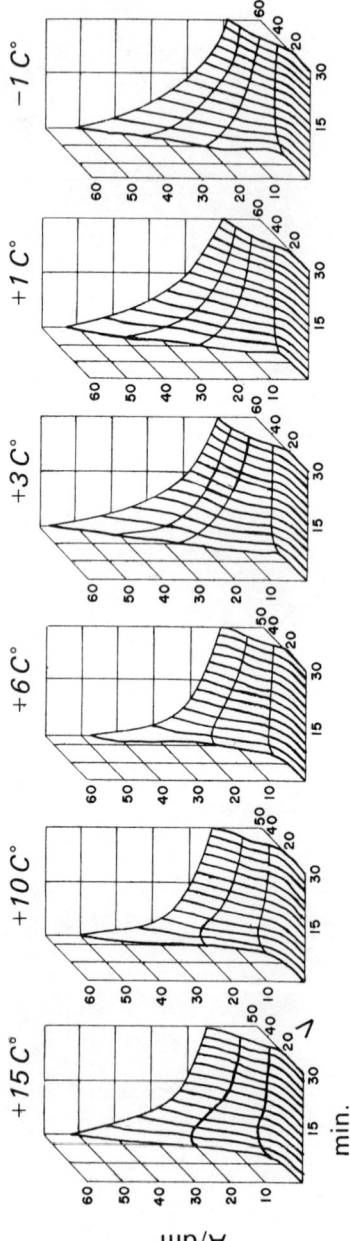

Fig. 57. The relationship between cell voltage and current density as a function of the time in 2–5% sulfuric acid at various temperatures.

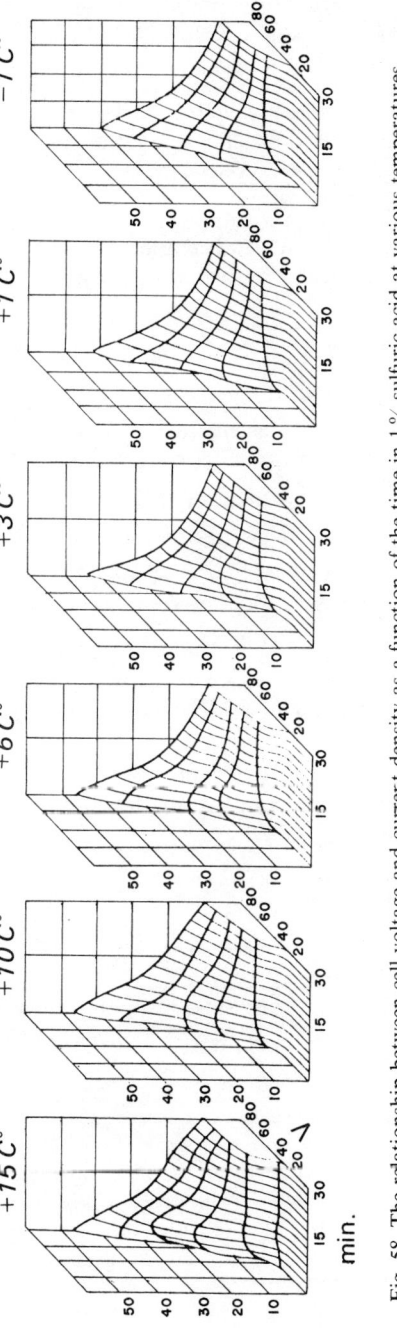

Fig. 58. The relationship between cell voltage and current density as a function of the time in 1 % sulfuric acid at various temperatures.

Franck[49] gives various values of m between 1 and 2 in various oxidizing systems. When an oxide with slight solubility is formed, m equals 1.5. Therefore, the rate of formation of the primary oxide layer is a function of the current passing through, which in turn depends on the concentration, temperature, and potential conditions. These relationships are shown on τ_p/I diagrams plotted with various parameters (Fig. 59).

Influence of the Chemical, Metallurgical, and Crystallographical Properties of the Base Metal on the Structure of the Oxide Layer: Epitaxy

The Chemical and Metallurgical Properties of the Base Metal

An anodic oxide coating is produced by a chemical or electrochemical conversion of the external crust of a base metal, unlike methods which produce a coating by the precipitation of a product from a solution onto a substrate surface (like galvanizing). The consequence of this fact is that the physical and mechanical properties, chemical composition, crystallographical and structural composition, state of energy, etc., of the surface layer of the base metal have a definite effect on the quality and structural properties of the metal oxide which replaces the external superficial metal layer; the thermal and mechanical history of the metal is included in this listing.

In consideration of the increased requirements for anodic oxidation coatings, the production of aluminum and its alloys suitable for "anodization" is increasing; excellent oxide coatings with a uniform structure, free of discoloration and stains, can be produced. The technical literature abounds with details for anodizing. In most industrially developed countries the

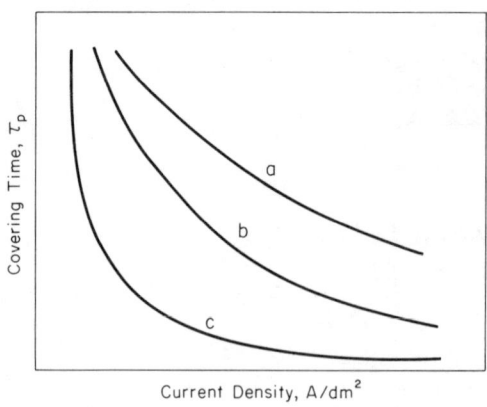

Current Density, A/dm^2

Fig. 59. Dependence of the covering time on the anodization current density using the Müller rule.[84] Curve (a) is in 1 % sulfuric acid, (b) in 5 % sulfuric acid, (c) in 20 % sulfuric acid.

technical instructions, test methods, and specifications for acceptance have been standardized; these details will be omitted.

The alloying elements influence the properties of the oxide layer. The purity and uniformity of the oxide layer are dictated principally by the silicon and heavy metal concentration in the base metal; the presence of significant concentrations of these components may make it difficult or even impossible to produce a coherent and adherent oxide layer. For instance, the oxide coating on the AlMgZnTi alloys has a higher hardness than on high-purity aluminum, AlMg, or AlMgSi alloys. However, on the metal alloyed with zinc and titanium, the hard oxide coating shows very little adherence and peels off after a slight impact, whereas the oxide coatings on aluminum, AlMg, or AlMgSi alloys adhere tightly to the base metal.

The incorporation of intermetallic compounds (e.g., aluminides, silicides) in the oxide layer depends on whether these substances dissolve in the electrolyte or remain unchanged.[85] Intermetallic substances soluble in sulfuric acid (e.g., Al_3Mg_2, Al_2Zn_3, $MgZn_2$, $Al_2Mg_3Zn_3$) may be partly incorporated in the oxide layer in the form of finely divided MgO and ZnO. Magnesium alloys can incorporate no more than 0.7% Mg in the form of MgO into the oxide layer[86-88]; the residual magnesium is dissolved out of the layer by the electrolyte. In alloys containing titanium, the titanium metal passes into the aluminum oxide layer as TiO_2. The intermetallic compounds $MnAl_6$, $FeAl_3$, $CuAl_2$, and Mg_2Si are insoluble in sulfuric acid, and therefore they are wedged into the structure of the oxide layer partly as oxides and partly unchanged. The distribution corresponds to their original sites in the microstructure of the base metal. In certain oxidizing processes pure silicon is embedded in the aluminum oxide without oxidation.

Aluminum products produced by hot or cold rolling, pressing, or other forming processes as well as in cast products have aluminum oxide stage inclusions which may have an irregular size and shape. Occasionally, sand grains or heavy metal oxides may appear wedged into a product surface. The arrangement of these inclusions in the product depends on the forming process. The small, sometimes submicroscopic inclusions can cause cloudy or discolored areas in the oxide layer during ordinary anodization; in special processes (e.g., hard oxide or direct colored oxide production) they may cause odd formations on the oxide layer.

We have previously shown that the oxide nuclei initiate on surface defects. Thus, foreign substances incorporated in the oxide layer influence not only the oxidation process but also the appearance, color, surface finish, and the physical, mechanical, and corrosion properties. The corrosion resistance is particularly sensitive to heavy metal compounds in the oxide (intermetallic iron compounds, for example); the corrosion resistance of these oxides is substantially less than that of coatings prepared on pure aluminum.

Further details on the metallurgical properties of the base metal on the quality of the oxide layers can be found in the chapter by Tajima in *Advances in Corrosion Science and Technology*, Vol. 1.[24]

The Influence of Crystallography and Microstructure on Oxide Formation

When single- or multicomponent coatings are formed on metals, epitaxy can play a role in development of the oxide crystal structure, its texture, and its macro and microformation. The effect of the microstructure and crystal orientation of the base metal can appear equally in all stages of the "guest" or oxide phase.

In epitactic systems the substrate influences nucleation and growth by substantially reducing the threshold value of the activation energy required to start the processes at certain sites of the metal surface and by inhibiting the start of the reaction at other sites.

The effect of crystallography and microstructure on oxide formation can be detected. The microstructure of aluminum casting can consist of either fine or coarse grains; the grain size will be determined primarily by the cooling rate. During slow cooling only a few solid nuclei appear at the outer mold surface. This causes dendritic solidification with large grains. At high cooling rates, however, a great number of nuclei are formed simultaneously in the melt, and after total solidification the metal matrix is usually uniform and fine-grained.

The dendritic solid has a decided effect on the oxide formation. Figure 60 shows a hard oxide layer produced on a very slowly cooled aluminum casting in dilute sulfuric acid, at a low temperature with a high cell voltage; here the dendritic arrangement of the oxide formations produced by a chemical transformation of the base metal is readily visible.

The coatings on AlSi alloys are also arranged in branching groups, as can be seen in Fig. 61. In this case not only is the effect of the dendritic structure noticeable, but also the effect of the silicide phase in the base metal. Both of these factors influence the growth and final appearance of the oxide.

Mott[2,89] has studied the influence of crystallography on the surface energy of the base metal and how it relates to the electrochemical process. He proposes that at the metal surface, a definite activation energy is required to permit metal ions to emerge from the crystal lattice. This activation energy must overcome the lattice energy which keeps the atoms in the lattice. Once

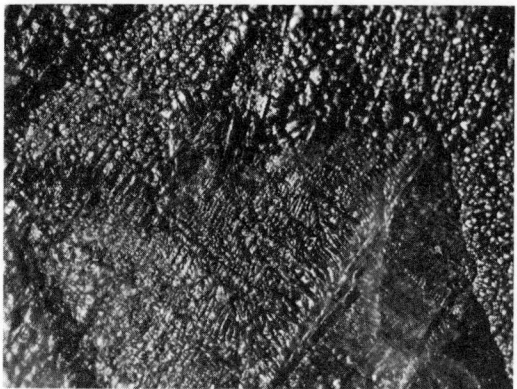

Fig. 60. Dendritic structure of an oxide layer prepared on an
aluminum casting in 1 % sulfuric acid with 50 V at ±1°C
(50 ×). (Reproduced at 65 %.)

the atoms are freed, they can collide with oxygen ions in the solution and form
aluminum oxide molecules. The aluminum ions continue to emerge from the
metal and diffuse through the thickening oxide layer toward the electrolyte.
The number of atoms removed from the lattice can be increased by increasing
the magnitude of the external energy sources (such as temperature or
potential).

The value of the energy required to remove the metal ions from their
places depends not only on the lattice structure but also on the sites they

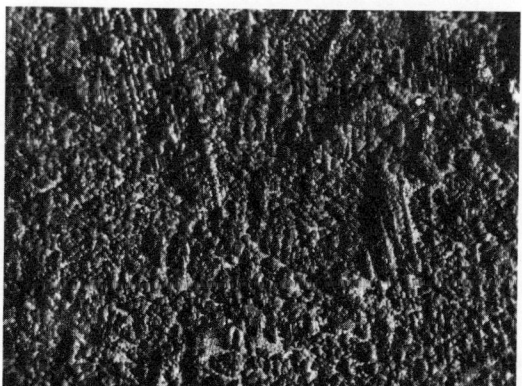

Fig. 61. Banded dendritic structure of an oxide layer
prepared on an AlSi (silumin) casting in 1 % sulfuric acid
with 50 V at ±1°C (50 ×). (Reproduced at 65 %.)

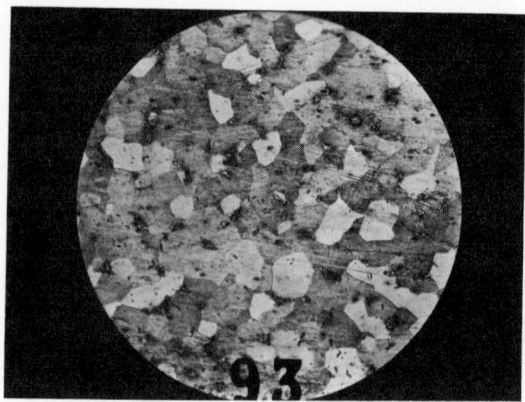

Fig. 62. Microstructure of hot-rolled 99.99Al consisting of
crystal grains of different orientation, after chemical etching
in a mixture of HCl, HNO$_3$, HF, and H$_2$O (150×).
(Reproduced at 65%.)

occupy in the crystal.[90,91] The removal of atoms arranged in crystal planes free
of defects usually requires more energy than does the removal of metal atoms
from edges, protruding peaks, or microstructural defects.

The surface of a chemically etched polycrystalline sample (Fig. 62) shows
that the total surface area due to roughness and pitting is larger than the
surface area of a sample of the same size which has been etched smooth. It can
be seen, then, that an applied potential can remove more atoms or ions from
the rougher surface than from the smooth surface. That is why the activities of
the individual surfaces may differ; they can etch differently and oxides or other
types of conversion compound phases can form at different rates.

The Relation between Overvoltage and the Geometrical Factors of the Base Metal

During anodization the difference between the equilibrium potential and
the actual potential of precipitation (i.e., the so-called overvoltage) represents
the activation energy required to start and to continue the electrode processes.

The activation energy (Q_h, Q_k, Q_l) required to remove a metal ion from a
crystal plane with (hkl) indices and the electrical overvoltages (η_h, η_k, η_l)
which liberate oxygen have different values. These values primarily determine
the planes on which the reaction products will appear; they determine the
morphological and structural difference that appear at the surface of the oxide
coating. If the voltage exceeds the upper limit, metal ions are able to leave the
crystal lattice structure simultaneously at different surfaces sites. The diffusion

of the ions through the oxide layer proceeds in a statistically uniform fashion. A cell voltage in excess of the local activation energy completely suppresses epitaxy, and oxide formation is uniform. However, under certain conditions, when oxide re-solution is repressed as in a dilute electrolyte at low temperature, the morphology of the base metal is readily visible on the final oxide coating.

Epitaxy in the Aluminum Aluminum Oxide Systems

Neuhaus and Gebhardt[92] have found that in aluminum/aluminum oxide systems the orientation relationships may differ; this depends on whether the "guest phase" (a natural oxide film, a coating formed from corrosion products, an artificially strengthened oxide film, etc.) has formed in a dry or moist atmosphere, at high temperatures, or in a aqueous solution through chemical or electrochemical reactions.

In thermal oxidation, Doherty[46] finds that oxide formation can be classified by the oxidation temperature: (1) oxidation below 500°C, where an amorphous oxide forms with a maximum thickness of 50 Å, and (2) oxidation above 500°C, where a crystalline oxide forms consisting of γ-A100H and γ'-A100H with a thickness of several 100 Å. In the latter case amorphous aluminum oxide is transformed to a regular crystal lattice by "devitrification."

The crystallinity of oxide phases formed in a moist or aqueous environment at low, medium, or boiling temperatures by chemical or electrochemical effects, and by corrosion or artificial conversion, is often completely obliterated since in the presence of water, aluminum oxide has a tendency to gel; therefore, the structure of the freshly formed oxide phase is amorphous. After some time (several days) an X-ray pattern can be generated; this indicates that the aluminum oxide phase is transforming from the gel to a crystalline structure by an internal rearrangement and by the loss of water. The types of crystal lattice which compose the anodic oxide layers are discussed by Tajima, who furnishes X-ray data.[24,76]

Below we shall present some examples of our results to illustrate the crystallographic relationships in the aluminum/aluminum oxide system. These results are based on experiments conducted on chemically etched high-purity (99.99%) thin aluminum sheets in 1% sulfuric acid at 0°C with 40 V applied.

The oxide that forms in the first stage of anodizing follows the lattice orientation of the individual grains (Fig. 62). Electron microscopy investigations of the oxide formations appearing on different crystal planes with the indices determined by X-ray diffraction have led to the following conclusions:

1. On planes with the indices (100), (110), and (111), 60 sec of anodizing formed primary oxides with characteristic arrangements, as shown in Figs. 63, 64, and 65. These primary formations are similar to the primary formations of electrocrystallization in the cathodic precipitation of copper, published by Hayashi et al.[93] (Fig. 66); and this seems to prove that the epitactic effect of lattice orientation is found in cathodic and anodic electrochemical processes.

2. Even after a long period of anodization, the oxide layer maintains the different orientations of the grains of the base metal (Fig. 67); the boundary between adjacent crystallites is readily visible in the electron photomicrograph (Fig. 68).

3. Prolonged anodizing gradually obliterates the boundaries until the oxide surface appears more or less uniform. This final surface resembles the pictures of oxide layers found in the technical literature.

Another interesting trait is that translucent layers of aluminum oxide exhibit optical birefringence. Franklin,[29] Huber,[94] Stirland and Bichsel,[95] Marge and Renouard[96] and others explain the origin of this phenomenon by assuming that the pore channel system of the oxide layer (KHR) is not perpendicular to the surface but lies at varying angles to it. This structural nonuniformity causes the birefringence, which may be small or great depending on the angle of the pore channels. Our experiments have shown

Fig. 63. Primary oxide on the crystal surface with the (100) orientation on chemically etched 99.99Al after anodizing for 60 sec in 1 % sulfuric acid with 40–45 V at ±1°C. Electron micrograph (14,000 ×). (Reproduced at 65 %.)

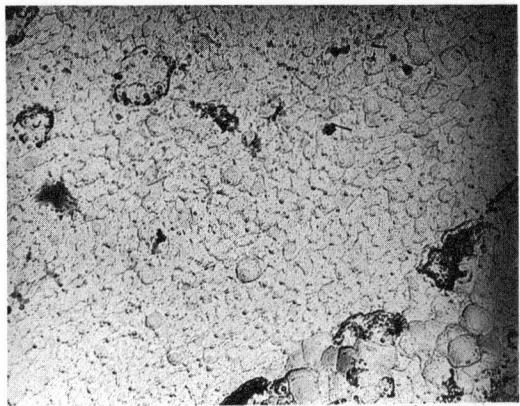

Fig. 64. Same as Fig. 63 but showing the primary oxides on the (110) surfaces. Electron micrograph (12,000 ×). (Reproduced at 65 %.)

that this phenomenon may not depend on the pore channel angles alone. Other effects are present.

When we anodized aluminum plate (Al 99.99) in 2–5% sulfuric acid at 0°C with 30–40 V, we found that the nuclei at certain points of the metal surface are more or less transparent under the optical microscope (Fig. 69). In polarized light these brightly transparent oxide formations appear as transparent or more-or-less darkened spots arranged according to the crystalline or subgrain arrangement of the base metal (Fig. 70). This phenomenon indicates epitaxy and controls the structure of the oxide particles formed in this fashion.

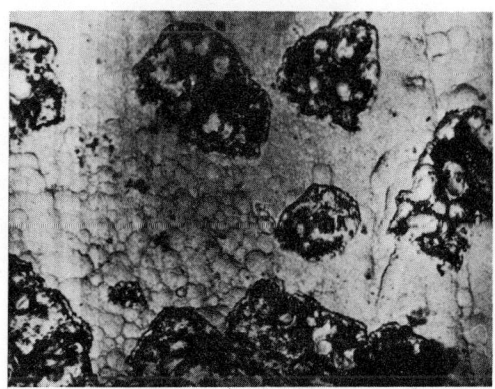

Fig. 65. Same as Fig. 63 but showing the primary oxides on the (111) surfaces. Electron micrograph (14,500 ×). (Reproduced at 65 %.)

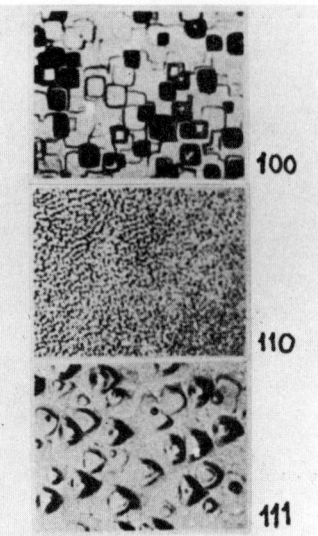

Fig. 66. Initial stage of cathodic copper deposition after anodizing in 0.5 M Cu (ClO$_4$)$_2$ solution at 25°C with a current density of 3 A/dm^2 on parts of the surface with different orientation. (Micrograph by Hayashi.[93])

Fig. 67. Oxide coating after 15 min on variously oriented surfaces of chemically etched 99.99Al in 1 % sulfuric acid at 40–45 V at ±1°C, in polarized light (250 ×). (Reproduced at 65 %.)

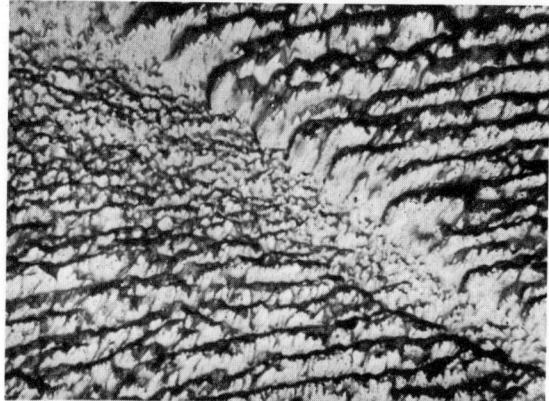

Fig. 68. Same as Fig. 67. Boundary between crystallites of different orientation. Electron micrograph (7700 ×). (Reproduced at 65%.)

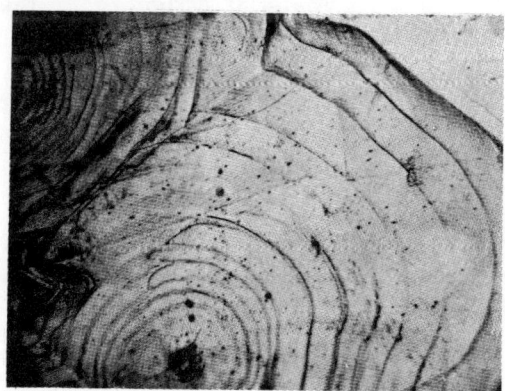

Fig. 69. Primary and secondary oxides obtained at the initial stage of anodic oxidation of chemically etched 99.99Al with 35 V at ±1°C, in ordinary illumination (250 ×). (Reproduced at 65%.)

Fig. 70. Oxide surface of Fig. 69 in polarized light (250 ×).
(Reproduced at 65%.)

THE RELATIONSHIPS BETWEEN THE STRUCTURE AND THE PHYSICAL, MECHANICAL, AND CHEMICAL PROPERTIES OF THE OXIDE LAYER

The Role of Material Structure in the Physical and Mechanical Properties

The Layer Thickness of Oxide Coatings

It is a fundamental fact that the thickness of oxide layers does not increase linearly with time of anodizing; in each process the thickening rate gradually decreases and attains a limiting value. The decrease is due partly to the reduction of the conductivity and partly to the shift in the equilibria of the reactions; the chemical re-solution and the nucleation process forming the oxide layer are competing reactions which shift the concentration of the reactants. There is a characteristic thickness reached by the oxide that is determined by the equilibrium between oxide formation and oxide re-solution. After attaining the characteristic boundary value, prolonging the time of oxidation does not increase the layer thickness. But while the oxidation of the base metal and the chemical re-solution of the oxide continue to decrease, a considerable number of dimensional defects can appear on the sample. Table 2 summarizes the optimum thickness values for oxide coatings produced by anodizing (page 305).

The Hardness and Wear Resistance of Oxide Coatings

The hardness of oxide coatings depends on several factors: (1) on the chemical composition and structural homogeneity of the base metal, (2) on the composition and temperature of the oxidizing electrolyte solutions, and (3) on the anodizing voltage and current density.

The hardness of oxide coatings on pure metals is usually higher than on alloys. Oxide layers prepared by dc current are harder than those produced by ac current. Oxide layers prepared in sulfuric acid, oxalic acid, or chromic acid baths are harder the lower the concentration and temperature of the electrolyte solution. Very hard oxide coatings (Vickers hardness 400–600 kg/mm^2) can be prepared on commercial-grade aluminum metal in 1–2% sulfuric acid at −1°C with a dc voltage and a high current density.[97–100] The hardness of ordinary oxides is greater near the base metal than near the external surface. The cross section in Fig. 71 shows the hardness distribution of an oxide layer.

The wear resistance reaches a maximum when the oxide layer has the optimum thickness as given in Table 2. Prolonged anodizing reduces the wear properties because of the resulting surface degradation.

But wear resistance is influenced by the alloying additions and their concentrations. Oxides prepared on pure Al and on AlMg alloys have the highest wear resistance; on alloys containing colored metals the wear resistance of oxide coatings is very slight.

The hardness and wear resistance required by industrial applications have been standardized, so such properties will not be listed here. Further information on this topic can be found in the work of Tajima.[24]

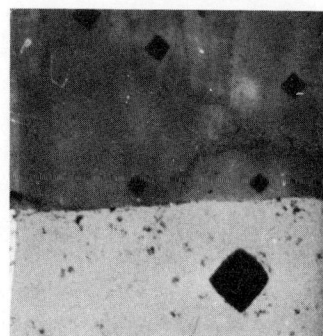

Fig. 71. Distribution of Vickers hardness values measured with the Hanemann tester on a polished section of an oxide coating (150 ×). (Reproduced at 75%.)

Tensile and Bending Strength of Oxide Coatings

Aluminum oxide coatings are generally brittle, and thus their tensile strength and elongation are low; however, they possess significant compressive strength. Brittleness is greater in the plane perpendicular to the direction of formation or thickening of the oxide layer; therefore, when anodized aluminum is subjected to bending, the oxide layer tends to crack. The degree of cracking depends on the anodizing conditions. Fine hairline fissures have little effect on the decorative or protective qualities of the coating after a sealing treatment, but coarser cracks in heavier oxide coatings caused by severe bending may lead to spalling of the coating. Malleable oxide coatings may be obtained in specially prepared anodizing solutions at higher temperatures with an ac current[101–103]; these methods are used to produce anodized aluminum wire for electrical coils.

It is impossible to directly measure the bending and tensile strength of oxide coatings, since the coating cannot be separated from the base metal. However, the oxide coating affects the strength of the base metal, and conclusions as to the mechanical properties of oxide coatings may be drawn from tensile, bending, or fatigue tests on anodized aluminum samples. To illustrate the relationships, we shall mention some of our results.[104–107]

The strength of anodized plates depends on the cross-sectional thickness of the oxide and the base metal. To investigate the change of strength which may occur, aluminum specimens were prepared and coated with an unusually heavy oxide layer; we prepared coatings of 100–150 μm and 150–200 μm (Vickers hardness = 460–500 kg/mm^2) on soft- (L) and hard- (K) rolled commercial (Al 99.5) aluminum plates (gauge 0.8, 1.2, and 2.0 mm) in cold dilute sulfuric acid with 50 V cell voltage for 30 and 60 min. The diagrams in Figs. 72 and 73 show the tensile and bending strength of the anodized aluminum specimens.

In tensile testing the metal core and the oxide layer are elongated together by the tensile stresses until the forming limit of the brittle oxide layer is reached. After passing this limit, widening hairline cracks form in the oxide layer (Fig. 74) along the direction of the zone boundaries of the porous and fibrous covering oxide layer. After the hairline cracks appear, the tensile load is supported exclusively by the base metal; and since the notch effect of the cracks weakens it, rupture occurs at lower tensile forces than in aluminum specimens of similar cross section with a smooth, unoxidized surface.

In bending tests, coarse cracks form after a certain amount of bending of the surface subjected to tensile stress, while on the opposite face, which is

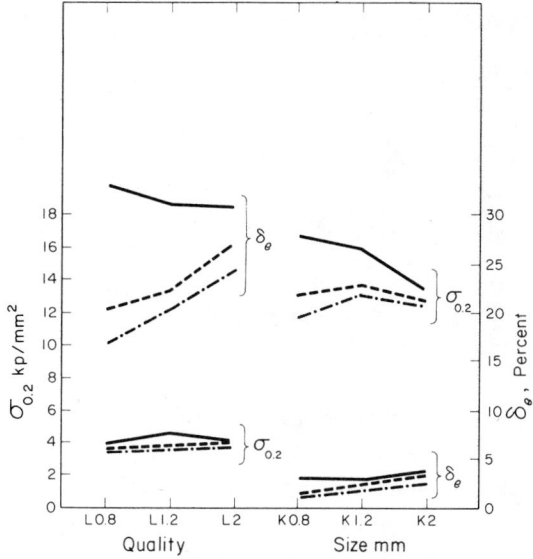

Fig. 72. Change of the yield point $(\sigma_{0.2})$ and uniform elongation (δ_e) of 0.8-, 1.2-, and 2-mm soft(L)- and hard (K)-rolled 99.999Al plates as a function of the thickness of the hard oxide layer formed on the surface: ——, untreated plate; ———, plate anodized for 30 min; —·—, plate anodized for 60 min.

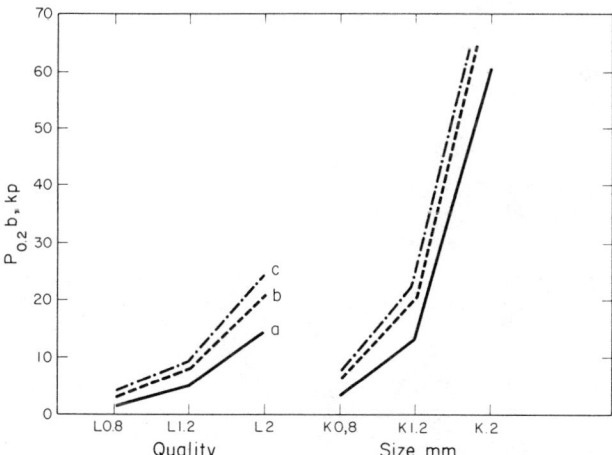

Fig. 73. Change of the bending limit $(P_{0.2}b)$ of 0.8-, 1.2-, and 2-mm soft (L)- and hard(K)-rolled 99.5 Al plate as a function of the thickness of the hard oxide layer formed on the surface: (a) untreated plate; (b) plate anodized for 30 min; (c) plate anodized for 60 min.

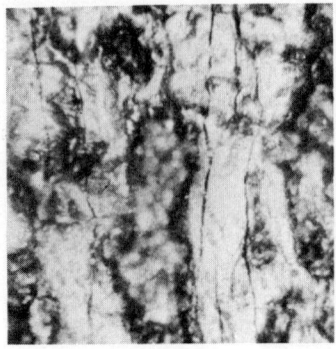

Fig. 74. Parallel hairline cracks in the hard oxide layer caused by tensile and bending stress (150 ×). (Reproduced at 75%.)

subjected to compression, the hard oxide with a high compressive strength exerts a "supporting" effect. As a consequence, the yield point in bending increases and deformation occurs at higher loads than in the base metal, and the oxide does not spall on the face subjected to tensile stress, even when the base metal itself has ruptured. Figure 75 shows the characteristics of oxide coatings in bending[108] and compares them with the cracking and peeling of other galvanic metal coatings after bending.

The resistance of oxide coatings to prolonged and repeated mechanical loading can be studied by fatigue tests. The fatigue tests carried out on light

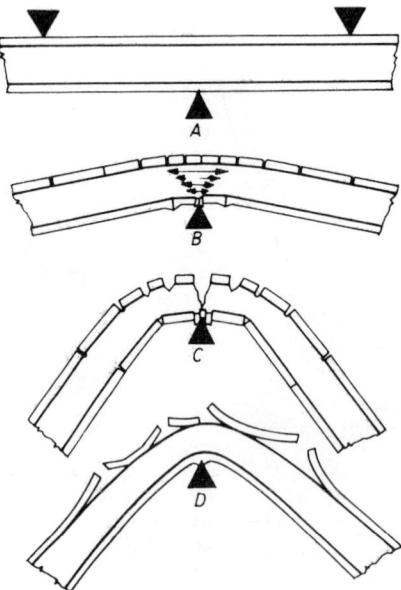

Fig. 75. Cracking of the anodic oxide coating during bending: A–C, oxide-coating aluminum plate; D, aluminum plate with galvanic metal coating. (From Schenk.[108])

metal plates with and without anodic oxidation and with an Amsler high-frequency fatigue tester with 5×10^6 tensile stress pulses showed that the decrease in fatigue life is related to the oxide thickness. Fatigue strength values measured with bending loads usually differ slightly from the values measured on plates without oxide coatings; these results are similar to those obtained in static bending tests.

Internal Mechanical Stresses in Oxide Coatings

The Development of Internal Stresses and Methods for Their Measurement. During anodization certain tensile or compressive stresses develop in the oxide coatings; their values depend on the material properties of the base metal, on the conditions of anodic oxidation, and on the conditions of anodic oxidation and on the layer thickness. These internal mechanical stresses influence the morphological and strength properties of the oxide coatings, and therefore they should be taken into account when considering the application of anodized products.

The effect of internal stresses in the layer during anodic oxide formation have been known for some time.[108] Elssner and Beyer[109] and Röhrig and Käpernick[110] have established a connection between spalling at rough edges and corners of aluminum objects and the differences of the specific volumes of aluminum metal and aluminum oxide. However, Davies *et al.*,[111] Vermilyea,[112] Bradhurst and Leach,[113] and others[114–117] have found that in addition to the ratio of specific volumes, mechanical effects due to structural changes in the oxide layer are also highly significant.

The dependence of the internal stresses, especially their direction and size, on the experimental conditions is interpreted partly by the volume increase caused by the reaction of $Al \rightarrow Al_2O_3$ and partly by the reduction of volume which accompanies the dissolution of the oxide by the reaction $Al_2O_3 \rightarrow Al^{3+}$. Theoretically one cannot calculate precisely the specific volume ratio of the base metal and the oxide layer (V/v) because the average molecular weight and density of the oxide, oxyhydrate, and hydroxide in the coating depend on the anodizing conditions; and the distribution and amount of foreign components incorporated in the layer can vary with the composition and microstructure of the base metal. The porosity of the oxide layers equals about $12-30\%$ of the total volume at the end of the anodizing process. The gradual change of the actual volume occupied by the oxide produces a transformation of the initial compressive stress in the oxide layer to tensile stress.

The variation in internal stresses can be detected by the Hoar–Arrowsmith[118] method or by its modified variant, the Stressograph, constructed by Stalzer.[119] In the electrolyzing cell of the Stalzer instrument an oxide layer is prepared on an anode consisting of a 10 mm × 75 mm × 0.5 mm aluminum strip bent into a ring. While anodizing the anode, the ring opens or closes depending on the stresses arising in the oxide layer; its displacement is recorded continuously as a function of time by a sensitive recorder. The internal stress, S (ponds/mm^2), depends on the current density, I (mA), on the thickness of the oxide layer, d (μm), and on the value of the coefficient K, which is defined by the instrumental parameters:

$$S = 20K\frac{I}{d} \tag{10}$$

Results of tests carried out with our measuring equipment have shown that during the formation of a porous oxide layer, a high compressive stress appears first. When the layer thickness increases, this compressive stress is gradually reduced; and at sufficiently high current densities, a tensile stress arises. Figures 76 and 77 show the relationships between the experimental factors.

In the oxide layers obtained by traditional anodizing processes, tensile stresses up to 6 kg/mm^2 may arise. In coatings obtained by special processes, such as hard oxide coatings prepared in dilute, cold sulfuric acid, compressive stresses of 9.5–10.5 kg/mm^2 may arise. In both cases the internal stresses are generally homogeneous if the oxide coatings were correctly prepared.

Relationships between the Oxide Structure and the Internal Stresses. In addition to volume considerations, the microstructure of the oxide can also

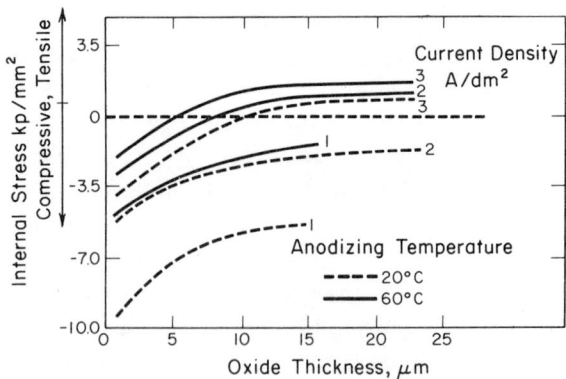

Fig. 76. Change of internal stress in an oxide coating of increasing thickness during anodizing in 20% sulfuric acid with 16–20 V at 20 and 60°C.

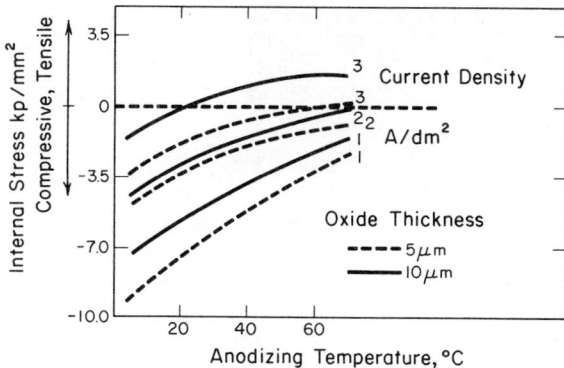

Fig. 77. Change of internal stresses in oxide coatings of various thicknesses prepared in 20% sulfuric acid with 16–18 V at 5–70°C.

play a part in the development of internal stresses. The structure of the oxide layer and the simultaneous development of stresses in the layer are controlled by the reaction mechanism. The compressive tensile stress transformations which occur by the relationships presented in Figs. 76 and 77 are due to the mechanical effects of the appearance, growth, and coalescence of primary oxide nuclei during oxide formation and during thickening of the coherent base layer. They are also due to the number, size, and distribution of pores, and therefore to the gradual reduction of the density of the oxide layer.

Next, we shall present some examples to indicate the relationships between the oxide structure and the heterogeneous internal stresses without attempting a complete discussion.

Structural effects which play a role in the development of internal stresses can be studied by optical and electron microscopic investigations in ordinary and polarized light of sections of aluminum samples anodized for various lengths of time. In the section of a sample plate anodized for 10 sec in 1% sulfuric acid with 40 V cell voltage (Fig. 78), the individual oxide nuclei and the coalesced oxide formations are already visible. Their external surface bulges out above the original plane of the metal surface, while the lower part penetrates the interior of the base metal with a curvature similar to the external surface. The V/v volume ratio can be measured on this figure by a simple geometric analysis.

The oxide layer produced in 5–10 min in 1–5% sulfuric acid, at -1°C with 50 V, is completely coherent but its thickness is not uniform; the external surface contains convex, coarse formations and some concave pits (Fig. 79a).

Fig. 78. Cross section of primary oxides obtained by anodizing for 10 sec in 1% sulfuric acid with 40 V at $\pm 1°C$ (150 ×). (Reproduced at 65%.)

Microscopic examination of the same section in polarized light shows that the mechanism of oxide formation actually controls the structural development of the oxide layer. In Fig. 79(b) it can be seen that the oxide layer is composed of individual fibrous blocks; their outer limits represent the extension of the primary oxide nuclei as well as the outlines of the grains of the base metal. Oxide nuclei created at various sites of the metal surface grow at different rates; thus, the height and mass of the oxide blocks are different in the given period of anodizing (5–10 min). The oxidation rate leads to a mutual influence on the growth of the oxide blocks and to their eventual mutual deformation; at some points they are wedged together and play an increased part in the evolution of internal mechanical stress in the oxide layer. Microscopic examination of samples anodized for longer periods (20–30 min) shows that the initial irregularities of the oxide layer are gradually obliterated and the fibrous blocks appear in close arrangement (Fig. 80).

Fig. 79. Cross section of hard oxide layer formed in 10 min of anodizing in 1–5% sulfuric acid with 50 V at $\pm 1°C$: (a) in ordinary light; (b) in polarized light (150 ×). (Reproduced at 65%.)

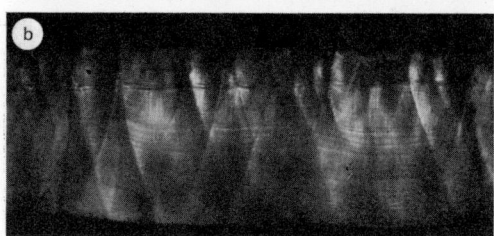

Fig. 80. Cross section of a hard oxide layer formed in 30 min in 1–5 % sulfuric acid with 50 V at ± 1°C: (a) in ordinary light; (b) in polarized light (150 ×). (Reproduced at 65 %.)

It should be stressed both from the theoretical and the practical points of view that an anomalous change of internal stresses, when the upper limits of the specified operational parameters are exceeded, may cause serious damage to the layered structure. In conventional anodization in sulfuric or oxalic acid, an excessive increase in the temperature of the electrolyte or in the current conditions produces a rapid dissolution of the oxide and coarse porosity, which leads to cracking and disintegration of the oxide layer.

When anodizing to produce a hard oxide (in 1 % sulfuric acid, +1°C to −1°C, and the cell voltage raised above the permitted value of 60 V), gap-like cracks appear due to internal stresses at sites where the oxide composition is more heterogeneous. Anodizing for prolonged periods at 80 V causes internal stresses which produce long horizontal ruptures in the oxide layer near the base metal; sometimes vertical cracks start from these ruptures in the direction of the external surface (Fig. 81). Cracks produced by compressive stresses tend to separate the layer from the metal surface; but the separation, curiously enough, never occurs at the metal surface but some distance from it. Shadows on the layer section indicate strong local heterogeneities, and the scatter in the values of the Vickers pyramid impressions at these areas also indicates a heterogeneous distribution of stresses.

When excessive voltages and low temperatures (such as anodizing in 1 % sulfuric acid at low temperature with 90–100 V) are used to produce hard coatings, the internal mechanical stresses may be very high and may produce

Fig. 81. Branching cracks in the hard
oxide layer obtained in 1 % sulfuric acid
with 80 V at ±1°C (150×). (Reproduced at 65%.)

serious damage and even peeling of the layers by cracks running parallel or
perpendicular to the metal surface (Fig. 82).

From these examples one may conclude that structural deformations of
various sizes and directions in the oxide are produced by both the isotropic or
anisotropic distribution of mechanical stresses at given sites of the layer.

*Oriented Structural Deformation in the Oxide Layer Induced by Internal
Stresses.* The following tests have yielded significant results concerning the
relationships between the structural changes and internal stresses in oxide
layers.

We anodized tubes of pure aluminum (50-mm) lengths with 10-, 20-, and
30-mm diameters with a wall thickness of 2 mm) in 0.1% sulfuric acid at −1°C
with 50 V. When the polished sections of the anodized tube samples were
examined under the microscope in polarized light, we found differences in the
structures of the oxide layers at the external tube surface and in the tube bore.

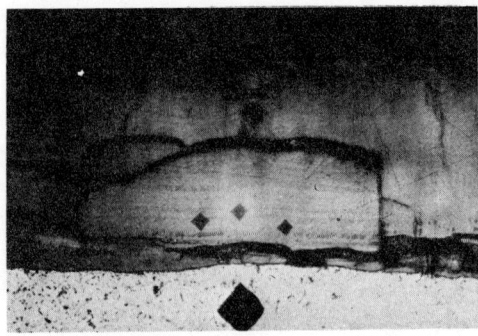

Fig. 82. Destruction of the oxide layer
by anodizing in 1% sulfuric acid with
100 V at ±1°C (150×). (Reproduced at
65%.)

On the internal and external surfaces of the tube with 30-mm diameter, the oxide layer was composed of conical blocks similar to the oxide layer shown in Fig. 80b. On the tube sample with 20-mm diameter, the external oxide layer was like the 30-mm tube, but within the bore the oxide contained indistinct and irregularly deformed shadows.

The 10-mm-diameter sample was like the other two samples on the external surface but contained cracks due to tensile stresses near the base metal (Fig. 83). In the oxide layer within the bore a curious crossbanded fibrous structure appears which was not previously observed. The fibres run at an angle of about 20–24° to the metal surface, and the fibres cross in an × form at an angle of about 40–44°. This oxide layer also contains internal cracks which are arranged near the surface of the layer. This shows that internal stresses were concentrated in this zone. The different strength conditions are also reflected by the fact that the external oxide layer has a Vickers hardness (VH) of 405 kg/mm^2, while the oxide layer in the bore has a VH of 470–476 kg/mm^2.

This phenomenon can be traced back to the difference in specific volumes (formulated by Schenk[108]), where $V/v = 1.8–1.5$. It can be explained by a rapid increase of internal compressive stresses in the tube bore, which is due to the increasing thickness and volume of the oxide layer; after the original limit of elasticity of Al_2O_3 has been exceeded, the continuously formed oxide products occupy sites in the gliding planes arranged according to the direction of compressive stress. The crossbanded fibrous structure in Fig. 83 resembles the gliding deformations found in crystalline solids which have been plastic-deformed. The fact that this structural change did not occur uniformly in the bores of the tube samples with 30-, 20-, and 10-mm diameters is probably due to the internal curvature of the surface.

There exists a critical radius limit beyond which internal stresses can create cracks or fractures in the thickening oxide layer. In the case of aluminum tubes with less than 10-mm diameter, the current distribution of the anodizing bath is reduced and oxide formation is less complete; therefore, there are restrictions to studying the internal stresses in oxide layers in samples of this size.

The Resistance to Electrical Breakdown of Oxide Layers

The aluminum oxide coatings are dielectric in nature; so a rinsed and dried oxide layer acts as a current insulator. Above a certain voltage, however, current breakdown occurs across an oxide layer placed between electrical

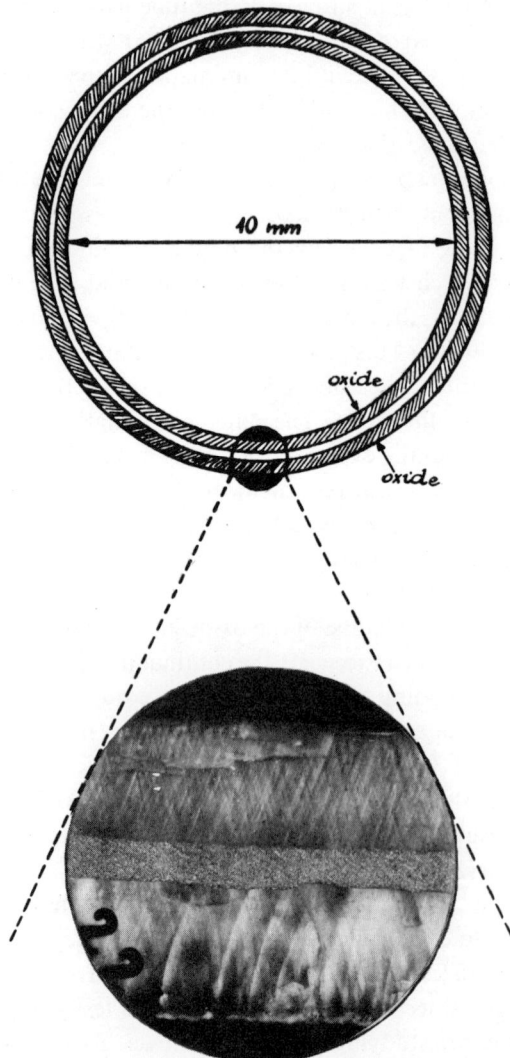

Fig. 83. Cross section of oxide formed on the external and internal surface of an aluminum pipe (O.D. 10 mm, wall thickness 2 mm) by anodizing for 30 min in 0.1% sulfuric acid with 50 V (150 ×). (Reproduced at 65%.)

poles. The breakdown voltage, or breakdown strength, depends on the material properties and thickness of the oxide layer. The electrical strength of ordinary thin oxide layers is low. The breakdown voltage of layers of more than 5–6 μm depends on the type of base metal and on the conditions of anodic oxidation.

The insulating capacity and breakdown strength of oxide coatings without any post-pore-sealing treatment changes slowly as a result of the aging processes (dehydration, formation of boehmite) in the oxide structure; in these cases the breakdown voltage should be measured shortly after the completion of anodic oxidation. The average values in Table 3 refer to the specific breakdown strengths of normal oxide layers prepared by usual anodizing processes without any post-pore-sealing treatment.

After a certain thickness, the value of $V/\mu m$ no longer increases linearly as the thickness increases; the rate of increase is reduced (Fig. 84). This behavior is due to microcracks forming in the heavy, hard oxide coatings, partly as a consequence of the nucleation mechanism and partly because of the differences in specific volume between the aluminum base and the aluminum oxide; when high peak voltages are applied, current breakthrough occurs relatively easily across these cracks.

In heavy oxide coatings, produced either with or without special methods, an increase of the silicon or colored metal content of the base metal can reduce the breakdown strength to 2–4 $V/\mu m$. After a pore-sealing treatment with oil, wax, or plastic, the breakdown voltage increases significantly and may be as high as 50–60 $V/\mu m$.

Since the electrical breakdown voltage depends not only on the thickness of the oxide coating but also on its structure, the measurement of breakdown strength can be used to characterize the oxide products obtained by various methods.

The Corrosion Resistance of Oxide Coatings

The corrosion resistance of oxide coatings prepared by different processes depends on the chemical effects of the atmosphere on the composition, physical structure, thickness, porosity, and cracking of the oxide layer. The type, amount, and distribution of foreign substances incorporated in the layer are also influential on the corrosion properties.

Table 3. Breakdown Strengths of Normal Oxide Layers

Oxidizing medium	Oxidizing current	Breakdown voltage, $V/\mu m$
Sulfuric acid	dc	35
Sulfuric acid	ac	25
Oxalic acid	dc	10–20
Oxalic acid	ac	5–8

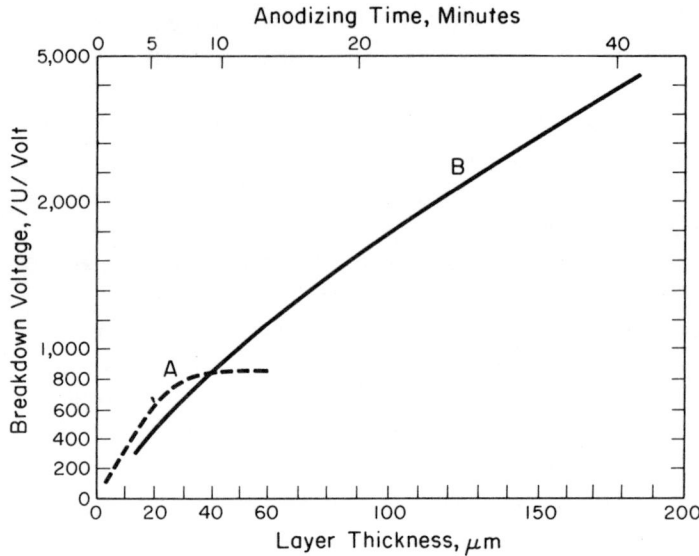

Fig. 84. Change in breakdown voltage as a function of layer thickness on 99.99 Al: A was anodized at 20° C in 15% sulfuric acid at 12.5 V; B was prepared at ±1° C in 1% sulfuric acid at 50 V.

An aggressive chemical medium or atmospheric moisture can penetrate through the pores and microcracks of the oxide layer to the barrier oxide layer and then to the base metal. As a result, little corrosion cells are formed very quickly at the bottom of the pores or cracks. This mechanism leads to a more rapid progress of the corrosive attack in the boundary zone of the base metal; a lamellar separation occurs under the oxide, and destruction of the base metal progresses.

The porous oxide coatings are treated by a pore-sealing post-treatment, as has been mentioned before. The increase of the resistance to chemicals and corrosion depends on the efficacy of the pore-sealing treatment; thus, it is as important as the anodization itself.

Corrosion is promoted by irregular thickness, the presence of holes and cracks, the incorporation of foreign substances, and structural discontinuities, because defect sites can be sources of corrosion. If one considers that anodic oxide formation and certain properties of the layer structure are dependent on the surface properties of the base metal, then clearly the chemical and corrosion resistance of anodized aluminum is also closely connected with the quality of the base metal. The combined effect of these factors determines the corrosion properties of the oxide coating, that is, whether corrosion appears on the whole oxide surface or only at definite defect sites on the surface, which may lead to pitting or intergranular corrosion.

The methods used in practice to measure corrosion resistance can be classified as follows:

1. Chemical solubility test: the Mylius method.
2. Testing the quality of porosity and of the pore-sealing treatment by the Kape method,[86–88] or the Tomashov–Bjalobsheski method.[19,120]
3. Accelerated laboratory corrosion testing with aggressive chemicals or electrochemical procedures (Cass, Corrodkote, Fact, or Kesternich methods, etc.).
4. Exposure testing in the natural environment.

The test for the chemical stability of the oxide layer (the Mylius method) consists of dissolving the oxide layer from the anodized specimen with 20% hydrochloric acid at room temperature. The volume of hydrogen gas evolved during dissolution in a specified period is measured. As shown by diagrams referring to our test series (Fig. 85), the Mylius test clearly indicates the differences in chemical stability and structure of the oxide layers prepared under various conditions.

The details of accelerated corrosion testing and of pore sealing treatments will be omitted since they can be found in the technical literature.

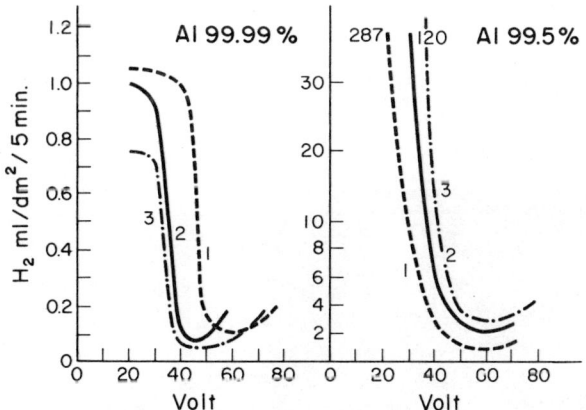

Fig. 85. Testing the solubility by the Mylius method of the oxide layers prepared in dilute sulfuric acid at 20–80 V. Oxides were prepared on 99.99% and 99.5% Al. Curve 1, 1% H_2SO_4; curve 2, 2.5% H_2SO_4; curve 3, 1% H_3BO_3 + 1% H_2SO_4.

Fig. 86. Appearance of corrosion cells on the surface of an oxide coating without pore sealing (at the points marked by arrows) during the Kesternich corrosion test (50 ×). (Reproduced at 65%.)

After the tests have been carried out, corrosion damage is assessed by optical chemical or electrochemical methods. Generally, corrosion starts at the edges and corners of specimens which have not been subjected to post-treatment (i.e., at discontinuities or even cracks). After some time, corrosion points appear irregularly distributed on the surface. The number and size of the corrosion cells increase rapidly, at which point metal and oxide dissolution occur simultaneously, as seen in Figs. 86 and 87.

Once the oxide layer has been penetrated, the corrosion of the base metal continues to spread (Figs. 88 and 89). The continuing production of corrosion products results in the breaking off of the oxide layer and the formation of open corrosion pits (Fig. 90). In some places the corrosion products penetrate the channels in the oxide layer and produce efflorescences on the surface; the oxide layer does not break up but bridges the pit in the base metal.

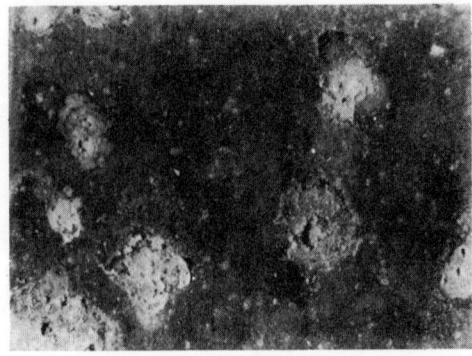

Fig. 87. Destruction of layer by pitting (50 ×). (Reproduced at 65%.)

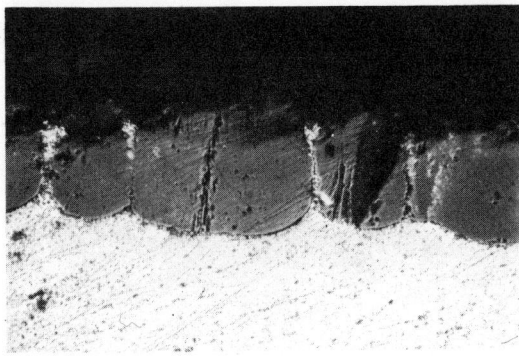

Fig. 88. Start of corrosion along the oxide cracks and at inclusions incorporated in the oxide (300 ×). (Reproduced at 65%.)

In all corrosion tests, whether accelerated or not, corrosion damage can be characterized by the weight loss of the specimen (in $g/mm^2 \cdot day$) and by the change of dimensions. These values can be employed to draw conclusions about the oxide formation as shown by the properties of the layer. These numerical values, as well as change in the properties of the oxide as observed through a microscope, can be employed to draw conclusions about the mechanism of oxide formation.

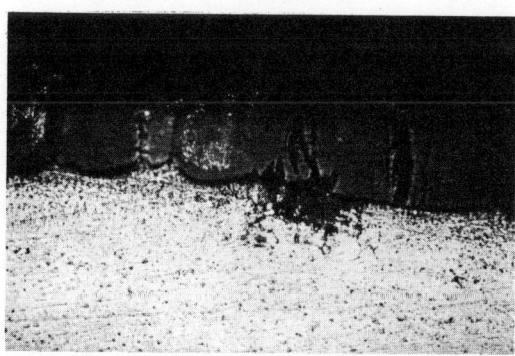

Fig. 89. Spreading of corrosion to the base metal along the gaps opened in the oxide layer (300 ×). (Reproduced at 65%.)

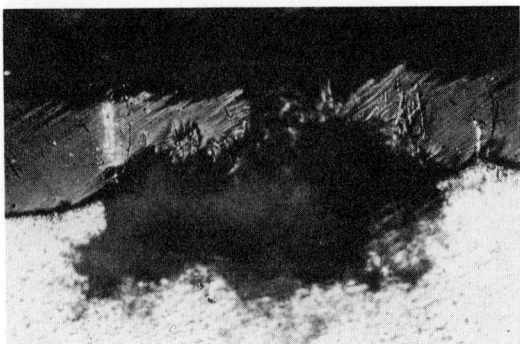

Fig. 90. Progress of corrosion attack and complete destruction of the oxide and the base metal (300 ×). (Reproduced at 65%.)

SUMMARY

From the results of investigations on the mechanism of atmospheric, thermal, or chemical oxidation in the metal/air, metal/oxygen, metal/water, or metal/electrolyte systems, it has generally been accepted that oxide formation starts on the metal surface with the appearance of oxide nuclei whose size and distribution depend on the experimental conditions. The coherent oxide coating is formed by the development of the first oxide nuclei, the appearance of further nuclei, and their coalescence and thickening. The structural composition and thickness of the oxide coatings always depend on the conditions of nucleus formation and on the subsequent physical and chemical processes.

In the anodic oxidation of aluminum, the reactions occurring at the metal/electrolyte interface under the influence of an electrochemical force field are explained by the KHR theory, by the formation of a homogeneous barrier layer, and by the subsequent oxidation reactions controlled by electrical breakdown.

There are several contradictions and deficiencies in the classical Keller–Hunter–Robinson theory, and it is suggested that a realistic basis for the mechanism of thickening of the covering oxide layer can better be found in the conclusions of Ginsberg, Wefers, Kaden, Murphy and Michelson, and Bogojavlenski, as well as results of our investigations in Hungary. In addition to the conventional test methods, we employed a special microcinematographic method which permits continuous study of the phenomena of oxide formation from the first moment of oxidation up to its completion. The information resulting from our studies is summarized as follows:

In conventional oxidation (i.e., in oxidizing electrolyte solutions at higher concentrations, with low or medium cell voltage), oxide formation starts simultaneously at each point of the aluminum surface and continues at the same rate. However, when anodizing is carried out in more dilute solutions, at higher voltages, oxide formation begins with the appearance of primary oxide nuclei only at highly active and energy-rich surface sites. Around these primary nuclei the process progresses by zones of secondary oxidation. The mechanical or chemical preparation of the surface of the base metal also influences the total reaction.

It can thus be seen that oxide formation does not start simultaneously at a uniform rate and does not propagate homogeneously over the surface; the oxide elements are formed separately in space and time. From our experimental observations and the results of the above-mentioned authors, we have evolved a model for explaining the mechanism of anodic oxide formation based on nucleation theory. The reactions, which occur partly simultaneously and partly in sequence, can be grouped into the following three stages, which assure the formation of a heavy covering oxide layer:

1. Initial stage. An oxygen or aluminum oxide film (barrier layer) appears on the aluminum surface, bonded by adsorption or chemisorption with a thickness of one to two molecular layers.
2. Nucleation stage (formation of oxide nuclei). At energetically favorable sites on the metal surface, nuclei are formed. Secondary zones of oxidation form at the edges of these oxide nuclei and progress laterally impinging and coalescing.
3. Forming stage (formation of the covering layer). Current conduction and material transport proceed perpendicularly to the base metal through the coherent oxide layer and also laterally parallel to the base metal.

The rate and extent of formation of the oxide layer depend mainly on the permeability of the oxygen film and oxide membrane formed in the initial stage by ion and electron diffusion. However, other processes also control oxide formation:

1. The selective chemical re-solution of the oxide products in the oxidizing electrolyte solutions.

2. The formation of micro and macro pores (the formation of which is characteristic of the anodization conditions) due to chemical resolution.

3. The transformation of the pores into long, hollow, fibrous pore channels due to the continuous thickening of the layer.

In the forming stage the material and current transport is supported not only by ion and electron diffusion but also by the exchange of materials in the chemical reactions of hydration and peptization as well as by the proton motion (protomery), which accompanies the changes of the hydrogen-bridge bonds in the less coherent oxide layer. After the forming stage, the oxide layer has all the characteristics of anodic oxidation discussed in the technical literature.

From optical and electron microscopic investigations of oxide layers in various orientations, it can be concluded that the oxide cells formed from primary oxide nuclei are dissimilar in size and shape and are arranged in larger units, or "bundles of cells." These bundles are also of different sizes; their shapes depend not only on the experimental parameters but also on the geometry of the reacting phase and on the sites occupied by the oxide elements. Studies under polarized light show that the oxide cell and pore channel bundles are grouped mostly in blocks within grain boundaries of the base metal. It can be seen clearly in polarized light that a differentiation parallel with the metal surface exists in the oxide layer; this results from the lateral, interlamellar propagation of material and current transport during oxidation.

It has been shown also that, although in conventional anodization it is observed only indirectly, the physical, chemical, and microstructural characteristics of the surface of the base metal play a decisive role in the morphology of oxide formation. In this respect, lattice deformations at crystal boundaries, lattice vacancies, dislocations, and foreign particles which penetrate to the surface of the metal object are very important. The epitaxy arising from such structural conditions and internal mechanical stresses can be explained by the nucleation mechanism. Metallic or nonmetallic (intermetallic) inclusions incorporated in the oxide layer, either with their original composition or as oxides, can influence not only the physical and mechanical properties of the layer but also the corrosion resistance of the oxide coating.

The explanation based on the nucleation mechanism and proposed for the partial processes of anodic oxide formation can be reconciled on several points with the KHR model.

ACKNOWLEDGMENTS

We have been studying the physical, chemical, and structural properties of thick and thin oxide membranes on aluminum surfaces in various media and under various energy conditions in Hungary for 20 years. This extended and fundamental work has been carried out at the Light Metals Research Institute (theoretical and practical research) and at the research laboratories of the General Design Institute for the Machine Industry (industrial plant testing).

I wish to express my appreciation for support by the managements of both Institutes. My thanks are also due to my colleagues, A. Domony, E. Lichtenberger-Bajza, G. Sinay, F. Kodó, M. Holló, T. Eöllös, and S. Egerházi, for the assistance they have given me in the course of my research.

In the technical realization of the microcinematographic investigations, I was assisted by M. Riedl and R. Karácsonyi from the Institute of Physical Chemistry at the Eötvös Lorand University of Sciences. My thanks to them for their valuable aid.

REFERENCES

1. W. Herrmann, *Wiss. Veroff. Siemens-Werkst. Sonderh.* **199**, 88 (1940).
2. N. Cabrera and N. F. Mott, *Rep. Prog. Phys.* **12**, 163 (1949).
3. G. Wyon, J. M. Marchin, and P. Lacombe, *Rev. Met.* **53**, 945 (1956).
4. M. Paganelli, *Alluminio* **27**, 159 (1958).
5. M. S. Hunter and P. Fowle, *J. Electrochem. Soc.* **103**, 482 (1956).
6. S. Tajima, Y. Tanabe, M. Shimura, and T. Mori, *Electrochem. Acta* **6**, 127 (1962).
7. R. W. Mott, *Electrochem. Ind.* **11**, 444 (1904).
8. W. J. Muller and K. Konopiczky, *Z. Phys. Chem.* **141**, 343 (1929).
9. S. Setoh and A. Miyata, *Phys. Chem. Res. (Tokyo)* **19**, 237 (1932).
10. Th. Rummel, *Z. Phys.* **99**, 581 (1936).
11. F. Keller, S. Hunter, and D. L. Robinson, *J. Electrochem. Soc.* **100**, 411 (1953).
12. R. C. Spooner, *J. Electrochem. Soc.* **102**, 156 (1955).
13. C. J. L. Booker, J. L. Wood, and A. Walsh, *Br. J. Appl. Phys.* **8**, 347 (1957); and *Nature*, **176**, 222 (1955).
14. G. Elssner, *Aluminum (Dusseldorf)* **35**, 374 (1959); and *Metall* **15**, 1194 (1961).
15. S. Wernick and R. Pinner, *Surface Treatment and Finishing of Aluminum and Its Alloys*, R. Draper Ltd., Teddington, England (1959).
16. S. Wernick, R. Pinner, E. Zurbrugg, and R. Weiner, *Die Oberflachen behandlung von Aluminum*, Eugen G. Leuze Verlag, Saulgau/Wurtt, Germany (1969).
17. J. D. Shreider, A. V. Bjalobsheski, Z. T. Zagricenko, and B. V. Serebrennikov, *Metalloved. Obrab. Met.* **11**, 14 (1956).
18. A. S. Nikosov, G. V. Kurganov, and N. J. Jarzemskaja, *Metaloved. Obrab. Met.* **12**, 66 (1957).
19. N. B. Tomashov and A. V. Bjalobsheski, *Tr. Inst. Fiz. Khim. Akad. Nauk SSSR* **3**, 17 (1953); and **4**, 5 (1957).

20. A Prati, F. Sacchi, and G. Paolini, *Alluminio* **30**, 467 (1961).
21. W. J. Campbell, Conference of The Institute of Metal Finishing, London, *J. Electrodepositors Tech. Soc.* **34**, 45 (1958).
22. H. Ginsberg, *Metall* **16**, 173 (1962).
23. E. F. Barkman, *Proc. Symp. Anodizing Aluminum*, Univ. Aston, Birmingham, England, p. 27 (1967).
24. S. Tajima, Anodic Oxidation of Aluminum, in *Advances in Corrosion Science and Technology* (M. G. Fontana and R. W. Staehle, eds.), Vol. 1, pp. 229–362, Plenum Press, New York (1970).
25. S. Tajima, N. Baba, and T. Mori, *Electrochim. Acta* **9**, 1509 (1964).
26. S. Tajima, T. Mori, and M. Shimura, *Alluminio* **31**, 191 (1962).
27. H. Neunzig and V. Rohrig, *Metall* **18**, 590 (1964).
28. W. Sautter, *Aluminum (Dusseldorf)* **41**, 760 (1965).
29. W. Franklin, *Nature* **180**, 1470 (1957).
30. W. Kaden, *Aluminum (Dusseldorf)* **39**, 33 (1963).
31. J. F. Murphy and C. E. Michelson, Proc. Conference Anodizing Aluminum, Univ. Nottingham, England, p. 83 (1961).
32. J. F. Murphy, Proc. Symp. Anodizing Aluminum, Univ. Aston, Birmingham, England, p. 3 (1967).
33. H. Ginsberg and K. Wefers, *Aluminum (Dusseldorf)* **37**, 19 (1961).
34. H. Ginsberg and W. Kaden, *Aluminum (Dusseldorf)* **39**, 33 (1963).
35. H. Ginsberg and H. Z. Neunzig, *Z. Metallkd.* **52**, 626 (1961).
36. A. F. Bogojavlenski, I. T. Vedernikov, *et al.*, *Zh. Prikl. Chim.* **31**, 310 (1958); **33**, 340 (1960); **35**, 1892 (1962); **38**, 952 (1965); **40**, 565 (1967).
37. A. F. Bogojavlenski and co-workers, Anodnaja Zashita Metalov Mashinostronije, M. 22, Moscow (1964).
38. A. F. Bogojavlenski, *Tr. Kaz. Aviats. Inst.* **90**, 3 (1966); *Kaz. Ord. Tr. Krasne Znameni Aviats. Inst.*, 59–95, (1968).
39. H. Akahori, *J. Electromicrosc. Jpn.* **10**, 175 (1961).
40. J. W. Diggle, T. C. Downie, and C. W. Goulding, *Chem. Rev.* **69**, 365 (1969).
41. J. Bénard, *Werkst. Korros.* **16**, 1044 (1965).
42. T. N. Rhodin, Reviews of International Conference, Bolton, Landing, N.Y., p. 508 (1959).
43. T. N. Rhodin, W. H. Orr, and D. Walton, *Mem. Sci. Rev. Met.* **62**, 157 (1965).
44. R. Darras and R. Dominget, *Mem. Sci. Rev. Met.* **62**, 148 (1965).
45. H. Mykura, *Mem. Sci. Rev. Met.* **62**, 167 (1965).
46. P. E. Doherty, *Mem. Sci. Rev. Met.* **62**, 196 (1965).
47. H. Fischmeister, *Mem. Sci. Rev. Met.* **62**, 27 (1965).
48. J. Paidassi and co-workers, *Mem. Sci. Rev. Met.* **62**, 271 (1965).
49. U. F. Franck, *Werkst. Korros.* **11**, 401 (1960); and **14**, 367 (1963).
50. D. A. Vermilyea, *Acta Met.* **2**, 482 (1954); and **5**, 492 (1957).
51. H. Fischer and F. Kurz, *Aluminum (Dusseldorf)* **32**, 126 (1956).
52. J. Osterwald, *Z. Elektrochem.* **66**, 492 (1962).
53. K. Hauffe, *Werkst. Korros.* **16**, 791 (1965).
54. C. Wagner, Symp. Diffusion and High Temperature Oxidation in Atom Movements, Cleveland, Ohio (1951); and *J. Appl. Phys.* **29**, 1262 (1958).
55. C. Wagner, *Z. Elektrochem.* **65**, 4 (1961).
56. P. Csokán, M. Riedl, and R. Karacsonyi, *Metalloberflaeche* **30**, (1970).
57. P. Csokán, Nucleation Mechanism of Oxide Formation of Aluminum during the Electrochemical Oxidation (Anodisation), lecture and projection of micro-kinematographical documentary film: *Corrosions-Week*, International Symposium,

European Federation of Corrosion, Budapest (1968); *Interfinish '68*, 7th International Conference, European Federation of Corrosion, Hannover, Germany (1968); Scientific-Technical Conference, Dom Techniky, Moscow (1971); Electrochemical Meeting of the Hungarian Scientific Academy, Mátrafüred, Hungary (1971); Annual Technical Conference, Institute of Metal Finishing, London (1972); 23rd Meeting of the International Society of Electrochemistry, Royal Swedish Academy of Engineering Sciences and The Swedish Corrosion Institute, Stockholm (1972); Annual Technical Conference, Institute of Metal Finishing, Edinburgh, Scotland (1973); Aluminum-Conference, Metal Finishing Society of Japan, Tokyo (1975); Meeting of Electrochemical Science, WASEDA University, Tokyo (1975).

58. P. Csokán, *DECHEMA-Monogr.* **39**, 229 (1960); *Metalloberflaeche* **15**, (1961); *Werkst. Korros.* **12**, 288 (1961); *Schleif- Poliertech.* (*Berlin*) **1**, 26 (1961); *Electroplat. Met. Finish.* **15**, 75 (1962); *Trans. Inst. Met. Finish.* **41**, 60 (1964); *Metalloberflaeche* **19**, 252 (1965); *Gep* **17**, 175 (1965); Tangungsber. Interfinish 68 Konf., Honnover p. 176 (1968); *Magy. Kem. Lapja* **24**, 501 (1969); *Metalloberflaeche* **23**, 326 (1969); *Gepgyartastechnologia*, **9**, 529 (1969); *Korroz. Figy.* **11**, 109 (1971); *Magy. Alum.* **8**, 203 and **8**, 337 (1971); *Galvanotechnik*, **63**, 132 (1972); *Magyar Alum.* **9**, 321 (1972); *Magyar Alum.* **10**, 8 (1973); *Electrodeposition Surf. Treat.* **1**, 24 (1973); *Trans. Inst. Met. Finish.* **51**, 6 (1973); *Trans. Inst. Met. Finish.* **52**, 92 (1974).

59. J. Glayman, *Galvano* **16**, 121 (1947).

60. W. Baumann, *Z. Phys.* **111**, 708 (1939).

61. W. Seith, *Diffusion in Metallen*, Springer-Verlag, Berlin (1955).

62. L. Vlasakova, *Werkst. Korros.* **9**, 537 (1958).

63. D. D. Eley and P. R. Wilkonson, *Proc. R. Soc. Ser. A.* **254**, 327 (1960).

64. S. Haruyama, *J. Electrochem. Soc. Jpn.* **33**, 181 (1965).

65. G. A. Dorsey, *Plating* **57**, 1117 (1970); and *J. Electrochem. Soc.* **114**, 466 (1969).

66. A. Domony, E. Lichtenberger-Bajza, and P. Csokán, *Werkst. Korros.* **11**, 701 (1960).

67. A. Domony and E. Lichtenberger-Bajza, *Metalloberflaeche* **15**, 134 (1961).

68. E. Rayola and A. W. Davidson, *J. Am. Chem. Soc.* **78**, 556 (1956).

69. T. Erdey-Gruz, *Elmeleti Fizikai-Kémia* (*Theoretical Physical Chemistry*), Tankönyvkiadó V. Budapest (1962).

70. A. Guntherschultze and H. Betz, *Elektrolyt-Kondersatoren*, Technisch Verlag A. Cranz, Berlin (1952).

71. G. Ulrich, *Wiss. Z. Elektrotech.* **2**, 201 (1956); and **3**, 651 (1957).

72. E. J. W. Verwey, *Physica* **2**, 1059 (1955).

73. A. Kordes, *Z. Krystallogr.* **91**, 193 (1935).

74. S. Tajima, *Metall.* **18**, 581 (1964).

75. A. Tajima, N. Baba, T. Mori, and M. Shimura, Proc. Symp. Anodizing Aluminum, Univ. Aston, Birmingham, England (1967).

76. H. Ginsberg and K. Wefers, *Metall* **16**, 173 (1962); and **17**, 202 (1963).

77. W. Kaden and H. Ginsberg, Proc. Conference Anodizing Aluminum, Univ. Nottingham, England (1961).

78. M. Eigen, Symposium on Hydrogen-Bonding, Lujubljana, Yugoslavia (1957).

79. A. Charlesby, *Proc. Phys. Soc.* **66**, 317 (1953).

80. T. P. Hoar and J. Yahalom, *J. Electrochem. Soc.* **110**, 614 (1963).

81. P. Winkel, G. Pistorius, and W. Ch. van Geel, *Phillips Rep.* **13**, 277 (1958).

82. A. W. Brace, *Metallurgia* **4**, 173 (1957); and **87**, 261 (1965).

83. P. Neufeld and H. O. Ali, *Trans. Inst. Met. Finish.* **48**, 175 (1970).

84. W. J. Müller, *Die Bedeckungstheorie der Passivität und die experimentelle Begrundung*, Akademische Verlagsgesellschaft, Berlin (1933).

85. H. Fischer, N. Budiloff, and L. Koch, *Korros. Metallwirtsch.* **18**, 62 (1942); and **20**, 115 (1944).
86. J. M. Kape, *J. Met. Finish.* **4**, 391 (1958).
87. J. M. Kape, *Electroplat. Met. Finish.* **12**, 59 (1959).
88. J. M. Kape, *Plating* **55**, 326 (1968).
89. N. F. Mott, *Trans. Faraday Soc.* **43**, 429 (1947).
90. J. N. Stranski and R. Kaishew, *Z. Phys. Chem.* **B26**, 317 (1934).
91. K. Hauffe, *Reaktionen in und an festen Stoffen*, Springer-Verlag, Berlin (1966).
92. A. Neuhaus and M. Gebhardt, *Werkst. Korros.* **17**, 567 (1966).
93. T. Hayashi, S. Higuchi, H. Kinoshita, and T. Ishida, *J. Electrochem. Soc. Jpn.* **37**, 64 (1964).
94. K. Huber, *Helv. Chim. Acta* **28**, 1416 (1945); and *J. Colloid Sci.* **3**, 197 (1948).
95. D. J. Stirland and R. W. Bichsel, *J. Electrochem. Soc.* **106**, 481 (1959); R. W. Bichsel, *Metall* **14**, 196 (1960).
96. A. Marge and M. Renouard, *Rev. Alum.* **38**, 1051 (1961).
97. P. Csokán, *Femip. Kut. Intez. Kozl.* **3**, 219 (1959).
98. P. Csokán, *Corros. Anticorr.* **8**, 158 (1960).
99. P. Csokán and Gy. Emöd, *Gep* **12**, 489 (1960).
100. P. Csokán and G. Sinay, *DECHEMA-Monogr.* **45**, 319 (1962).
101. P. Csokán, *Femip. Kut. Intez. Kozl.* **6**, 267 (1962).
102. P. Csokán, Ved.-Tech. Spolocnost DOM TECH., Bratislava, Czech. **3**, 8–1 (1962).
103. P. Csokán, *Werkst. Korros.* **15**, 307 (1964).
104. P. Csokán and G. Sinay, *Werkst. Korros.* **11**, 224 (1960).
105. P. Csokán and G. Sinay, *Gep* **14**, 126 (1962).
106. P. Csokán, *Elektrie* **8**, 245 (1962).
107. P. Csokán and G. Sinay, *Blech* **9**, 650 (1962).
108. M. Schenk, *Werkstoff Aluminum und seine anodische Oxydation*, A. Francke Verlag, Bern (1949).
109. G. Elssner and A. Beyer, *Arch. Metallkd.* **2**, 120 (1948).
110. H. Röhrig and E. Käpernick, *Z. Metallkd.* **31**, 101 (1949).
111. J. A. Davies, *et al.*, *J. Electrochem. Soc.* **109**, 999 (1962); and **110**, 849 (1963).
112. D. A. Vermilyea, *J. Electrochem. Soc.* **109**, 999 (1962); and **110**, 849 (1963).
113. D. H. Bradhurst and J. W. L. Leach, *Trans. Br. Ceram. Soc.* **62**, 793 (1963); and *J. Electrochem. Soc.* **113**, 1245 (1966).
114. S. F. Bubar and D. A. Vermilyea, *J. Electrochem. Soc.* **113**, 892 (1966); and **114**, 882 (1967).
115. L. Young, *J. Electrochem. Soc.* **110**, 589 (1963).
116. R. S. Alwitt, *J. Electrochem. Soc.* **114**, 843 (1967).
117. D. J. Arrowsmith, E. A. Culpan, and R. J. Smith, Proc. Symp. Anodizing Aluminum, Univ. Aston, Birmingham, England, p. 17 (1967).
118. T. P. Hoar and D. J. Arrowsmith, *Electroplat. Met. Finish.* **10**, 141 (1957).
119. M. Stalzer, *Metalloberflaeche* **18**, 263 (1964); and *Galvanotech. + Oberflaechenschutz* **7**. 39 (1966).
120. N. D. Tomashov and A. V. Bjalobsheski, *Tr. Inst. Fiz. Khim. Akad. Nauk SSSR* **5**, 114 (1957).
121. R. M. Barrer, *Diffusion in and through Solids*, Cambridge University Press, New York (1952).

INDEX